Recent Advances in Numerical Simulations

Edited by Francisco Bulnes and Jan Peter Hessling

Published in London, United Kingdom

IntechOpen

Supporting open minds since 2005

Recent Advances in Numerical Simulations
http://dx.doi.org/10.5772/intechopen.91589
Edited by Francisco Bulnes and Jan Peter Hessling

Contributors
Mouna Merzougui, Julia N. Soulakova, Trung Ha, Gilberto Rodrigues Liska, Luiz Alberto Beijo, Marcelo Ângelo Cirillo, Flávio Meira Borém, Fortunato Silva De Menezes, Myron Polatayko, Hani Akbari, Sabine Bauer, Ivanna Kramer, Ivaldo Leão Ferreira, José Adilson de Castro, Amauri Garcia, Mounir Khelladi, James Henry Ricketts, Roger Jones, Siliang Yang, Francisco Bulnes, Ali Erfani Agah, Gerald Tendayi Marewo, Yury Vasilyevich Yanilkin, Vadim Kolobyanin, Vladimir Shmelev, Domenico De Luca, Alessandro Petruzzi, Marco Cherubini, Simone Di Pasquale, Giovanni Bruna, Juan Carlos García-Limón, Víctor Sánchez-Suárez, Luis Alfredo Ortiz-Dumas, Francesco Fiorito, Deo Prasad, Alistair Sproul

First published in London, United Kingdom, 2021 by IntechOpen
IntechOpen is the global imprint of INTECHOPEN LIMITED, registered in England and Wales, registration number: 11086078, 5 Princes Gate Court, London, SW7 2QJ, United Kingdom
Printed in Croatia

British Library Cataloguing-in-Publication Data
A catalogue record for this book is available from the British Library

Additional hard and PDF copies can be obtained from orders@intechopen.com

Recent Advances in Numerical Simulations
Edited by Francisco Bulnes and Jan Peter Hessling
p. cm.
Print ISBN 978-1-83968-168-4
Online ISBN 978-1-83968-169-1
eBook (PDF) ISBN 978-1-83969-315-1

For EU product safety concerns:
IN TECH d.o.o., Prolaz Marije Krucifikse Kozulić 3, 51000 Rijeka, Croatia, info@intechopen.com or visit our website at intechopen.com.

Meet the editors

Dr. Francisco Bulnes has a Ph.D. in Mathematical Sciences from the Instituto de Matemáticas, Universidad Nacional Autonoma de México (IM/UNAM). He is the director of Advanced International Research in Mathematics and Engineering (IINAMEI) and editor in chief of the journal Mathematics. Dr. Bulnes has authored more than 100 journal papers and several books in mathematics and physics research. He is a well-known pioneer and creator of theories, theorems, and math objects. He has received various honors and awards (Doctorates Honoris Causa) from universities and both governmental and non-governmental organizations. He has two post-doctorates in mathematics from Cuba and Russia. His scientific work has been published by different countries including Spain, China, the United States, Russia, Cuba, United Kingdom, México, and India where he has even been honored through tribute in a British publishing house.

Jan Peter Hessling earned a Ph.D. in Theoretical Physics from Chalmers University of Technology, Gothenburg, Sweden, in 1996, and an MSc in Physics from the University of Massachusetts, USA, in 1991. Since then, he has been devoted to novel mathematical and statistical concepts, recently focusing on modeling uncertainty and digital filtering. Dr. Hessling is the original proposer of Deterministic Sampling for uncertainty quantification and Dynamic Metrology for the analysis of dynamic measurements utilizing custom digital filtering. He has authored four book chapters for InTechOpen and about twenty journal articles. Since 2016, these concepts are further developed in the privately held company Kapernicus AB, which is dedicated to applied mathematical R&D, offering statistical model sampling (SavvySampler®) and digital filtering of road surfaces (RoadNotes®) worldwide.

Contents

Preface

A numerical simulation is a computing calculation following a program that develops a mathematical model for a physical, social, economic, or biological system. Numerical simulations are required for analyzing and studying the behavior of systems whose mathematical models are very complex, as in the case of nonlinear systems. Likewise, a numerical simulation is a branch of numerical analysis studied in approximation theory inside the functional analysis.

This book examines different methods used in numerical simulations, including adaptive and stochastic methods as well as the finite element analysis research developed for the implementation and improvement of nature models studied in physics, biology, chemistry, and engineering processes. It also discusses applications of numerical simulations in demography, dynamical economics, and social behavior. The book also highlights the importance of using 2D and 3D models to capture the essence of a concept in terms of a graphic image of a process, which has as its mathematical model either a differential, integral, or functional equation and runs on a numerical calculation through a computer program. Likewise, the book discusses developments in simulation environments for physics and engineering such as virtual prototype analysis, which optimizes and develops accurate bases for engineering work and creation of new technology, minimizing and improving the cost of developing real prototypes. Forecasting and fluid models are necessary for predicting phenomena such as atmospheric phenomena. Also in dynamical systems with big complexity, such as interstellar trips and sidereal exploration have big utility inside dynamics aerospatial simulation, which can establish and plan a whole trip helping to the aero-spatial engineering for a program the flying computer and control instruments in the optimization of fuel, homework programming in the space and behavior of the interstellar spaceship.

In dynamics aerospatial simulation and interstellar trips, the numerical simulation is widely used with the purpose of developing space programs whose measurements give certainty and optimization to all systems and servo-systems of a spaceship.

Dr. Francisco Bulnes
Professor,
IINAMEI Director,
Head of Research Department in Mathematics and Engineering,
TESCHA,
Chalco, Mexico

Jan Peter Hessling, Ph.D.
Theoretical Physics,
Independent Scientist Focusing on Novel Methods of Uncertainty Quantification,
Sweden

Section 1

Numerical Simulation in Advanced Signal Analysis

Chapter 1

Femtosecond Laser Pulses: Generation, Measurement and Propagation

Mounir Khelladi

Abstract

In this contribution some basic properties of femtosecond laser pulse are summarized. In sections 2.1–2.5 the generation of femtosecond laser pulses via mode locking is described in simple physical terms. In section 2.6 we deal with measurement of ultrashort laser pulses. The characterization of ultrashort pulses with respect to amplitude and phase is therefore based on optical correlation techniques that make of the short pulse itself. In section 3 we start with the linear properties of ultrashort light pulses. However, due to the large bandwidth, the linear dispersion is responsible for dramatic effects. To describe and manage such dispersion effects a mathematical description of an ultrashort laser pulse is given first before we continue with methods how to change the temporal shape via the frequency domain. The chapter ends with a paragraph of the wavelet representation of an ultrashort laser pulse.

Keywords: femtosecond laser pulse, ultrashort pulse, autocorrelation, characterization, dispersion, spectral phase, wavelet

1. Introduction

Propagation of ultrashort optical pulses in a linear optical medium consisting of free space [1–5], dispersive media [6, 7], diffractive optical elements [8–10], focusing elements [11] and apertures [12, 13] has been extensively studied analytically, though only a few isolated attempts have been made on numerical simulation. However, analytical methods have the limitations of not being able to handle arbitrary pulse profiles.

2. Ultrashort laser pulses generations

The central aim of this section is to give a concise introduction to nonlinear optics and to provide basic information about the most-widely used tunable femtosecond laser sources, in particular tunable Ti:sapphire oscillators and Ti:sapphire amplifiers or optical parametric amplifiers.

2.1 Titane sapphire oscillator

In 1982, the first Ti:sapphire laser was built by Mouton [14]. The laser tunes from 680 nm to 1130 nm, which is the widest tuning range of any laser of its class1.

Nowadays Ti:sapphire lasers usually deliver several watts of average output power and produce pulses as short as 6.5 fs (**Figure 1**) [14].

At high intensities, the refractive index depends nonlinearly on the propagating field. The lowest order of this dependence can be written as follows:

$$n(r) = n_0 \frac{1}{2} n_2 I(r) \quad (1)$$

n_0: linear index refractive.

where n_2 is the nonlinear index coefficient and describes the strength of the coupling between the electric field and the refractive index n. The intensity is:

$$I(r) = e^{-gr^2} \quad (2)$$

The refractive index changes with intensity along the optical path and it is larger in the center than at the side of the nonlinear crystal. This leads to the beam self-focusing phenomenon, which is known as the Kerr lens effect (see **Figure 2**).

Consider now a seed beam with a Gaussian profile propagating through a nonlinear medium, e.g. a Ti:sapphire crystal, which is pumped by a cw radiation. For the stronger focused frequencies, the Kerr lens favors a higher amplification. Thus, the self-focusing of the seed beam can be used to suppress the cw operation, because the losses of the cw radiation are higher. Forcing all the modes to have equal phase (mode-locking) implies that all the waves of different frequencies will interfere (add) constructively at one point, resulting in a very intense, short light pulse.

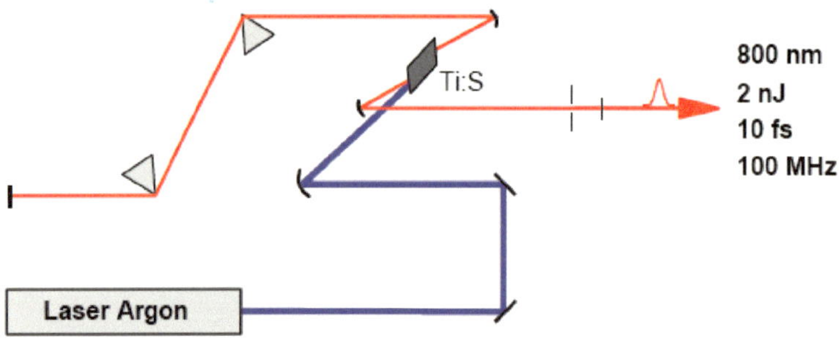

Figure 1.
A Ti: Sapphire oscillator.

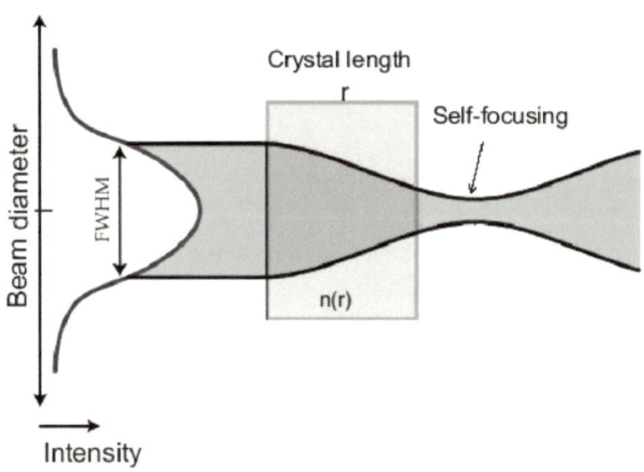

Figure 2.
The Kerr lens effect and self-focusing.

The pulsed operation is then favored, and it is said that the laser is mode-locked. Thus, the mode-locking occurs due to the Kerr lens effect induced in the nonlinear medium by the beam itself and the phenomenon is known as Kerr-lens mode-locking.

2.1.1 Examples

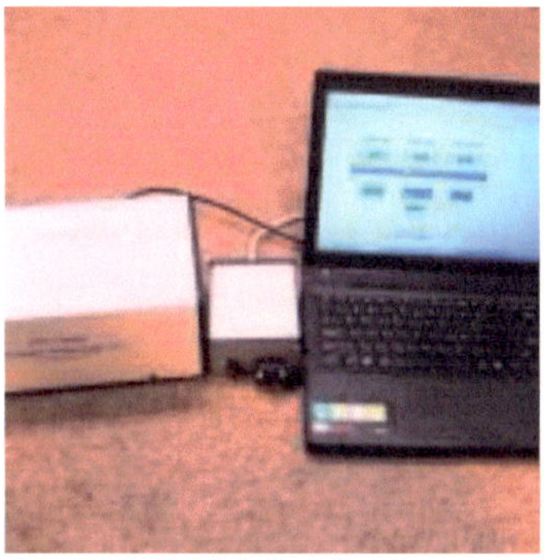

Auto TPL Tripler for laser oscillator.

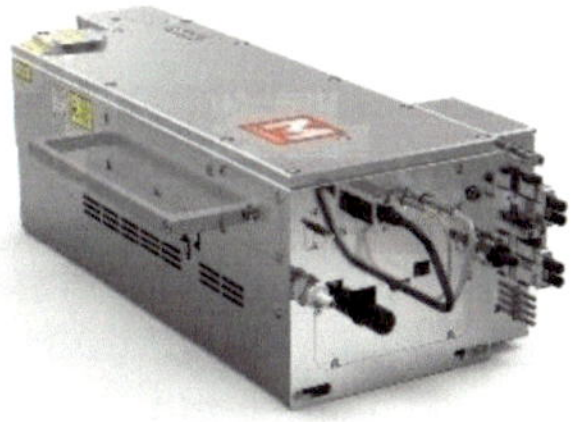

Sprite XT: Tunable ultrafast Ti: sapphire laser (**Figure 3**).

The modes are separated in frequency by $\nu = c/2L$, L being the resonator length, which also gives the repetition rate of the mode-locked lasers:

$$\tau_{rep} = \frac{1}{T} = \frac{c}{2L} \tag{3}$$

Where L is length of cavity and T is period.

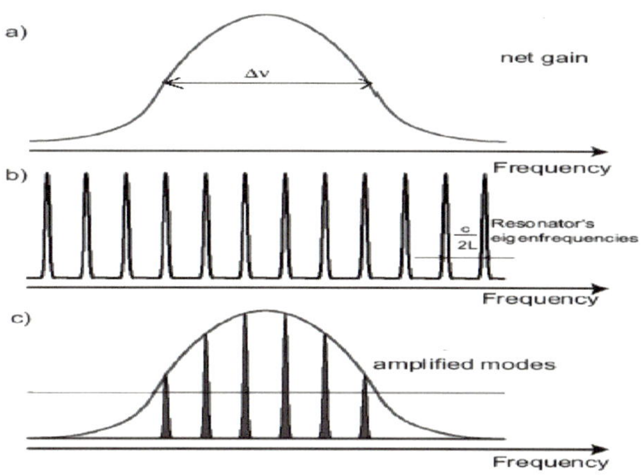

Figure 3.
The Kerr lens mode-locking (KLM) principle. (a) the net gain curve (gain minus losses). In this example, from all the longitudinal modes in the resonator (b), only six (c) are forced to have an equal phase.

Moreover, the ratio of the resonator length to the pulse duration is a measure of the number of modes oscillating in phase. For example, if $L = 1\,m$ and the emerging pulses have 100 *fs* time duration, there are 10^5 modes contributing to the pulse bandwidth. There are two ways of mode-locking a femtosecond laser: passive mode-locking and active mode-locking. In a laser cavity, these modes are equally spaced (with spacing depending on the cavity length). The electric field distribution with N such modes in phase (considered to be zero, for convenience) can be written as:

$$E(t) = \sum_{n}^{N-1} E_n e^{iw_0 + n\Delta wt} \propto \frac{e^{iN\Delta wt} - e^{iw_0 t}}{e^{i\Delta wt} - 1} \qquad (4)$$

Where w_0 is the central frequency and Δw is the mode spacing, this appears as a carried wave with frequency domain.

n: is integer from 1 to N.

The laser intensity is given by

$$\mathrm{I(t)} \propto [\mathrm{E(t)}]^2 = \frac{\sin^2[(2n+1)\Delta w.t/2]}{\sin^2(\Delta w.t/2)} \qquad (5)$$

Where E(t): electrical field.

This is series of pulses with width inversely proportional to the number of modes that are locked in phase of the mode spacing. The concept of mode-locking is easier said than done.

Figure 4. shows how the time distribution of a laser output depends upon the phase relations between the modes. **Figure 4a** is the resultant intensity of two modes in phase **Figure 4b**, is the resultant intensity of five modes in phase and a period repetition of a wave packet from the resultant constructive interference can be seen.

2.2 CPA laser system

CPA is the abbreviation of chirped pulse amplification. Chirped pulse amplification is a technique to produce a strong and at the same time ultrashort pulse. The concept behind CPA is a scheme to increase the energy of an ultrashort pulse while avoiding very high peak power in the amplification process.

In the CPA technique, ultrashort pulses are generated typically at low energy $\sim 10^{-9}$ J, with a duration around 10^{-12}–10^{-14} seconds and at a high repetition rate of about 10^8 1/s in an oscillator (**Figure 5**).

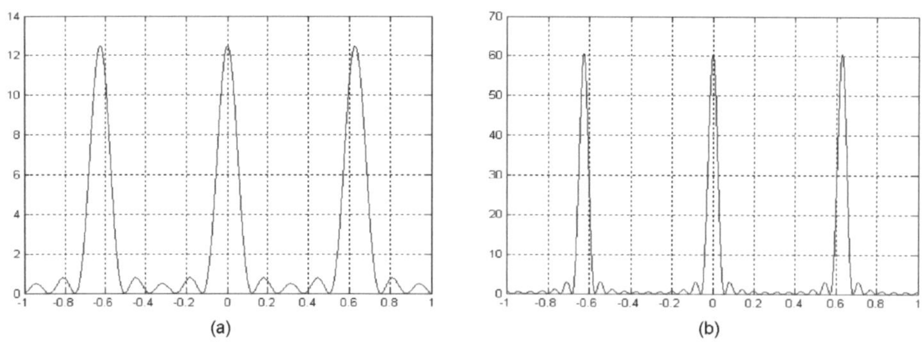

Figure 4.
The influence of the phase relation between oscillating modes on the output intensity of the oscillation. (a) Two modes in phase, and (b) five modes in phase.

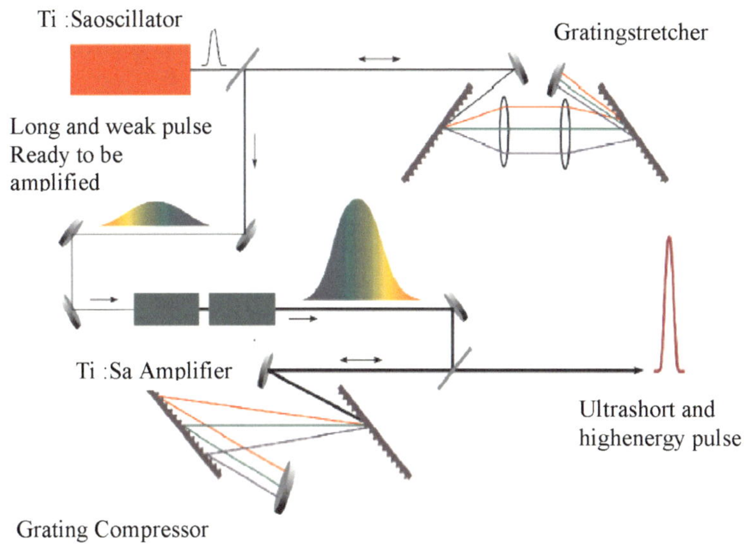

Figure 5.
Schematics of a chirped pulse amplification system, showing the duration and energy level of the signal at the different stages of the system.

2.3 Multipass and regenerative amplification

Two of the most widely used techniques for amplification of femtosecond laser pulses are the multipass and the regenerative amplification. In the multipass amplification different passes are geometrically separated (see **Figure 6a**).

The regenerative amplification technique implies trapping of the pulse to be amplified in a laser cavity (see **Figure 2b**). Here the number of passes is not important. The pulse is kept in the resonator until all the energy stored in the amplification crystal is extracted. Trapping and dumping the pulse in and out of the resonator is done by using a Pockelcell (or pulse-picker) and a broad-band polarizer [15].

Of all potential amplifier media, titanium- doped sapphire has been the most wide spread used. It has several desirable characteristics which make it ideal as a high-power amplifier medium such as a very high damage threshold (~8-10 J/cm^2), a high saturation fluence, and high thermal conductivity.

2.4 Stretcher-compressor

By using a dispersive line (combination of gratings and/or lenses), the individual frequencies within a femtosecond pulse can be separated (stretched) from each other in time (see **Figure 7a**).

In normal materials, low frequency components travel faster than high frequency components; in other words, the velocities of large wavelength components are higher than that of shorter ones. These materials induce a positive group velocity dispersion on a propagating pulse. To compensate the positive GVD (Group Velocity Dispersion) and rephase the dephased components a setup which produces negative group velocity dispersion is needed [15].

2.4.1 Grating compressor

Four identical gratings in a sequence as shown in **Figure 8** make up a grating compressor. A pulse impinges on the first gratings with an angle of θ. Then from the second grating the spectral components in the spectrum travel together in parallel

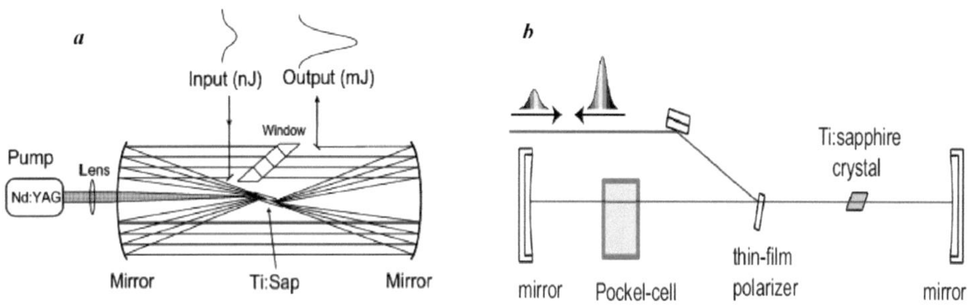

Figure 6.
(a) The amplifier configuration uses two spherical mirrors in a multi-pass confocal configuration to make the signal pass eight times through the amplifying medium, (b) schematic principle of a regenerative amplifier.

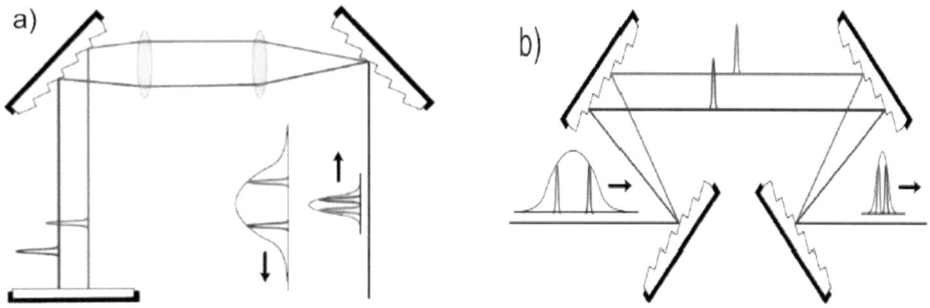

Figure 7.
Principle of a stretcher (a) and a compressor (b). The stretcher setup extends the temporal duration of the femtosecond pulse, whereas the gratings' arrangement in the compressor will compress the time duration of the pulse. Both setups are used in femtosecond amplifiers.

directions but with wavelength dependent position (spatially chirped). The gratings are set in such a way that their wavelength dispersions are reversed which implies that the exiting ray from the second grating is parallel to the incident ray to the first grating.

The group delay induced by the grating compressor is given

$$T_g = \frac{2L}{c.\sqrt{1-\left(\frac{\lambda}{d}-\sin\gamma\right)^2}}\left[1+\left(\frac{\lambda}{d}-\sin\gamma\right)\sin\gamma\right] \tag{6}$$

where, is the light wavelength, L is the distance between the gratings, d is the grating's constant and γ is the incidence angle of the beam to the first grating. Dispersion of the group delay is obtained as:

$$GDD = \frac{d^2\varnothing}{dw^2} = -\frac{\lambda^3 L}{\pi c^2 d^2}\left[1-\left(\frac{\lambda}{d}-\sin\gamma\right)^2\right]^{-3/2} \tag{7}$$

GDD is Group Delay dispersion.

The third order dispersion produced in the grating compressor will be:

$$TOD = \frac{1}{L}\frac{d^3\varnothing}{dw^3} = -\frac{d^2\varnothing}{dw^2}\frac{6\pi\lambda}{c}\left[\frac{1+\frac{\lambda}{d}\sin\gamma-\sin^2\gamma}{1-\left(\frac{\lambda}{d}-\sin\gamma\right)^2}\right] \tag{8}$$

TOD is Third Order Dispersion.

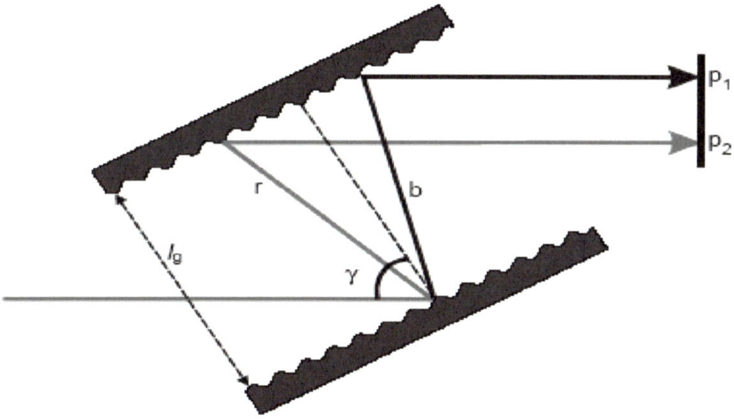

Figure 8.
Grating pairs used in the control of dispersion. R and b indicate the relative paths of arbitrary long- and short-wavelength rays. γ is the (Brewster) angle of incidence at the prism face. Light is reflected in the plane (p1–p2) in order to remove the spatial dispersion shown.

2.4.2 Prism compressor

A prism compressor is built of four sequentially arranged identical prisms used in a geometry similar to **Figure 9**; often at their minimum deviation (to decrease geometrical (spatial) distortion of prisms on the beam) and in their Brewster angle (to minimize power loss). Because of the symmetry in the arrangement, it is possible to place a mirror after the second prism (as we did in the grating compressor setup) perpendicular to the beam propagation direction. The first prism spreads the pulse spectral components out in space. In the second prism the red frequencies of the spectrum must pass through a longer length in the glass than the blue frequencies.

The optical path of a ray propagating in the compressor is defined as [15]: GDD will result

$$GDD = \frac{4L\lambda^3}{2\pi^2c^2}\left\{sin\beta\left[\frac{d^2n}{d\lambda^2}+\left(\frac{dn}{d\lambda}\right)^2\left(2n-\frac{1}{n^3}\right)\right]-2\cos\beta\left(\frac{dn}{d\lambda}\right)^2\right\} \tag{9}$$

β: is angle.

The third order dispersion can also be evaluated in the same manner to that used above

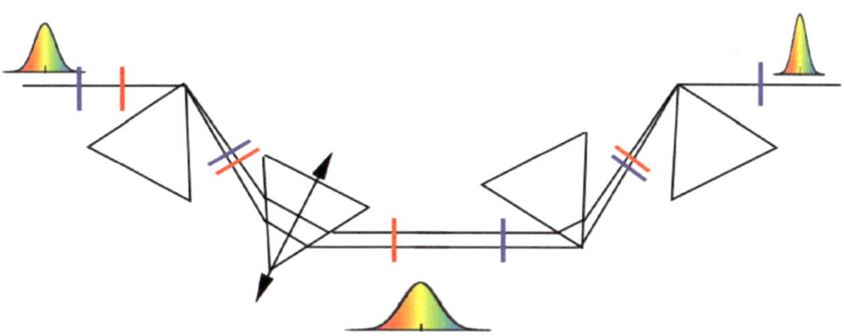

Figure 9.
Pulse compressor.

$$TOD = -\frac{L^4}{4\pi^2 c^3 L}\left[3\frac{d^2 n}{d\lambda^2} - \lambda\frac{d^3 n}{d\lambda^3}\right] \tag{10}$$

n: is refractive index.

2.5 Mathematical description of laser pulses

In order to understand the behavior of ultrashort light pulses in the temporal and spectral domain, it is necessary to formulate the relation between the two domains mathematically. It is important to introduce the concept of the amplitude and the phase of the electric field because the generation, measurement, and shaping of ultrashort laser pulses is based on measuring and influencing these properties. The electric field in the time-domain is invariably connected with its counterpart $E(w)$ in the frequency-domain via a Fourier transform:

2.5.1 Pulse duration and spectral width

The statistical definitions are usually used in theoretic calculations and given as

$$\langle t^2 \rangle = \frac{\int_{-\infty}^{+\infty} t^2 |E(t)|^2 dt}{\int_{-\infty}^{+\infty} |E(t)|^2 dt}; \langle w^2 \rangle = \frac{\int_{-\infty}^{+\infty} w^2 |E(w)|^2 dw}{\int_{-\infty}^{+\infty} |E(w)|^2 dw} \tag{11}$$

In case of Gaussian pulse is easy to determine pulse duration and spectral width by applying FWHM (Full Width Half at Maximum) of intensity. One can show that these quantities are related through the following universal inequality.

$$\Delta w \Delta \tau \geq 1/2 \tag{12}$$

Therefore, one defines the pulse duration $\Delta\tau$ as the Full Width at Half Maximum (FWHM) of the intensity profile and the spectral width Δw as the FWHM of the spectral intensity. The Fourier inequality is then usually given by

$$\Delta w \Delta \tau \geq K \tag{13}$$

where K is a numerical constant, depending on the assumed shape of the pulse.

2.5.1.1 Gaussian pulse

The Gaussian pulse, which is most commonly used in ultrashort laser pulse characteristics. The pulse is linearly chirped and represented by

$$A(t) = A_0\, exp\left(\frac{-(1+i\alpha)t^2}{\tau_g^2}\right) \text{ with } \Delta\tau_p = \sqrt{2\, ln\, 2\tau_g} \tag{14}$$

A_0: amplitude, α: chirp,
τ_g: pulse duration at FWHM.
$\Delta\tau_p$:pulse duration at FWHM after propagation.

The instantaneous frequency is given as

$$w(t) = w_0 + \frac{d\varphi(t)}{dt} = w_0 - \frac{2\alpha}{\tau_g^2}t \tag{15}$$

$W_{0:}$ central pulsation.
$w(t)$: instantaneous pulse.
$\varphi(t)$is instantaneous phase.

The spectral intensity can be derived by taking the Fourier-transform of Eq.14, it also has the Gaussian shape [16].

The shortest possible pulse, for a given spectrum, is known as the *transform-limited pulse duration*. It should be noted that Eq. (13) is not equality, i.e. the product can very well exceed K. If the product exceeds K the pulse is no longer transform-limited and all frequency components that constitute the pulse do not coincide in time, i.e. the pulse exhibits frequency modulation is very often referred to as a *chirp* (**Figure 10**).

2.5.2 Time domain description

Since in this paper the main emphasis is on the temporal dependence, all spatial dependence is neglected, i.e., $E(x, y, z, t) = E(t)$. the electric field $E(t)$, is a real quantity and all measured quantities are real. However, the mathematical description is simplified if a complex representation is used:

$$\tilde{E}(t) = \tilde{A}(t).e^{-iw_0 t}. \tag{16}$$

where $\tilde{A}(t)$ is the complex envelope, usually chosen such that the real physical field is twice the real part of the complex field, and w_0 is the carrier frequency, usually chosen to the center of the spectrum. In this way the rapidly varying is separated from the slowly varying envelope $\tilde{A}(t)$. $\tilde{E}(t)$ can be further decomposed into:

$$\tilde{E}(t) = \left|\tilde{E}(t)\right|.e^{i\varphi_0}.e^{-i\varphi(t)} = \left|\tilde{E}(t)\right|.e^{i\varphi_0}.e^{-i(\varnothing(t)-w_0)}. \tag{17}$$

$\varphi(t)$ is often to as the temporal phase of the pulse and φ_0 the absolute phase, which relates the position of the carrier wave to the temporal envelope of the pulse (see **Figure 11**). In $\varnothing(t)$ the strong linear term due to the carrier frequency, wt, is omitted.which means that a nonlinear temporal phase yields a time-dependent frequency modulation- the pulse is said to carry a chirp (illustrated in **Figure 11b**).

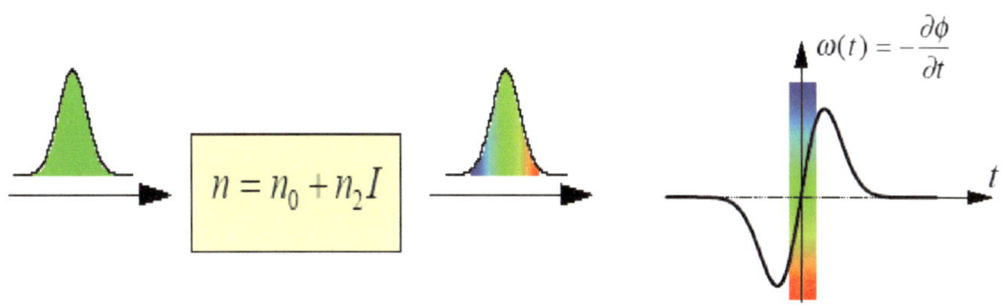

Figure 10.
Self-phase modulation. Variation in the instantaneous frequency $w(t)$ of the transmitted pulse after the propagation through a nonlinear medium with a positive nonlinear index of refraction n_2.

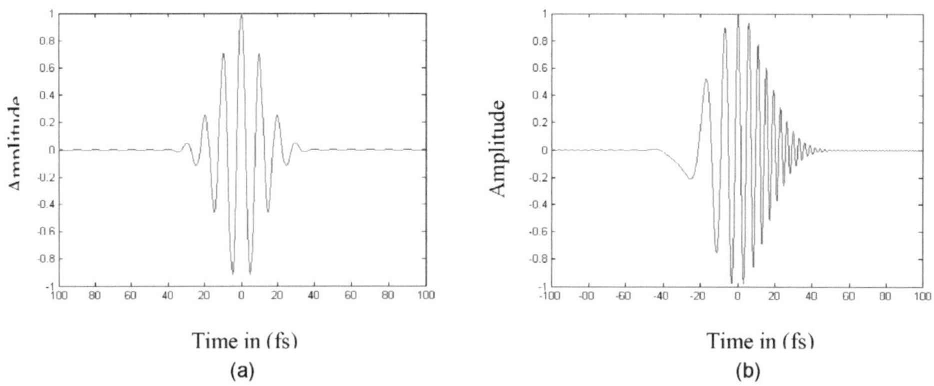

Figure 11.
(a) The electric field of an ultra-short lasers pulse, (b) the electric field of an ultrashortlasers pulse with a strong positive chirp.

An ultrashort pulse of light will lengthen after it has passed through glass as the index of refraction, which dictates the speed of light in the material, depends nonlinearly on the wavelength of the light. The wavelength of an ultrashort pulse of light is formed from the distribution of wavelengths either side of the center wavelength with the width of this distribution inversely proportional to the pulse duration.

2.5.2.1 Phase and chirp

Instantaneous phase function of $E(t)$ can be described as the sum of temporal phase andproduct of carrier frequency with time by the relation $w(t) = \frac{d}{dt}\varnothing(t) + w_0 t$.

Carrier frequency w_0 has been choosen by minimizing of temporal variationof phase $\varnothing(t)$. The first deriviation of $w(t)$ is defined by temporally-dependentcarrier frequency as the result of applying the derivation we receive relation expended in series. Then carrier frequency time denotes quadratic chirp.

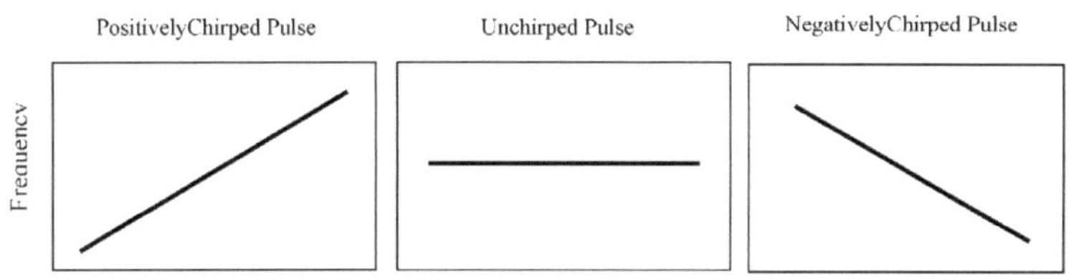

Positive chirp is when leading edge of pulse is red-shifted in relation to central wavelength and trailing edge is blue-shifted. Negative chirp happens in opposite case. Linear chirp, instantaneous frequency varies linearly with time. The presence of chirp results in significant different delays between the spectrally different components of laser pulse causing pulse broadening effect and leading to a duration-bandwidth.

Chirps always appear when ultrashort laser pulses propagate through a medium such as air or glass, where the spectral components of the pulse are subject to a different refractive index. This effect is called *Group Velocity Dispersion* (GVD).

2.5.3 Lens frequency domain description

The frequency representation is obtained from the time domain by a complex Fourier transform,

$$E(w) = \frac{1}{\sqrt{2\pi}} \int_{-\infty}^{+\infty} E(t).e^{-iwt}.dt. \tag{18}$$

Just as in the time domain, $\tilde{E}(w)$ can be written as:

$$\tilde{E}(w) = |\tilde{E}(w)|e^{i\varphi(w)}. \tag{19}$$

where $\varphi(w)$ now denotes the spectral phase. An inverse transform leads back to the time domain,

$$\tilde{E}(t) = \frac{1}{\sqrt{2\pi}} \int_{-\infty}^{+\infty} \tilde{E}(t).e^{iwt}.dw. \tag{20}$$

From Eq. (20) it is clear that $\tilde{E}(t)$ can be seen as a superposition of monochromatic waves. A common procedure is to employ Taylor expansion

$$\varnothing(w) = \varnothing_0 + \sum_{n=1}^{\infty} \frac{1}{n!} a_n (w - w_0)^n \text{ with } a_n = \left.\frac{d^n \varnothing}{dw^n}\right|_{w=w_0}. \tag{21}$$

$\varnothing(w)$ is spectral phase, w is pulsation, and a_n is variation of the spectral phase with pulsation.

It can be seen by inserting this Taylor expansion into Eq. (21) that the first, two terms will not change the temporal profile of the pulse.

2.6 Ultrashort pulse measurement techniques

The most commonly used technique of ultrashort laser pulse width measurement is concerned with the study of the temporal intensity profile $I(t)$ through its second-order correlation function that is obtained by the second-harmonic generation. Ultrashort - pulse characterization techniques, such as the numerous variants of frequency resolved optical gating (FROG) and spectral phase interferometry for direct electric-field reconstruction (SPIDER), fail to fully determine the relative phases of wellseparated frequency components. If well-separated frequency components are also characterized gate pulses are used [17].

2.6.1 Non-interferometric techniques

2.6.1.1 Intensity autocorrelation

Complex electric field $E(t)$ corresponds to intensity $I(t) = |E(t)|^2$ and an intensityautocorrelation function is defined by

$$A(\tau) = \int_{-\infty}^{+\infty} I(t)I(t+\tau)dt \tag{22}$$

$A(\tau)$ is quadrature detection

I(t) is intensity and E(t) is electrical field

Two parallel beams with a variable delay are generated, then focused into a second-harmonic-generation crystal to obtain a signal proportional to $E(t) + E(t+\tau)$. Only the beam propagating on the optical axis, proportional to the cross product $E(t)E(t-\tau)$ is retained. This signal is then recorded by a slow detector, which measures

$$I(\tau)=\int_{-\infty}^{+\infty}|E(t)E(t-\tau)|^{2}dt=\int_{-\infty}^{+\infty}I(t)I(t-\tau)dt \tag{23}$$

$I(\tau)$ is exactly the intensity autocorrelation $A(\tau)$.

This numerical factor, which depends on the shape of the pulse, is sometimes called the deconvolution factor. If this factor is known, or assumed, the time duration (intensity width) of a pulse can be measured using an intensity autocorrelation. However, the phase cannot be measured [17, 18].

2.6.1.2 FROG

The technique of Frequency-Resolved Optical Gating (FROG) has been introduced by Trebino and coworkers. In FROG technique signal E_1 has been temporally shifted about τ through time-delay element in respect with signal E_2. Then, two signals have been in nonlinear medium non-interferometrically overlapped. As the result of SFG or DFG process (at the efficient phase matching conversion) one receive the FROG signal.

In construction with process II also collinear geometry is possible.

$$E_{FROG}(\Omega,\tau)\propto\int_{-\infty}^{+\infty}E_1(t-\tau)E_2(t)\exp{(i\Omega t)}dt \tag{24}$$

$$\propto\int_{-\infty}^{+\infty}\tilde{E}_1(w)\tilde{E}_2(\Omega-w)\exp{(iw\tau)}dw \tag{25}$$

$$\propto\int_{-\infty}^{+\infty}dt\int_{-\infty}^{+\infty}\tilde{E}_1(w)E_2(t)\exp{(-iwt)}\exp{(i\Omega t+iw\tau)}dw \tag{26}$$

The spectral intensity

$$I_{FROG}(\Omega,\tau)\propto|E_{FROG}(\Omega,\tau)|^{2} \tag{27}$$

Where:

τ is delay between two temporal electrical field $E_1(t)$ and $E_2(t)$

Ω is delay between two spectral electrical field $E_1(w)$ and $E_2(w)$

$I_{FROG}(\Omega,\tau)$ is called FROG-trace. Relations (25), (26) are only the mathematical representation of Eq. (24). The task of receiving unknown complex signals E_1 and E_2 from measured FROG trace is known as the FROG reconstruction problem.

The resulting trace of intensity versus frequency and delay is related to the pulse's spectrogram, a visually intuitive transform containing both time and frequency information. Using phase retrieval concepts that the FROG trace yields the full intensity $I(t)$ and phase $\varnothing(t)$ of an arbitrary ultrashort pulse. As has been already mentioned, several schemes and methods exist for frequency-resolved optical gating as a technique for the full characterization of ultrashort optical signals as complex electric fields [14, 19].

2.6.1.2.1 Examples

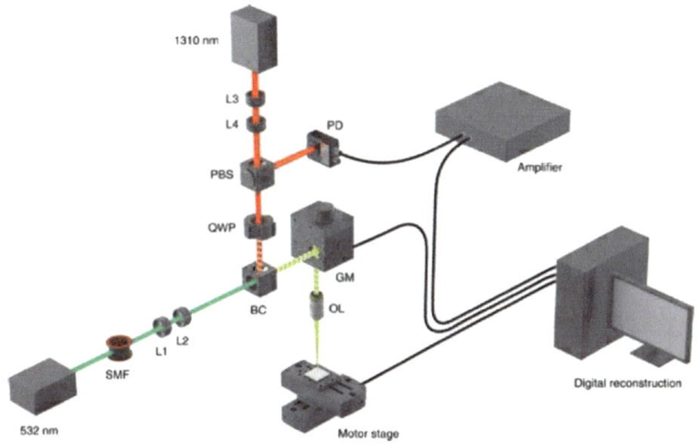

Experimental setup. PARS microscopy with 532-nm excitation and 1310-nm integration beams. BC, beam combiner; GM, galvanometer mirror; L, lens; OL, objective lens; PBS, polarized beam splitter; PD, photodiode; QWP, quarter wave plate; SMF, single mode fiber.

2.6.2 Interferometric techniques

2.6.2.1 Interferometric autocorrelation

Setup for an interferometric autocorrelator is similar to the field autocorrelator above, with the following optics added: L: converging lens, SHG: secondharmonic generation crystal, F: spectral filter to block the fundamental wavelength. A nonlinear crystal can be used to generate the second harmonic at the output of a Michelson interferometer in a collinear geometry. In this case, the signal recorded by a slow detector (**Figure 12**).

$$I(\tau) = \int_{-\infty}^{+\infty} |E(t) + E(t-\tau)|^2 dt \tag{28}$$

$I(\tau)$ is called the interferometric autocorrelation. It contains some information about the phase of the pulse: the fringes in the autocorrelation trace wash out as the spectral phase becomes more complex (**Figure 13**).

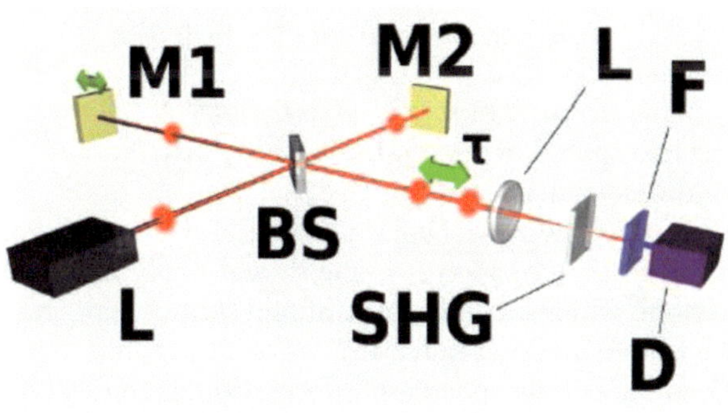

Figure 12.
Setup for an interferometric autocorrelator, similar to the field autocorrelator above, with the following optics added: L: Converging lens, SHG: Second-harmonicgeneration crystal, F: Spectral filter to block the fundamental wavelength.

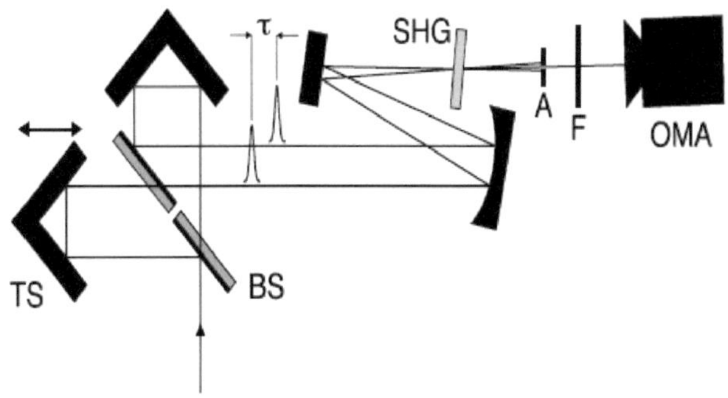

Figure 13.
FROG reconstruction scheme. When both E_1 and E_2 have been unknown, then we dealwith blind-FROG problem. When $E_1 = E_2$ then we have to do with SHG-FROG problem.

2.6.2.2 Spider

The Spectral Phase Interferometry for Direct Electric-field Reconstruction technique (SPIDER) is based on spectral interferometry and needs no components which has to be shifted over the measurement process. From the signal $E(t)$ that should be characterized, the copy-signal is being generated by beam splitter

$$E(t-\tau)\,exp\,[iw_0\tau] \tag{29}$$

The time between the signal and copy itself has been established through fixed position at the optical delay-line. Then the copy of signal goes through phase filter (dispersive medium, for instance SF10 glass), so arises the signal

$$E_M(t) = F^{-1}\{\tilde{E}(w)\,exp\,[i\varnothing_M(w)]\} \tag{30}$$

E_M electrical field at point M
$\varnothing_M$: phase of electrical field
Through the phase filter electric field $\tilde{E}(w)$ get additional spectral phase $\varnothing_M(w)$, which corresponds to temporal extension $E(t)$. From the signals

$$E_M(t)\ and\ E(t-\tau)\,exp\,[iw_0\tau] + E(t) \tag{31}$$

SFG-Signal can be created

$$E_{SFG}(t) \propto E_M(t)\{E(t-\tau)\,exp\,[iw_0\tau] + E(t)\} \tag{32}$$

$$= E_M(t)E(t-\tau)\,exp\,[iw_0\tau] + E_M(t)E(t) \tag{33}$$

As square law detectors are not sensitive to the phase, the measurement of the intensity (whether it is spatial or spectral) is an easy task but the measurement of the phase needs indirect solutions (**Figure 14**).

Spectral interferometry allows us obtain difference between two spectral phases. To the spectral interferometry spectrum, one should apply fast Fourier transform and as the product one will achieve in form of one center peak and two sidebands lower peaks in the time domain (**Figure 15**).

Centered peak contains only spectrum information. One filter out two peaks and from existing one can receive spectral phase difference by applied inverse fast Fourier transform. To the main advantages of the SPIDER method can be counted following properties: pulse retrieval is direct (non-iterative), minimal data are required: only one spectrum yields spectral phase [20].

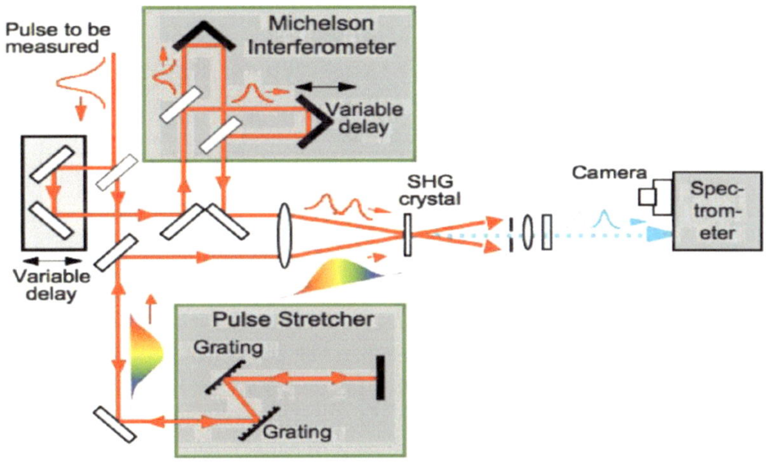

Figure 14.
Main steps of SPIDER technique.

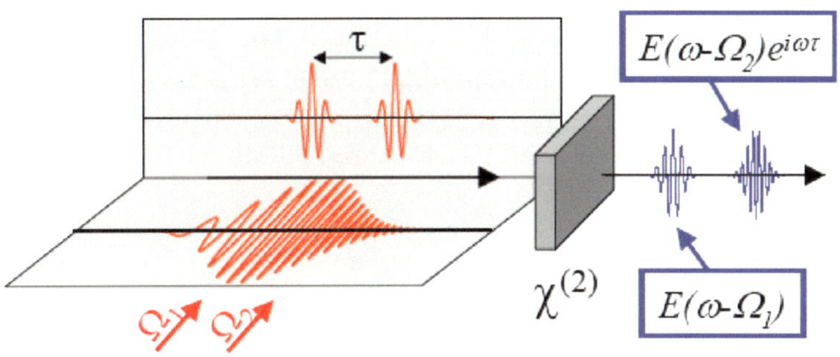

Figure 15.
Generation of two sheared replicas of the input pulse by non-linear interaction with a chirped pulse.

2.6.2.2.1 Examples

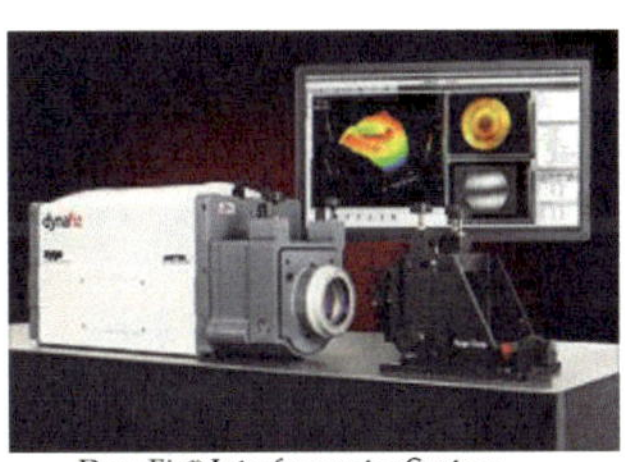

DynaFiz® Interferometer System

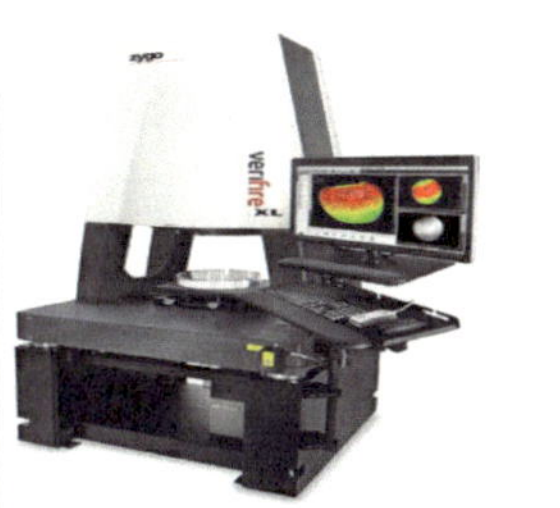

Verifire™ *XL* interferometer system

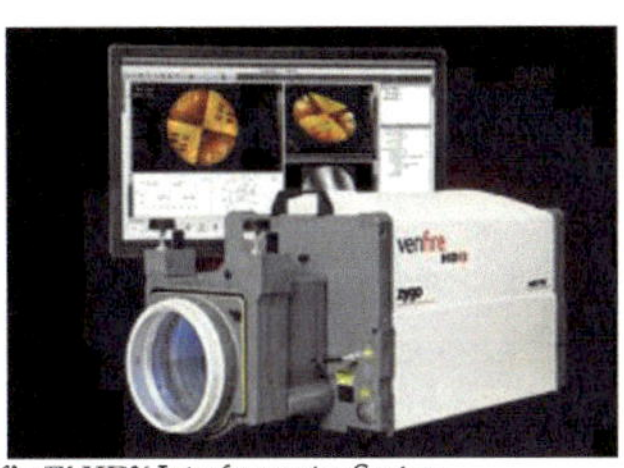

Verifire™ HDX Interferometer System

3. Propagation of a light pulse in a transparent medium

The frequency Fourier transform of a Gaussian pulse has already been given as

$$E(w) = exp\left(\frac{-(w - w_0)^2}{4.\Gamma}\right). \tag{34}$$

$\Gamma = \frac{2 log 2}{\Delta_0^2}$ is coefficient of function Gaussian

An ultrashort Fourier limited pulse has a broad spectrum and no chirp; when it propagates a distance through a transparent medium, the medium introduces a dispersion to the pulse inducing an increase in the pulse duration. To investigate and determine the dispersion, we assume a Gaussian shape for the pulse. The electric field of the pulse is given as Eq. (35).

After the pulse has propagated a distance z, its spectrum is modified to

$$E(w,z) = E(w)\, exp\,[\pm ik(w)z], k(w) = \frac{n(w).w}{c}, \tag{35}$$

where k(w) is now a frequency-dependent propagation factor. In order to allow for a partial analytical calculation of the propagation effects, the propagation factor is rewritten using a Taylor expansion as a function of the angular frequency, assuming that $\Delta w \ll w_0$ (this condition is only weakly true for the shortest pulses). Applying the Taylor expansion to Eq. (37), the pulse spectrum becomes.

Δw is bandwidth of the pulse and w_0 is central pulsation

$$k(w) = k(w_0) + k'(w - w_0) + \frac{1}{2}k''(w - w_0)^2 + \ldots, \tag{36}$$

where $k' = \left(\frac{dk(w)}{dw}\right)_{w_0}$. and $k'' = \left(\frac{d^2k(w)}{dw^2}\right)_{w_0}$,

$$E(w,z) = exp\left[-ik(w_0)z - ik'z(w - w_0) - \left(\frac{1}{4\Gamma} + \frac{i}{2}k''\right)(w - w_0)^2\right]. \tag{37}$$

The time evolution of the electric field in the pulse is then derived from the calculation of the inverse Fourier transform of Eq. (39),

$$e(t,z) = \int_{-\infty}^{+\infty} E(w,z).e^{-iwt}dw \tag{38}$$

so that

$$e(t,z) = \sqrt{\frac{\Gamma(z)}{\pi}}.exp\left[iw_0\left(t - \frac{z}{V_\varnothing(w_0)}\right)\right] \times exp\left[-\Gamma(z)\left(t - \frac{z}{V_g(w_0)}\right)^2\right] \tag{39}$$

Where

$$V_\varnothing(w_0) = \left(\frac{w}{k}\right)_{w_0}, V_g(w_0) = \left(\frac{dw}{dk}\right)_{w_0}, 1/(\Gamma(z) = 1/\Gamma + 2ik''z. \tag{40}$$

In the first exponential term of Eq. 40, it can be observed that the phase of the central frequency w_0 is delayed by an amount $\frac{z}{V_\varnothing}$ after propagation over a distance z. Because the phase is not a measurable quantity, this effect has no observable consequence. The phase velocity $V_{\varnothing}(w_0)$ measures the propagation speed of the plane wave components of the pulse in the medium. These plane waves do not carry any information, because of their infinite duration.

The second term in Eq. 40 shows that, after propagation over a distance z, the pulse keeps a Gaussian envelope. This envelope is delayed by an amount z/V_g, V_g being the group velocity. The second term in Eq. 40 also shows that the pulse envelope is distorted during its propagation because its form factor $\Gamma(z)$, defined as

Depends on the angular frequency w through $k''(w)$,

$$k'' = \left(d^2k/dw^2\right)_{w_0} = \frac{d}{dw}\left(\frac{1}{V_g}\right)_{w_0}. \tag{41}$$

This term is called the "Group Velocity Dispersion". The temporal width of the pulse at point z:

$$\Delta\tau_z = \Delta\tau_0\sqrt{1 + 4.(\Gamma.k''z)^2}. \tag{42}$$

with $k'' = \frac{\lambda^3}{2.\pi.c^2}\frac{d^2n}{d\lambda^2}\Gamma = \frac{2\log 2}{\Delta_0^2}$,

$\Delta\tau_0$ initial pulse before propagation inside the medium

3.1 Application in litharge index SF56 medium

In optical materials, the refractive index is frequency dependent. This dependence can be calculated for a given material using a Sellmeier equation, typically of the form

$$n^2(w) = 1 + \sum_{i=1}^{m}\frac{B_i w_i^2}{w_i^2 - w^2} \tag{43}$$

B_i: is the coefficients of glass see in **Table 1**

The index of litharge SF56 is given by the following expression (43):

where w_i is the frequency of resonance and B_i is the amplitude of resonance. In the case of optical fibers, the parameters w_i and B_i are obtained experimentally by fitting the measured dispersion curves to Eq. (44) with $m = 3$ and depend on the core constituents [21].

3.2 Parameter of dispersion

An ultrashort Fourier limited pulse has a broad spectrum and no chirp; whenit propagates a distance through a transparent medium, the medium introduces dispersion to the pulse inducing an increase in the pulse duration. We consider dispersions of orders two. The pulse broadens on propagation because of group velocity dispersion (GVD).

In summary, the propagation of a short optical pulse through transparent medium results in a delay of the pulse, a duration broadening and a frequency chirp. All these phenomena are increase with distance of propagation. We shown in

B_i	***1.81651732***	***0.428893631***	***1.07186278***
$\lambda_i(\mu m)$	*0.0143704198*	*0.0592801172*	*121.419942*

Table 1.
Parameters for litharge SF56 glasses.

Figure 16 that the duration broadening is not linear for ultrashort pulse. Specially under 70 fs or less. The Eq. (43) is not applicable for pulse less than 70 fs. For minimize these parameters we introduce the nonlinear phenomena as Self phase modulation (SPM), Soliton pulse, dispersion compensate fiber.

3.3 Group velocity dispersion

The Group Velocity Dispersion (GVD) is defined as the propagation of different frequency components at different speeds through a dispersive medium. This is due to the wavelength-dependent index of refraction of the dispersive material [22].

$$\varphi(w) = \varphi(w_0) + (w - w_0)\frac{d\varphi}{dw}\bigg| + \frac{1}{2!}(w - w_0)^2\frac{d^2\varphi}{dw^2}\bigg|_{w=w_0} + \ldots + \frac{1}{n!}(w - w_0)^n\frac{d^n\varphi}{dw^n}\bigg|_{w=\Omega} \tag{44}$$

$$\begin{cases} \varphi(\lambda) = \dfrac{2\pi}{\lambda}n(\lambda)z \\ \dfrac{d\lambda}{dw} = -\dfrac{\lambda^2}{2\pi c} \\ \dfrac{d\varphi}{dw} = -\dfrac{z}{c}\left[\dfrac{dn}{d\lambda} - n\right] \\ \dfrac{d^2\varphi}{dw^2} = +\dfrac{\lambda^3}{4\pi^3c^2}\dfrac{d^2n}{d\lambda^2}z \\ \dfrac{d^3\varphi}{dw^3} = -\dfrac{\lambda^4}{4\pi^2c^3}\left[3\dfrac{d^2n}{d\lambda^2} + \lambda\dfrac{d^3n}{d\lambda^3}\right]z \\ \dfrac{d^4\varphi}{dw^4} = +\dfrac{\lambda^5}{8\pi^3c^4}\left[12\dfrac{d^2n}{d\lambda^2} + 8\lambda\dfrac{d^3n}{d\lambda^3} + \lambda^2\dfrac{d^4n}{d\lambda^4}\right]z \\ \dfrac{d^5\varphi}{dw^5} = -\dfrac{\lambda^6}{16\pi^4c^5}\left[60\dfrac{d^2n}{d\lambda^2} + 60\lambda\dfrac{d^3n}{d\lambda^3} + 15\lambda^2\dfrac{d^4n}{d\lambda^4} + \lambda^3\dfrac{d^5n}{d\lambda^5}\right]z \\ \dfrac{d^6\varphi}{dw^6} = +\dfrac{\lambda^7}{32\pi^5c^6}\left[360\dfrac{d^2n}{d\lambda^2} + 480\lambda\dfrac{d^3n}{d\lambda^3} + 180\lambda^2\dfrac{d^4n}{d\lambda^4} + 24\lambda^3\dfrac{d^5n}{d\lambda^5} + \lambda^4\dfrac{d^6n}{d\lambda^6}\right]z \end{cases} \tag{45}$$

Itseemsto methatwe can write $\phi^{(i)} = \frac{d^i\phi}{dw^i}$ asarecurrence, giving $\phi^{(i)}$ basedonder-ivativesoforderi, the index of refraction. Matrix form, wecan writet

$$\begin{bmatrix} \varnothing^{(2)} \\ \varnothing^{(3)} \\ \varnothing^{(4)} \\ \varnothing^{(5)} \\ \varnothing^{(6)} \end{bmatrix} = (-1)^n 2.\pi.z\left[\frac{\lambda}{2.\pi.c}\right]^n \begin{bmatrix} 1 & 0 & 0 & 0 & 0 \\ 3 & 1 & 0 & 0 & 0 \\ 12 & 8 & 1 & 0 & 0 \\ 60 & 60 & 15 & 1 & 0 \\ 360 & 480 & 180 & 24 & 1 \end{bmatrix} \tag{46}$$

The various terms of the Taylor expansion to order n can be written in the shape of a matrix [A], which's we can express various terms A $_{ij}$.

$$\varnothing(w) = \varnothing(w_0) + (w - w_0)\varnothing^{(1)} + \sum_{i=2}^{p}\frac{1}{i!}(w - w_0)^i\varnothing^{(i)}\big|_{w=w_0}. \tag{47}$$

$\varnothing(w)$is Taylor series of phase and $\varnothing^{(p)}$ is the orders of Taylor series

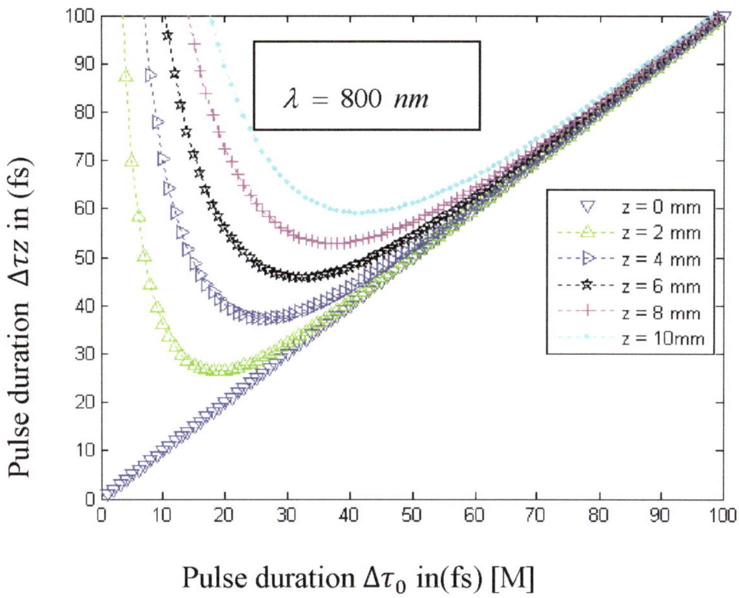

Figure 16.
Temporal broadening of the transform- limited pulse for differentvalues of the propagation distance z.

$$\varnothing^{(p)} = (-1)^p.2\pi.z\left[\frac{\lambda}{2.\pi.c}\right]^p \sum_{j=2}^{p} \lambda^{j-1} A(p-1, j-1) n^{(j)} \text{ with } p > 2 \tag{48}$$

Analytically known and experimentally observed propagation affects such as spectral shift, pulse broadening and asymmetry in dispersive media can be easily brought out in the simulation using formalism presented here. In addition, such studies can be extended to pulses of arbitrary temporal shape without any further algorithmic complexity by numerical simulation. Higher order dispersion effects can be handled easily in the numerical simulation unlike in the case of analytical calculation (**Figure 17**) [22].

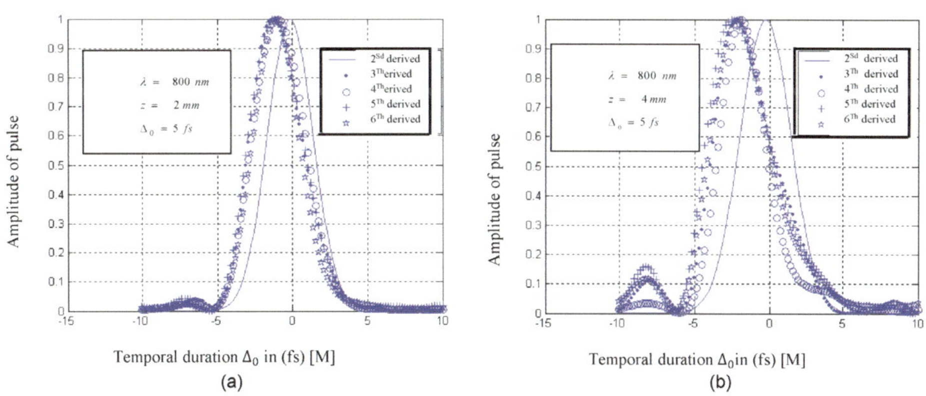

Figure 17.
(a) The pulse broadens on propagation as a result of group velocity dispersion (GVD) (b) the pulse shape is no longer Gaussian, and it becomes asymmetric due to higher order dispersions.

4. Time-frequency decomposition

4.1 Wavelet theory

The wavelets are very particular elementary functions, these are the shortest vibrations and most elementary that one can consider. One can say that the wavelet east carries out a zooming on any interesting phenomenon of the signal which place on a small scale in the vicinity of the point considered [23].

4.2 Wavelet techniques

Starting with a signal $e(t)$, in plane $z = 0$, we define wavelet centered at Ω by (**Figure 18**):

$$\theta(\Omega) = E(w).exp\left[-\frac{(w-\Omega)^2}{4\gamma}\right], \text{with } E(w) = \frac{E_0}{2.\pi}\sqrt{\frac{\pi}{\Gamma}}exp\left[\frac{(w-w_0)^2}{4.\Gamma}\right], \quad (49)$$

$$\theta(t, z=0) = TF\{\theta(\Omega, z=0)\} \quad (50)$$

TF it is the Fourier Transform equation
z is the distance of propagation

$$\theta(t, z=0) = E_0\sqrt{\frac{\gamma}{\gamma+\Gamma}}.exp\left[\frac{-(w_0-\Omega)^2}{4(\gamma+\Gamma)}\right].exp\left[-\frac{\gamma\Gamma}{\gamma+\Gamma}t^2\right].exp\left[j\frac{\gamma w_0+\Gamma\Omega}{\gamma+\Gamma}t\right] \quad (51)$$

- In time, the pulse is also Gaussian, of parameter $\frac{\gamma\Gamma}{\gamma+\Gamma}$.
- The maximum of amplitude of the wavelet $\theta(t, z=0)$ vary with Ω, center frequency of analysis on Gaussian of parameter $\gamma + \Gamma$.
- The signal propagates in the positive z direction in a linear dispersive and transparent medium, which fills the half space $z > 0$ and whose refractive index is $n(w)$. After propagation, the wavelet $\theta(\Omega, x)$ may be written as

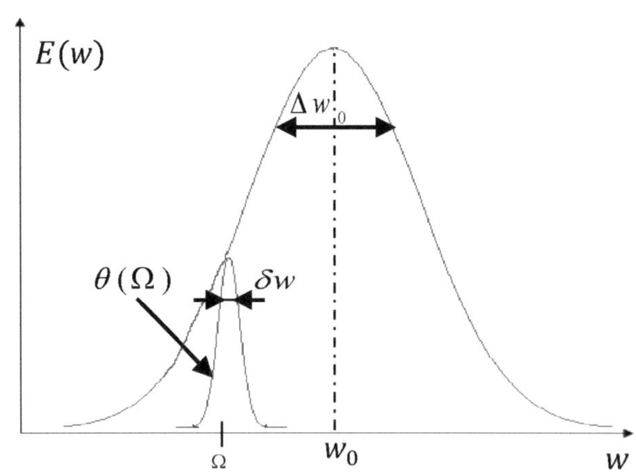

Figure 18.
Gaussian envelope decomposed on a number of waveletwecalculates the electric field associated with the wavelet $\theta(\Omega, z = 0)$.

$$\theta(\Omega, z) = \frac{E_0}{2.\sqrt{\pi\gamma}} E(w).exp\left[-\frac{(w-\Omega)^2}{4\gamma}\right].exp\left[j\varnothing(w)\right]. \tag{52}$$

As already mentioned, $\tau_{wavelet}$ is large enough to ensure that analyzing function has only non negligible values over a spectral range lying in the neighborhood of Ω in (**Figure 19**). Under these circumstances, we have

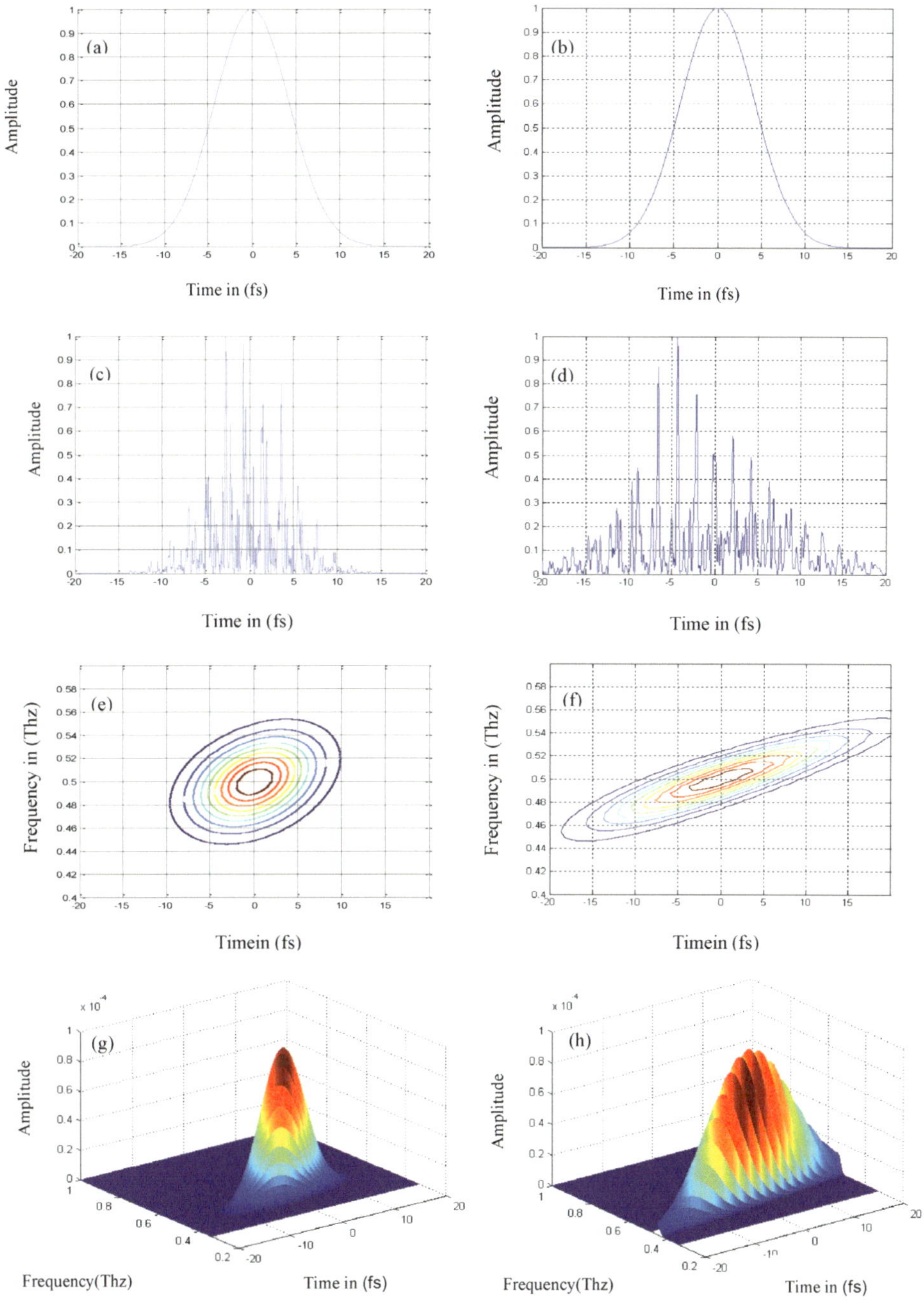

Figure 19.
(a) Initial pulse, (c) pulse after propagation of the 10 cm in glass SF56 (e) contour of the wavelet, (g) the wavelet representation, (b) initial pulse, (d) pulse after propagation of the 10 cm in the silica medium, (f) contour of the wavelet, (h) the wavelet representation [22].

$$\varnothing(w) = \varnothing(\Omega) + (w - \Omega)\frac{d\varnothing}{dw}\bigg| + \frac{1}{2!}(w - \Omega)^2\frac{d^2\varnothing}{dw^2}\bigg|_{w=\Omega} + \dots + \frac{1}{n!}(w - \Omega)^n\frac{d^n\varnothing}{dw^n}\bigg|_{w=\Omega} + \theta(w)._{w=\Omega} \tag{53}$$

Neglecting the higher terms in Eq. (49):

$$\varnothing(w) = \varnothing(\Omega) + (w - \Omega)\frac{d\varnothing}{dw}\bigg|_{w=\Omega} + \frac{1}{2!}(w - \Omega)^2\frac{d^2\varnothing}{dw^2}\bigg|_{w=\Omega} + \theta(w). \tag{54}$$

$$\theta(\Omega, z) = \frac{E_0}{2.\sqrt{\pi\gamma}}\sqrt{\frac{\pi}{\Gamma}}exp\left[-\frac{(w - w_0)^2}{4\Gamma}\right] .\,exp\left[-\frac{(w - \Omega)^2}{4\gamma}\right].\,exp\left[j\varnothing^{(0)} + j(w - \Omega)\varnothing^{(1)} + \frac{1}{2}j(w - \Omega)^2.\varnothing^{(2)}\right] \tag{55}$$

$$\theta(t, z) = \frac{1}{2\pi}\int_{-\infty}^{+\infty}\theta(\Omega, z).\,exp\,(jwt)dw \tag{56}$$

We calculates the temporal electric field associated with the wavelet $\theta(\Omega, z)$.

$$\theta(t, z) = \frac{1}{2\pi}\frac{E_0}{2.\sqrt{\pi\gamma}}\sqrt{\frac{\pi}{\Gamma}}e^{\left[-\frac{(\Omega - w_0)^2}{4\Gamma}\right]}e^{\left(j\varnothing^{(0)}\right)} \times e^{-\left[\frac{1}{4\Gamma} + \frac{1}{4\gamma} - \frac{1}{2}j\varnothing^{(2)}\right]\Omega^2}.e^{\left[\frac{(\Omega - w_0)}{2\Gamma} - j\varnothing^{(1)}\right]\Omega} \times\left(\int_{-\infty}^{+\infty}e^{-\left[\frac{1}{4\Gamma} + \frac{1}{4\gamma} - \frac{1}{2}j\varnothing^{(2)}\right]w^2}.e^{\left[\frac{1}{4\Gamma} + \frac{1}{4\gamma} - \frac{1}{2}j\varnothing^{(2)}\right]2\Omega w} \times e^{\left[-\frac{(\Omega - w_0)^2}{2\Gamma} - j\varnothing^{(1)}\right]}.e^{jwt}\right)dw \tag{57}$$

The amplitude of the incident Ω wavelet is given from Eq. (58) by

$$\theta(t, z) = \frac{E_0}{2.\sqrt{\pi\gamma}}\sqrt{\frac{\Gamma(z)}{\Gamma}}.\,exp\left(j\varnothing^{(0)}\right)exp\left(-\Gamma(z)\left[t + \frac{z}{V_g(\Omega)}\right]^2\right) \times exp\left(-\frac{(\Omega - w_0)^2}{4\Gamma}\left[1 - \frac{\Gamma(z)}{\Gamma}\right]\right).\,exp\left[j\left(1 - \frac{\Gamma(z)}{\Gamma}\right)\Omega + \frac{\Gamma(z)}{\Gamma}w_0\right]\left(t + \frac{z}{V_g(\Omega)}\right). \tag{58}$$

This wavelet is characterized by a Gaussian envelope. This decomposition is valid only for the values of Δw much larger than δw ($\Delta w \gg \delta w$).

The delay of group of the wavelet $\left[t + \frac{z}{V_g(\Omega)}\right]$ is characterized by a Gaussian envelope which is the temporal width.

The delay of group of the wavelet is inversely proportional to the velocity of group its envelope propagates without deformation [22].

4.3 Simulations

4.3.1 Parameters of the simulations

Pulse initial: $\Delta\tau_0 = 20\,fs$
(Wavelength)$\lambda = 800\ nm$

Pulse of the wavelet: $\Delta\tau_{wavelet} = 1000\,fs$

Longer of the medium: $z = 10\ cm$

To describe the propagation of the pulse, we only consider the propagation of the maximum of each wavelet in a three dimensional representation:

Figure 19(c) shows that when pulse propagate inside SF56 glass present minor dispersion and distortion compared to the silica fiber in **Figure 19d**.

Figure 19(e) shows that the SF56 glass resist in temporal domain than silica fiber as shown in **Figure 19(f)**. but, in frequency domain the both SF 56 glass and silica fiber are the same modification.

Figure 19(g) and **(h)** shown the amplitude of the wavelet

5. Conclusion

Generating ultrashort light pulses requires a laser to operate in a particular regime, called mode-locking which many be illustrated either in the frequency or the time domain. Depending on the particular case, one description is much more intuitive than the other and we have chosen to present the simpler approach. The generation of femtosecond laser pulses via mode locking is described in simple physical terms. As femtosecond laser pulses can be generated directly from a wide variety of lasers with wavelengths ranging from the ultraviolet to the infrared no attempt is made to cover different technical approaches.

The ability to accurately measure ultrashort laser pulses is essential to creating, using, and improving them, but the technology for their measurement has consistently lagged behind that for their generation. The result has been a long and sometimes quite painful history of attempts—and failures—to measure these exotic and ephemeral events. The reason is that many pulse-measurement techniques have suffered from, and continue to suffer from, a wide range of complications, including the presence of ambiguities, insufficient temporal and/or spectral resolution and/or range, an inherent inability to measure the complete pulse intensity and/or phase, an inability to measure complex pulses, and misleading results due to the loss of information due to idiosyncrasies of the technique or multishot averages over different pulses.

Finally, we have demonstrated here the possible decomposition of an ultrashort pulse into an infinite number of longer Fourier transform limited wavelets which propagate without any deformation through a dispersive medium. After propagation through the medium, the pulse may be visualized in a three-dimensional representation by the locus of the wavelet maxima.

Author details

Mounir Khelladi
Abou-bekrBelkaid University of the Tlemcen, Algeria

*Address all correspondence to: mo.khelladi@gmail.com

References

[1] Florența A.C, Dynamics of Ultra-short Laser Pulse Interaction with Solids at the Origin of Nanoscale Surface Modification" geboren am 16.11.1971 in Berca, Rumänien Cottbus (2006)

[2] Chritov, I.P., Propagation of femtosecond light pulses, *Opt. Commun.* 53, 364–366.

[3] Sheppard, C.R, Xianosong, G, Free-space propagation of femtosecondligthpulses, *Opt. Commun.* 133, 1–6. (1997)

[4] Agrawal,G.P, Spectrum-induced changes in diffraction of pulsed beams, *Opt. Commun.* 157, 52–56. (1998).

[5] Kaplan, A.E., Diffraction-induced transformation of near cycle and subcycle pulses, J. *Opt. Soc. Am. B 15*, 951–956. (1998)

[6] Agrawal, G.P, Far-Field diffraction of pulsed optical beams in dispersive media, *Opt. Commun.* 167, 15–22. (1999).

[7] Porras,M.A, Propagation of single-cycle pulse light beams in dispersive media, *Phys. Rev. A* 60, 5069–5073. (1998).

[8] Ichikawa, H, Analysis of femtosecond-order optical pulses diffracted by periodic structure, *J. Opt. Soc. Am.* A 16, 299–304. (1999).

[9] Pietstun, R., Miller, D.B., Spatiotemporal control of ultrashort optical pulses by refractive-diffractive-dispersive structured optical elements, *Opt. let. 26*, 1373–1375. (2001).

[10] Kempe, M., Stamm, U., Wilhelmi, B., and Rudolph, W., Spatialand temporaltransformation of femtosecond laser pulses by lenses systems, J. *Opt. Soc. 38*, 1058–1064. (1999).

[11] Matei,G.O,Gili, M.A., Spherical aberration in spatial and temporal transformation of femtosecond laser pulses by lenses and lens systems, *J. Opt. Soc. Am. B9*, 1158–1165. (1992).

[12] Fuchss,U., Zeintner, U.D., and Tunnermenn (A.),Ultrashort pulse propagation in complex optical systems, *Opt. Express 13*, 3852–3861. (2005).

[13] Leonid.S., Ferrari.A, and Bertolotti. M, Diffraction of a time Gaussian-shaped pulsed plane wave from a slit, *Pure. Appl. Opt5*, 349–353. (1996).

[14] Moulton.P.f, Quant. Electron. Conf., Munich, Germany, June (1982).

[15] Holzwarth,"Measuring the Frequency of Light using Femtosecond Laser Pulses" aus Stuttgart den 21. December (2000).

[16] Rulliere, C, Femtosecond laser pulses: principles and Experiments, *Springer,* (1998).

[17] Joanna.M, "Seutp of very advanced for phase and amplitude reconstruction of electric field (VAMPIRE)"Master Thesis submitted to the Faculty for the Natural Sciences and for Mathematics of the Rostock University Germany (2007).

[18] Abdolah.M.K, "Manipulation and characterization of femtosecond laser pulses for cluster spectroscopy" vorgelegt von aus Teheran (Iran) July (2007).

[19] Greg.T, Andy. R Margaret.M, Measurement of 10 fs Laser Pulses. 1077-260X,IEEE (1996).

[20] Hofmann.A, Bestimmung des elektrischen Feldesultrakurzer Laserpulsemit SPIDER, Teil1: Theorie, Projektpraktikumsbericht, PhysikalischesInstitut, EP1, Universität Würzburg (2005).

[21] Jong Kook K. Investigation of high nonlinearity glass fibers for potential

applications in ultrafast nonlinear fiber devices, Dissertation submitted to the Faculty of the Virginia (2005).

[22] Khelladi.M, Seddiki.O, and Bendimerad.T, Time-Frequency Decomposition of an Ultrashort Pulse: Wavelet Decomposition, *Radioengineering ISSN*, April (2008) Vol.17, pp.1210–2512, N1

[23] Meyer,Y, Jaffort, S., Riol,O, waveletanalysis, *edition Française de scientifique American* (1987)

Chapter 2

Numerical Simulations of Detections, Experiments and Magnetic Field Hall Effect Analysis to Field Torsion

Francisco Bulnes, Juan Carlos García-Limón, Víctor Sánchez-Suárez and Luis Alfredo Ortiz-Dumas

Abstract

Field torsion models are considered from the experiments realized in electronic-dynamical devices and magnetic censoring of a Hall Effect sensor to detect torsion under electrical restricted conditions and space geometry. In this last point, are obtained 2D and 3D-models of torsion energy, which enclose the field theory concepts related with torsion, and open several possibilities of re-interpretations that can be useful in technological applications in the future. Likewise, are considered some measurements that evidence the torsion as field observable. Through geometrical models obtained from theorems and other results are demonstrated the conjectures required to understanding of torsion, as a geometrical and physics invariant most important in the deep study of the Universe. Also, applications in astrophysics and cosmology of these geometrical models are obtained to show Universe phenomena understudy of torsion.

Keywords: field torsion models, Hall effect sensor, spectral torsion, torsion detection, torsion energy, magnetic field, spining, spinors

1. Introduction

A deep study of torsion carry us to determine the spectral form of the torsion explored through energy signals that evidence the torsion as an primordial field born from the spins and fermion interaction of the matter-space interaction when the agitation of the space produces the fundamental material particles which create through duality particle/wave the matter in the Universe. This new spectral form that have been defined from the curvature energy concept [1, 2], and expressed as the value of integrals on cycles of a space, come from a generalization of curvature in analysis and integral geometry called integral curvature [3, 4], and obtained through co-cycles (values of the integrals) in a tempered distributions topological space.

After, and applying this new form of curvature to explore and measure several phenomena in the space–time, furthermore of establish a Universe theory through integrals, have been obtained curvature of the space–time in the macroscopic levels as well as microscopic levels, where in this second item, have been obtained results referent to the creation of gravity and matter in the Universe [5], considering the causes and origin of these and the evolution of the Universe, being the curvature energy $\kappa(\omega_1, \omega_2)$, an theoretical and practical element to consider in all modern Universe studies, and also the technological applications derived of these as corollaries of the great curvature energy theory.

As has been mentioned, the microscopic phenomena in the universe are the causes and are the things that give origin to the gravity and thus to the matter such as is known, through a large process in the Universe development. Likewise, an field observable that result of the microscopic interactions from particles spin level is the field torsion, which let in evidence much other space–time phenomena as the inflation, possibly the role of neutrinos, the baryongenesis, the proliferation of **H**-particles and the form of evolution of all sidereal bodies, which in their different steps show mechanism where curvature and torsion are geometrical invariants that give meaning to the evolution of these sidereal objects and possibly to explain the role of the Higgs boson, the meaning of the dark matter and more.

The present chapter will explain and expose some theoretical models, numerically obtained in 2D and 3D-dimensions of torsion and its meaning in field theory and the intrinsic study of torsion as field observable, as geometrical invariant and possible central concept to the new and future technologies that will be required to the human survival.

2. Numerical mathematical models: field torsion, torsion energy, spinors

The torsion is a double curvature of a space or body, which is result of interaction of two fields, where one field is the electromagnetic field and other a field relative to the matter, which is the gravitational field in a wide sense (even considering the quantum gravity).

Likewise, as mentioned in the introduction, the torsion born from the matter spins which under the action of the electromagnetic field produce spin-waves whose geometrical invariants are spinors in the invariant theory [6, 7] (see **Figure 1**).

This has been formulated in a first conjecture (**Figure 2**).

Conjecture 2.1. The curvature in spinor-twistor framework can be perceived with the appearing of the torsion and the anti-self-dual fields [7].

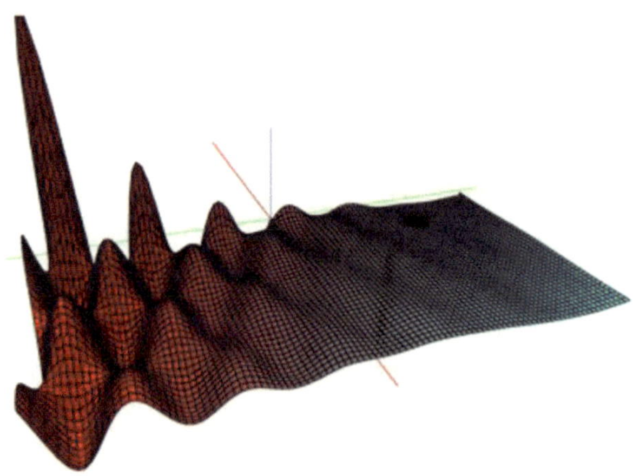

Figure 1.
Spinor image of spin-waves in torsion. This is a 2-dimensional model of spin-waves generated for a magnetic dilaton in a cylindrical-spiral trajectory of movement [8, 9]. In this 2-Dimensional surface $s(\omega_1, \omega_2)$ *have been considered a mesh of 100 points of density and the corresponding torsion model is obtained as cylindrical solution* Z = exp(0.5x)BesselY(0.1x, 2y)sin(4x) + BesselJ(0.6x, 6), *of the corresponding field equation* $(P + \chi T)\varphi = 0$.

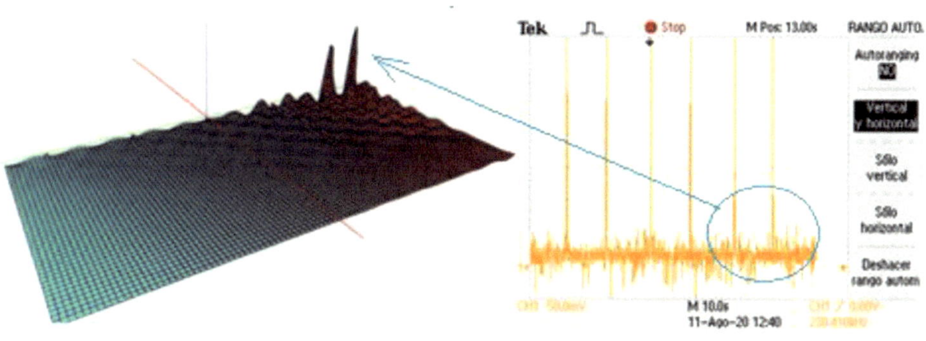

a) b)

Figure 2.
*(a). The corresponding solution to the field equation in cylindrical regime is Z = exp(−0.5(x + 1))BesselY (0.5x, 8y)sin(7x)BesselJ(0.3y,20). The rotation of cylinder was realized b). (b) Corresponding signals detected in the torsion detection and showed in the wave generator, under electronics measuring conditions: frequency of 235 kHz, wave sample of 13 seconds, voltage of 4 Volts (**Figure 3a**)).*

From a point of view of an electronics study [10] (using a magnetic Hall sensor[1]) was obtained an evidence of the existence of torsion as field observable, where was used a magnetic particle as dilaton [5, 6] moving through of a trajectory whose torsion is constant in all space [4]. Likewise, was obtained the signal[2]:

$$\tau(\omega) = \frac{1}{2\pi}\frac{b}{l^3}\int_{-0.5s}^{+0.5s} \Pi\left(\frac{t}{1}\right) e^{-j\omega t} dt = \frac{1}{2\pi}\left[\frac{(2.5V - \tilde{\alpha})b}{l^3}\right]\frac{\sin(0.5\omega)}{0.5\omega}, \tag{1}$$

which belongs to the signals set that evidence the torsion under permanent torsion conditions:

$$\left\{\frac{\sin\omega L}{\omega L}, \frac{\sinh\omega L}{\omega L}; \frac{\cos\omega L}{\omega L}, \frac{\cosh\omega L}{\omega L}\right\}, \forall L = \frac{n2\pi t}{\omega T}, T > 0 \tag{2}$$

---

1 **Lemma** [10]. We consider a sensor Hall device $\mathcal{L}^{H}_{Hall}$. The current deflection detected for the magnetic field change in the sensor produces per volume unit a torsion energy:

$$\tau = \frac{V}{2\pi}\frac{b}{(a^2 + b^2)}\frac{1}{l}\left(= \frac{\text{Volts}}{(\text{meter})^3}\right).$$

2 The torsion is detected with conditions of movement. Likewise, by the lemma 3. 1 [10], the produced magnetic field in the dilaton must be corresponding to

$$H = \frac{I}{2\pi}\frac{a^2}{l^3},$$

whose magnetic field of the dilaton must be decreasing in the cycles for seconds of the turns for that these are detectable by the Hall type sensor (with low velocity). Then is conditioned the signal the initial constant voltage signal

$$V_0 = \Pi\left(\frac{t}{13}\right) = \begin{cases} 2.5V - \alpha & |t| \leq 0.5s \\ 0 & |t| \geq 0.5s \end{cases},$$

which is a rectangular signal of conditioning. Here $\tilde{\alpha}$, is an amplifier factor of the voltage for be detectable.

In our experiments and for our case the spin/waves detection are very little, and need signals filters to obtain a signal in a time range of the 10 seconds, at least. The corresponding screen and 2-dimensional spinor surface $s(\omega_1, \omega_2)$ can be viewed in the **Figures 3a** and **4b**. The oscillations are little and the dilaton or particle reveals the torsion in the space–time [2, 4] under action of an electromagnetic field considering the kinematic invariance. Then chooses a trajectory in our electronic device of constant torsion (**Figures 3c** and **4**) [11].

Conjecture 2.2 (F. Bulnes) [9, 10]. The torsion is the geometrical invariant of the interaction between energy and space.

The appearing of waves in the space–time agitation is due the energy agitation of the particles that from their spins, are interchanging by the screw effect generated by the force $\delta(t_1, t_2)$ (see the **Figure 4**).

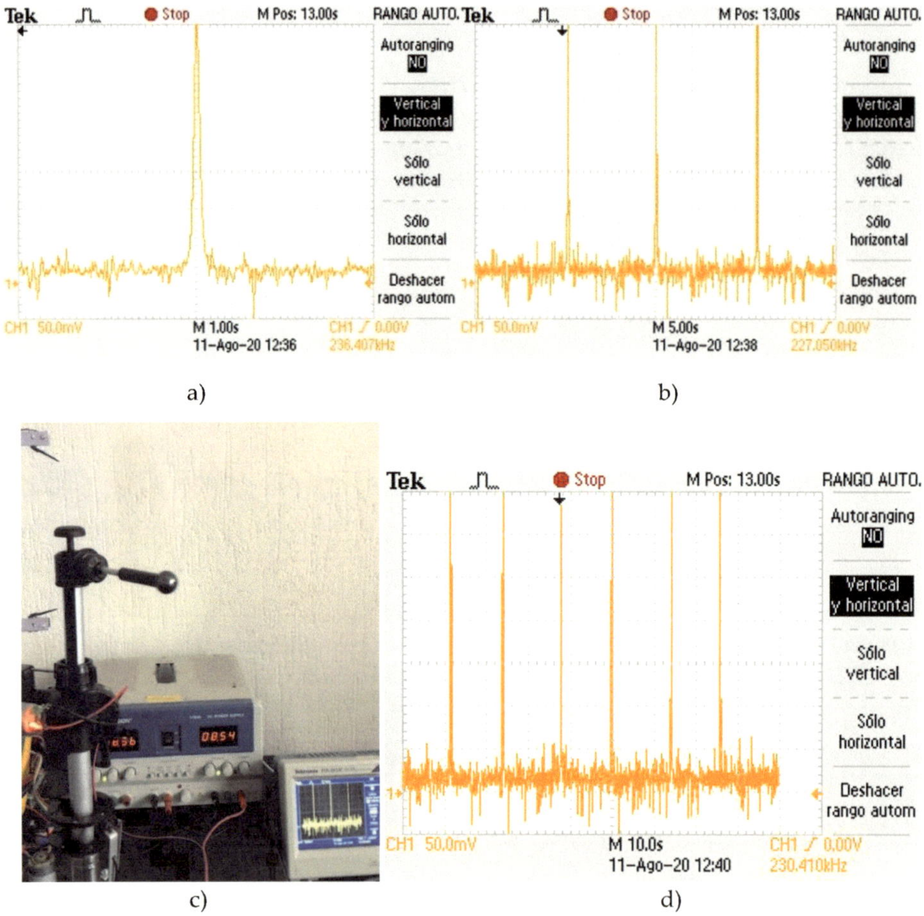

Figure 3.
(a) Energy spectra obtained for the conditioning of voltage signals. (b) This represents a quasi-periodic signal where in each 14 seconds to risk a pico-voltage V_p*, when the dilaton spins to a velocity of 12 rpm, where is observed that their amplitude is very major than the subsequent amplitudes until to risk a new maximum* V_p*. (c). The frequency* ω*, is maintained in a 227 to 236 kHz range. In the sensing process was realized a closer of the energy spectra 1 sec/div (figure a)) to 5 sec/div (figures b, c) and 10sc/div (figure d)) to 50 mV for the three cases. We consider a conditioning of the signal through the operational amplifier of instrumentation AD620, which permit us view an output voltage proportional to 1.4 mV per Gauss detected.*

DOI: http://dx.doi.org/10.5772/intechopen.96779

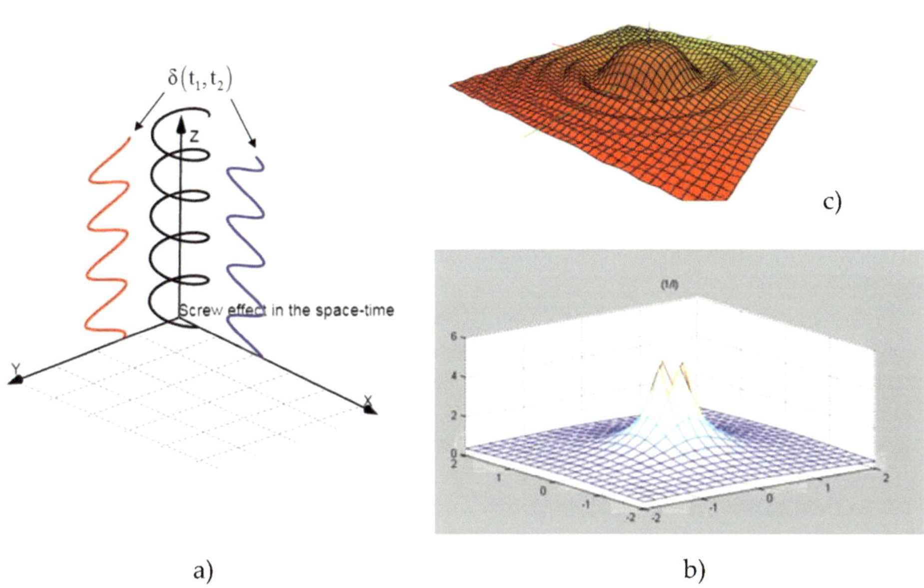

Figure 4.
(a). The screw effect in the space–time. (b). Mono-pole of electromagnetic field. This represents only $\delta(t_1, t_2)$*, of a field source in the space–time, whose torsion is established in the Z-axis. (c). 2-dimensional surface shows the field source with torsion effect waving the 2-dimensional surface around the source.*

3. Other cosmological phenomena re-interpretations related to 2D and 3D-torsion geometrical models and the signal analysis realized

In the studies of the Universe the creation of geometrical models or numerical simulation of the space–time phenomena is more complicated and obeys to a re-interpretation that obeys a modern field theories and cohomology theory of topological spaces, even the incorporation of microscopic theories of the Universe, because here are cause of all phenomena in the Universe.

However, as has been said the most important geometrical invariant is the curvature, being the torsion a second curvature.

For example, a re-interpretation that obeys a modern field theories and cohomology theory of topological spaces, with the incorporation of microscopic theories of the Universe can be established as follows.

We can create a 2D-numerical model of the screw effect and re-interpretation in the cosmology objects as black holes or sources as stars or behavior intersidereal magnetic alignment of galaxies, using 2-dimensional complex surfaces considering the Morera's and Cauchy-Goursat's theorems to be evaluate them and can be applied in an numerical program. For example, on singularities or poles in the space–time, considering the space–time a complex Riemannian manifold with singularities. This could represent the surface of the real part of the function $g(z) = \frac{z^2}{z-1}$. The moduli space of this point is less than 2 and thus lie inside one contour. Likewise, the contour integral can be split into two smaller integrals using the Cauchy-Goursat theorem having finally the contour integral [12] $\oint_C g(z)dz = \oint_C \left(0-\frac{1}{z-z_1}\right)dz = 0-2\pi i = -2\pi i$, (see the **Figure 5a**)). Likewise this value is a traditional cohomological functional element of $H^f(\Pi-\ell', \Omega^r) = \mathbb{C}$. This element is a contour around the singularity as can be viewed in the **Figure 5a**).

Form a cosmological evolution, the torsion play a fundamental role in the forming of all interstellar objects, even the Universe itself. The macroscopic density fluctuations are echoes from big-bang until today “Large CMB” (where quantum field fluctuations had place (**Figure 5**)).

Likewise, from a point of purely cosmological view, the existence of these singular points reveals the existence of sidereal objects where big quantities of flow (using Poincarè arguments) of energy are expulsed or/and attracted along the space–time as twistors satisfying the geometrical Penrose model of a black hole. The torsion as field observable always is present. The spins s and –s, corresponds to different field interactions whose images can correspond to the twistor spaces $\mathbb{P}^+$, and $\mathbb{P}^-$.

Here we can incorporate the following cohomology space that re-interprets field theory objects with geometrical objects considering the Universe as complex Riemannian manifold (topological space) to a field source:

$$\mathrm{H}^1(\mathbb{P}\mathcal{T}, \mathcal{O}(-2h\text{-}2)) \cong \ker\left(\mathrm{U},_{h(k)}\right) = \left\{\varphi \in \mathrm{C}^2(\mathrm{U})\big|_{h(k)}\varphi = 0, \text{in } \mathrm{U} \subset \mathbb{M}\right\}, \qquad (3)$$

The field torsion results evident with some work on twistor-spinor framework [7, 9], and using the electronics interpretations studied in other additional experiments that extend the design of experiments in electronics and realized in our researches (**Figure 6**).

A study realized from a point of view of the mathematical physics and particle physics carries to the following conjecture, considering the particle fermion with boson gauge for torsion and called broson.

Conjecture 3.1 (F. Bulnes, M. Ramírez, L. Ramirez, O. Ramírez). The macroscopic image of the broson actions must be a flow electro-gravitational energy

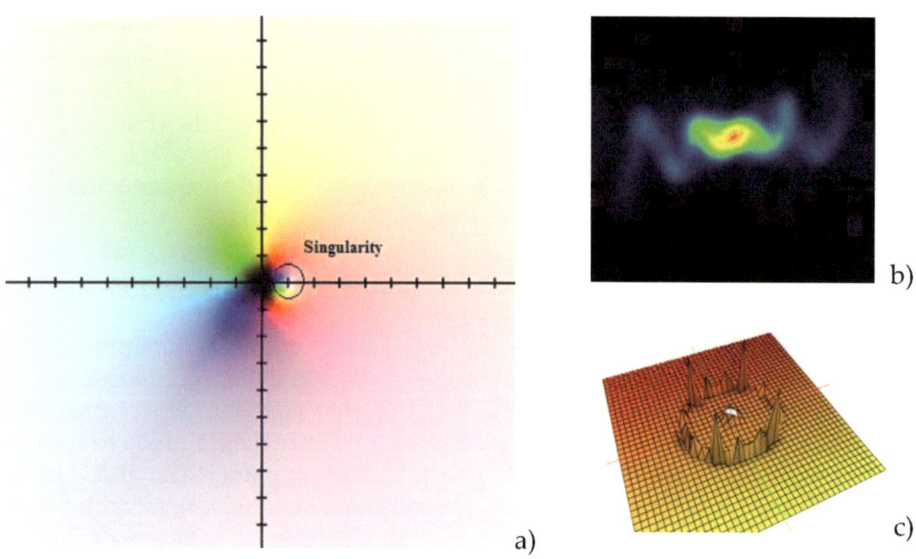

Figure 5.
(a). Pole or singularity of the complex function $g(z) = \frac{z^2}{z-1}$. *(b). The black hole with astronomic catalogue SS 433 is a giant black hole that wobbles. With the VLA and the VLBA, we have watched the wiggle of its jets over time. Here spectral image, we can see its corkscrew appearance obtained for the screw effect. This appearance obeys to the agitation produced by the gravitational field and electromagnetic fields due the particles around (right-handed neutrinos keV) of the black hole in the vacuum space. (c). Source 2-dimensional model considering the little perturbations in its horizon detected in neighborhood of the source (also can be considered as singularity type. For example a peak or cusp). The surface is Z = (0.2coth (−0.2ln (0.3x^2 + 0.2y^2)))/(x^2 + y^2) where happens the perturbation phenomena. This surface represents a quantum field fluctuations in the beginning of the Universe from big-bang stage until any stage of the space–time evolution before of the baryongenesis.*

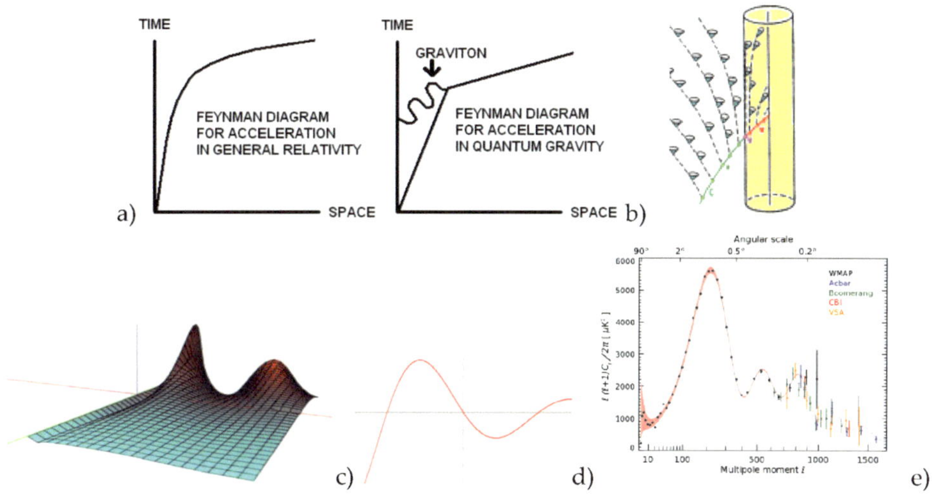

Figure 6.
(a) The curvature can be measured to quantum level [8] as distortions given by the link-wave between a hypothetical particle as dilaton (gauged graviton) and the trace on relativistic Feynman diagram followed in quantum gravity [5, 8]. (b) The torsion in the case of the cylindrical trajectory, for different times and considering the causality structure of the space–time, can be determined by different deviations to the world lines in each case having as microscopic oscillations predicted by the Majorana fermions [13]. (c) [14]. Microscopic perturbation on a cylindrical surface. Also is considered the causal structure given by light cones. The red segment in the Figure 6b), corresponds to the surface model given in 9c). (d). Curve of the cylindrical waves. (e). Observe that similar waves of the spectrum power of CMB radiation temperature anisotropy in terms for 90° until 0.4° to the multi-pole moment. Remember the relation ofh singularities with the field torsion in the aspect of big production or decaying of energy.

whose micro-states begin from the actions of non-Abelian fields that are measured by the gauge fields.

Definition 3.1. A broson is a hypothetical particle that is a fermion and that come from the Branes, being this hypothetical particle wrapped by gauge bosons in the space–time [15].

The value of broson is the intrinsic geometry of the torsion as field observable, associated to a bundle. In field theory can be considered a framework operator belonging to certain algebras (can be quantum algebras) that comply with certain conditions to solve field equations (as Dirac equation for example $(P + \chi \mathrm{T})\mathrm{o}_{\mu}^{a} = 0$) using **H**-states (states or densities in a Hamiltonian manifold, are not **H**-particles necessarily) as basis [16]. Likewise, the Dirac equation can take the following form:

$$\mathrm{d}\,\mathbf{h} = 0, \quad \forall \mathbf{h} \in \mathbf{H}, \tag{4}$$

The gauge bosons produce torsion in the microscopic space due the electromagnetic characteristic of these bosons that are photons [14] realizing backreaction with the covered space by the gravity. As a result pending for prove will be:

Theorem (F. Bulnes) 3.1. We consider the space–time with CPVT effects. The **H**-torsion is the deformation energy between neutrinos and antineutrinos or curvature/fermion spin energy (space–time-curvature/spin couplings between fermions/anti-fermions).

4. Conclusions and future research

The way of the evolution of the Universe depends of the mechanism between the vacuum space, where live fields of particles and the energy derived from the field interaction. This mechanism due the conjecture 2.1, in the section 2, explain the possible field interacting in the space as oscillations born from the microscopic space–time characteristics.

However, considering the limitations of our experiments with electronic devices only we can see and interpret with arguments of geometry, certain traces of electronic signals that evidences the torsion under a magnetic field determined in certain voltage range and a movement of cylindrical trajectory, which as we know, is the constant torsion. However, this verifies the conjecture 1.2., and some theorems established in other studies in theoretical physics and mathematical physics.

One of the future goals is obtain advanced Hall magnetic sensor designed inside the quantum electronics, which permits an evidence of torsion more clear and no so depending of the geometry restrictions as a trajectory of constant torsion, without the managing of a dilaton and more sensible of the microscopic environment and the presence of gravity. This will have that to be through the fermions differencing in the non-harmonic analysis that appear in the anti-symmetric behavior of the curvature energy measured from field interactions [8].

The applications of torsion further of Universe understanding, are diverse and much, with vanguard technological developments:

- Advanced vehicles of anti-gravitation and electromagnetic impulse,
- Total control of the mind, conscience and brain,
- Quantum Communication,
- Nanomedicine: Spintronics and radionics devices of total cure,
- Artificial intelligence: advanced positron brains,
- More understanding of the Universe: deep understanding

If we consider the multi-poles as the sources of the electromagnetic nature of the space–time (of fact their moduli stack is obtained by equivalences in field theory using some *gerbes* of derived categories as has been mentioned in field theory in some before works [17]), we can to use the loops around of these poles as contours of the cohomological functionals $H^1(\Pi - \coprod, \mathcal{O}(V)) = \mathbb{C}$, [18, 19] to evaluate through the residue theorem their energy and these values that are amplitudes, can be used in spinor waves, of fact we can consider the partial wave expansions of the space–time suggested by the conformal actions in 4-dimensional and 2-dimensional spaces.

The Conway integrals can be considered in axisymmetric boundaries and also non-axisymmetric cases where matter is confined within axisymmetric boundaries, for example in a galaxy. If we consider the electromagnetic nature of the iso-rotations for magnetic intersidereal fields in galaxies, we need other formalism based in twister geometry, where elliptic integrals are analogues in the space $H^1_{\mathcal{L}}(U'', \mathcal{O}(\text{-}2))$.

The cohomological analogous are “poles” which can be interpreted as “sources” of electromagnetic radiative energy (see the **Figure** 7).

Finally, is opportune to sign that the methods and results of the research are numerical simulations (much 2D and 3D-dimensional geometrical models and some analogies with the sidereal objects), on themes parallel and related to the gravity (no gravity precisely) considering the torsion methods and experiments of our torsion theory as analogous to detect gravity waves, but in this case detect waves of torsion in an indirect way.

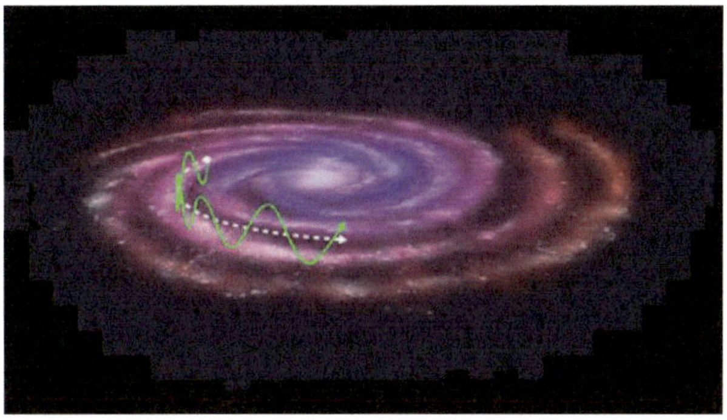

Figure 7.
The responsible electromagnetic energy of the accretion and iso-rotation of a galaxy through its spinor representation [18, 19]. This responsible electromagnetic energy can be proved that is torsion energy.

Technical notation and singles

$\delta(t_1, t_2)$	2-dimensional impulse function. Here $t_1 = t_2$, where $\delta(t_1, t_2) = \delta(t)\delta(t) = \delta^2(t)$.
H	Magnetic field produced in the dilaton during the sensing process of the Hall torsion sensor.
$\mathbf{H}$	Hopf algebra which is an operators algebras of quantum field states. This also can be identifying as a Hamiltonian densities manifold. In mathematical physics and derived geometry is interpreted as Higgs states algebra.
$H^1_{\mathcal{L}}(U'', \mathcal{O}(-2))$	Cohomological group or space of integrals of the field equations whose sheaf holomorphic vector bundle of helicity -2, has algebraic representation through a lines bundles whose polynomials have zeros in the poles or singularizes evaluated by Conway integrals, for example.
$\mathbb{M}$	Complex space–time.
τ	Torsion. Here in our research is torsion energy.
$\tau(\omega_1, \omega_2)$	Spectral torsion. Torsion energy.
mV	Mili-volts.
$h(k)$	Differential operator of the field equation with helicity h(k). This differential operator appears in the wave equation of electromagnetic field.
VLBA	Very Long Baseline Array. Term in astrophysics referred to a system of ten radio telescopes which are operated remotely from their Array Operations Center located in Socorro, New Mexico.
keV	Kilo-electron-volts. Energy unity that corresponds to $1{,}6 \times 10^{-19}$ joules.
$H^1(\Pi - \coprod, \mathcal{O}(V)) = \mathbb{C}$	Cohomological functional. Contents all complex values of cohomological contours to singularities which as poles are evaluated by Cauchy integrals, Conway integrals and extensions of these. This cohomological equality of spaces represents all these integrals.
VLA	Radio-astronomical observatory (called Karl G. Jansky Very Large Array (VLA)) is located to an altitude 2124 meters over sea level.
Large CMB	Cosmic Microwave Background of Large range.

$s(\omega_1, \omega_2)$	Spectral spinor surface. This in the quantum case models the gravitational waves related with the field torsion.
CPTV	Violation of the CPT.
H-Torsion	Torsion of the **H**-states.
ω	Angular frequency given for $\omega = \frac{2\pi}{T}, T > 0$.
CPT	In particle physics means Charge-Parity-Time. It is a symmetry type which establish rules of symmetry or invariance of physical laws that permit the simultaneous transformation between these. For example, the spatial inversion can be achieved with spin inversion, which is a parity inversion in particles.
o_μ^a	Cartan tetrad symmetric (or anti-symmetric) tensor component for unify the gravity with electromagnetism in a simply way. This represents a 4-metric tensor component use to the torsion of the gravity actions. With movement this could be the kinematic tensor required in the equivalences between twistors with spinors, in twistor geometry.
SS 433	Binary system of X-rays SS433 located in the Eagle constellation. The singles and number means Star Spectrum (intersidereal object observed only with radiotelescopes). The number is the corresponding number of the observed object of this type. Also known as V1343 Aquilae, located at 5.5 kpc in the galactic plane (l = 39.7° and b = −2.2°).
$(P + \chi T)$	Differential Dirac operator of the Dirac field equation.
P	D'Alambert operator equal to $\frac{\partial^2}{\partial t^2} - \left(\frac{\partial^2}{\partial x^2} + \frac{\partial^2}{\partial y^2} + \frac{\partial^2}{\partial z^2}\right)$.
H-states	States or Higgs fields of **H**. In general these are boson or fermion states.

Author details

Francisco Bulnes[1,2*], Juan Carlos García-Limón[2], Víctor Sánchez-Suárez[2] and Luis Alfredo Ortiz-Dumas[2]

1 IINAMEI, Research Department in Mathematics and Engineering, TESCHA, Mexico

2 Electronics Engineering Division, TESCHA, Research Department in Mathematics and Engineering, TESCHA, Mexico

*Address all correspondence to: francisco.bulnes@tesch.edu.mx

References

[1] F. Bulnes, I. Martínez, A. Mendoza and M. Landa, "Design and Development of an Electronic Sensor to Detect and Measure Curvature of Spaces Using Curvature Energy," Journal of Sensor Technology, Vol. 2 No. 3, 2012, pp. 116–126. doi: 10.4236/jst.2012.23017.

[2] Bulnes F, Martínez I, Zamudio O. Fine curvature measurements through curvature energy and their gauging and sensoring in the space. In: Yurish SY, editor. Spain: Advances in Sensors Reviews 4, IFSA; 2016.

[3] F. Bulnes, "Curvature Spectrum to 2-Dimensional Flat and Hyperbolic Spaces through Integral Transforms," *Journal of Mathematics*, Vol. 1 (1), pp17–24.

[4] F. Bulnes, *Research of Curvature on Homogeneous Spaces*, TESCHA, State Government of Mexico, 2010.

[5] Bulnes F. Gravity, curvature and energy: Gravitational field intentionality to the cohesion and union of the universe. In: Zouaghi T, editor. Gravity and Geoscience Applications, Industrial Technology and Quantum Aspect. London, UK: IntechOpen; 20 December 2017. DOI: 10.5772/intechopen.71037. Available from: https://www.intechopen.com/books/gravitygeoscience-applications-industrialtechnology-and-quantum-aspect/gravity-curvature-and-energygravitational-field-intentionality-to-thecohesion-and-union-of-the-uni

[6] R. Penrose, W. Rindler (1984) Volume 1: Two-Spinor Calculus and Relativistic Fields, Cambridge University Press, United Kingdom. https://doi.org/10.1017/CBO9780511564048

[7] Bulnes, F. , Stropovsvky, Y. and Rabinovich, I. (2017) Curvature Energy and Their Spectrum in the Spinor-Twistor Framework: Torsion as Indicium of Gravitational Waves. Journal of Modern Physics, 8, 1723–1736. doi: 10.4236/jmp.2017.810101.

[8] Bulnes, F. (2017) Detection and Measurement of Quantum Gravity by a Curvature Energy Sensor: H-States of Curvature Energy, Recent Studies in Perturbation Theory. In: Uzunov, D., Ed., InTech, https://www.intechopen.com/books/recent-studies-in-perturbation-theory/detectio n-and-measurement-of-quantum-gravity-by-a-curvature-energy-sensor-h-states-of -curvature-ener https://doi.org/10.5772/68026

[9] Bulnes, F. Martínez, I. Zamudio, O. Navarro, E. (2020) Kinematic-Energy Measurements of the Torsion Tensor in Space-Time. In: Bulnes, F., Ed., Intech, https://www.intechopen.com/online-first/kinematic-energy-measurements-of-the-torsion-tensor-in-space-time DOI: 10.5772/intechopen.92815

[10] Dr. Francisco Bulnes, A J. C. García-Limón, L. A. Ortiz-Dumas and V. A. Sánchez-Suarez. Detector of Torsion as Field Observable and Applications. American Journal of Electrical and Electronic Engineering. 2020; 8(4):108–115. doi: 10.12691/ajeee-8-4-2

[11] F. Bulnes, J. C. García-Limón: https://youtu.be/0s5H8w6JfyA

[12] L. V. Ahlfors, *Complex Analysis*, Mac Graw Hill Inc. 1966.

[13] S. Gorbuis, E. Guivenchy, Torsion Matter and their Corresponding Dirac Equations on Neutrinos and Matter/anti-Matter Asymmetries (Part II: gravitations insights), Journal on Photonics and Spintronics, Vol. 4 (1), 2015, pp15–25.

[14] F. Bulnes, "Electromagnetic Gauges and Maxwell Lagrangians Applied to the Determination of Curvature in the

Space-Time and their Applications," Journal of Electromagnetic Analysis and Applications, Vol. 4 No. 6, 2012, pp. 252–266. doi: 10.4236/jemaa.2012.46035.

[15] M. A. Ramírez. I. Ramírez, O. Ramírez, F. Bulnes, Field Ramifications: The Energy-Vacuum Interaction that Produces Movement, Journal on Photonics and Spintronics, Vol. 2 (4), 2013, pp. 4–11.

[16] Prof. Dr. Francisco Bulnes, "Mukai-Fourier Transform in Derived Categories to Solutions of the Field Equations: Gravitational Waves as Oscillations in the Space-Time Curvature/Spin IV," Int. Sci. Conf. Algebrain and Geometry Methods of Analysis, May 30–June 4, 2018, Odesa, Ukraine.

[17] F. Bulnes, Extended d−Cohomology and Integral Transforms in Derived Geometry to QFT-equations Solutions using Langlands Correspondences, Theoretical Mathematics and Applications, Vol. 7 (2), pp51–62.

[18] Francisco Bulnes. Mathematical Electrodynamics: Groups, Cohomology Classes, Unitary Representations, Orbits and Integral Transforms in Electro-Physics, American Journal of Electromagnetics and Applications. Vol.3, No. 6, 2015, pp. 43–52. doi: 10.11648/j.ajea.20150306.12

[19] Francisco Bulnes, Ronin Goborov, Integral Geometry and Complex Space-Time Cohomology in Field Theory, *Pure and Applied Mathematics Journal*. Special Issue: Integral Geometry Methods on Derived Categories in the Geometrical Langlands Program. Vol. 3, No. 6–2, 2014, pp. 30–37. doi: 10.11648/j.pamj.s.2014030602.16

Section 2

Advanced Numerical Modelling

Chapter 3

A Monotonic Method of Split Particles

Yury Yanilkin, Vladimir Shmelev and Vadim Kolobyanin

Abstract

The problem of correct calculation of the motion of a multicomponent (multimaterial) medium is the most serious problem for Lagrangian–Eulerian and Eulerian techniques, especially in multicomponent cells in the vicinity of interfaces. There are two main approaches to solving the advection equation for a multicomponent medium. The first approach is based on the identification of interfaces and determining their position at each time step by the concentration field. In this case, the interface can be explicitly distinguished or reconstructed by the concentration field. The latter algorithm is the basis of widely used methods such as VOF. The second approach involves the use of the particle or marker method. In this case, the material fluxes of substances are determined by the particles with which certain masses of substances bind. Both approaches have their own advantages and drawbacks. The advantages of the particle method consist in the Lagrangian representation of particles and the possibility of" drawbacks. The main disadvantage of the particle method is the strong non-monotonicity of the solution caused by the discrete transfer of mass and mass-related quantities from cell to cell. This paper describes a particle method that is free of this drawback. Monotonization of the particle method is performed by spliting the particles so that the volume of matter flowing out of the cell corresponds to the volume calculated according to standard schemes of Lagrangian–Eulerian and Eulerian methods. In order not to generate an infinite chain of spliting, further split particles are re-united when certain conditions are met. The method is developed for modeling 2D and 3D gas-dynamic flows with accompanying processes, in which it is necessary to preserve the history of the process at Lagrangian points.

Keywords: Eulerian method, PIC method, numerical simulation, gas-dynamic

1. Introduction

Correct calculations of multi-material flows is the greatest challenge for ALE and Eulerian CFD codes, especially those using mixed cells at interfaces. There are two basic approaches to solving the advection equation for the multi-material case. In the first (grid-based) approach, interfaces are identified, and their position on the grid is tracked at each time step. The interface can be identified both explicitly, or it can be recovered based on the field of volume fractions. The latter algorithm serves as a basis for widely used methods, like the VOF method [1] (concentration method [2]). The second approach involves material particle methods first proposed by Harlow (the PIC method [3]). In this case, material fluxes from cells, including mixed ones, are controlled by particles, to which certain material masses are assigned.

Both approaches have their advantages and drawbacks. The advantages of the particle method consist in the Lagrangian representation of particles and the possibility of assigning material information to them. This minimizes the errors of solving the advection equation by the grid-based Eulerian methods. A number of modifications of the PIC method have been developed to improve its accuracy and extend the range of its applications [4–9]. An overview of these methods is provided in [10].

The central drawback of the particle method is the highly non-monotonic character of the solution caused by the discrete transfer of mass and mass-related quantities from cell to cell. The corresponding error can be reduced in the most straightforward manner by increasing the number of particles in cells, but such an increase limits the method's performance, especially in the 3D case. To minimize this drawback, a number of method modifications are employed. In [11, 12], for this purpose, the authors use particles having different masses. This approach, however, does not eliminate the need of involving a large number of particles. In a different approach, particles are used only in a limited part of the integration domain, for example, near interfaces [13]. As a result, only a small number of cells contain large quantities of particles. The rest part of the domain in this case is treated by the grid-based methods. Such a selective use of particles, however, does not eliminate the error of solving the advection equation by the grid-based methods and the need of remembering the history of a particular process in a large volume of the material.

This paper proposes a particle method that minimizes these drawbacks. Monotonization of the particle method is performed by particle splitting, so that the material volume flowing out of the cell corresponds to the volume calculated by schemes based on the grid approach. In order to prevent endless splitting, such split particles are further recombined under certain conditions. This approach allows us to do with a small number of particles in the cell, while delivering a monotonic solution.

2. Problem statement

The split-particle (SP) method has been implemented in a code called EGAK in the 2D approximation. In the source code, the major quantities for numerical solution of the multi-material gas dynamic equations include node-centered velocity vector components u_x and u_y and cell-centered thermodynamic quantities: density ρ_ξ, specific (per unit mass) internal energy e_ξ, and volume fractions $\beta_\xi = V_\xi/V$ of the constituent materials.

Particles can also be specified for some materials (in a particular case, these can be all materials). Each particle (with index p) has its coordinates in space $x_p(t)$, $y_p(t)$ and velocity vector components $u_{xp}(t)$, $u_{yp}(t)$ (these are used in interim calculations in the Lagrangian step). In addition, all particles represent thermodynamic states of the corresponding material (density, specific internal energy, volume): $\rho_{\xi p}$, $e_{\xi p}$, $V_{\xi p}$. Note that densities and volumes of particles can also give us their masses. Also note that in the method proposed particle velocities are obtained by interpolation between nodal velocities rather than "remembered" like in the classical PIC method.

Approximation of the corresponding equations is performed in two steps using a decomposition procedure. The **first** (Lagrangian) step involves calculations of the gas dynamic equations without convective members, i.e. gas dynamic equations in Lagrangian variables. In the **second** (Eulerian) step, a new grid is constructed (it generally coincides with the grid of the previous timestep), and the quantities are remapped onto it. As inputs, this step takes the outputs of the Lagrangian step. In turn, this step is divided into two sub-steps, i.e., approximation of the advection equation is done with decomposition in directions.

The difference equations below are presented in as much detail as needed to understand the algorithm of particle introduction; please refer to [14, 15] for a more detailed description of EGAK's basic difference scheme. In what follows, if no confusion is possible, no subscripts or subscript *n* are used to denote the outputs of the previous timestep, and subscripts n+1/2 and n + 1 denote the outputs of the Lagrangian and the Eulerian step, respectively.

3. Lagrangian step

3.1 Approximation of the cell-centered quantities

The Lagrangian gas dynamic equations are approximated using EGAK's standard scheme. As outputs, the Lagrange step delivers updated node-centered velocities, as well as densities, energies and volume fractions of each constituent material. This also applies to cells containing particles.

3.2 Definition of particle-specific quantities

In addition to the material-specific quantities, some particle-specific quantities are also defined for particles in the cells containing particles.

3.3 Updating of particle coordinates

Updated particle coordinates are found in two steps:

Step 1. At the Lagrangian step, particles are assumed to move together with the cell and inside the cell, without crossing its boundaries. The relative change in the particle position in the cell is associated with the difference in divergences (compression ratios) of different materials as a result of employing one closing model or another for the mixed-cell gas dynamic equations. In this study, we use only one assumption that the materials have equal divergences. This means that the sub-cell motion of particles does not change their position relative to the grid nodes.

Particle coordinates, $\tilde{x}_p^{n+1/2}$, $\tilde{y}_p^{n+1/2}$ are updated by bilinear interpolation between coordinates of cell nodes $x^{n+1/2}$, $y^{n+1/2}$, just as at time t^n.

Step 2. It is easy to show that the calculations of particle velocities by bilinear interpolation violate the law of conservation of momentum in the particle-containing cell. To ensure its conservation, the calculated particle velocities are corrected as follows:

1. Components of cell momentum are calculated:

$$P_{cx} = \frac{1}{4}\left(u_{x0}^{n+1/2} + u_{x1}^{n+1/2} + u_{x2}^{n+1/2} + u_{x3}^{n+1/2}\right) \cdot M, \quad (1)$$

$$P_{cy} = \frac{1}{4}\left(u_{y0}^{n+1/2} + u_{y1}^{n+1/2} + u_{y2}^{n+1/2} + u_{y3}^{n+1/2}\right) \cdot M.$$

Here, M is the cell mass and u_{xi}, u_{yi} (i = 0, 1, 2, 3) are the velocity vector components at four cell nodes.

2. Components of the total momentum of particles belonging to the cell are calculated:

$$P_{px} = \sum_p \tilde{u}_{xp}^{n+1/2} \cdot m_p,\; P_{py} = \sum_p \tilde{u}_{yp}^{n+1/2} \cdot m_p. \tag{2}$$

Here, the particle velocities are calculated using the particles' coordinates determined by bilinear interpolation and previous particle coordinates:

$$\tilde{u}_{xp}^{n+\frac{1}{2}} = \frac{\tilde{x}_p^{n+1/2} - x_p^n}{\tau},\; \tilde{u}_{yp}^{n+\frac{1}{2}} = \frac{\tilde{y}_p^{n+1/2} - y_p^n}{\tau}, \tag{3}$$

were $\tau = t^{n+1} - t^n$.

3. Coefficients $\lambda_x = P_{cx}/P_{px}$, $\lambda_y = P_{cy}/P_{py}$ are calculated.

The particles' velocities and coordinates are updated using the resulting weights:

$$u_{xp}^{n+1/2} = \lambda_x \cdot \tilde{u}_{xp}^{n+1/2},\; u_{yp}^{n+1/2} = \lambda_y \cdot \tilde{u}_{yp}^{n+1/2}; \tag{4}$$

$$x_p^{n+1/2} = x_p^n + u_{xp}^{n+1/2} \cdot \tau,\; y_p^{n+1/2} = y_p^n + u_{yp}^{n+1/2} \cdot \tau.$$

3.4 Determination of particle velocity, density and energy

Changes in the relative density and energy of particles of a given constituent material are assumed to be equal to the corresponding relative changes in these quantities calculated for the respective material on average. This gives the following formulas:

$$\rho_{\xi p}^{n+1/2} = \rho_{\xi p}^n + \left(\rho_\xi^{n+1/2} - \rho_\xi^n\right)\rho_{\xi p}^n/\rho_\xi^n, \tag{5}$$

$$e_{\xi p}^{n+1/2} = e_{\xi p}^n + \left(e_\xi^{n+1/2} - e_\xi^n\right), \tag{6}$$

$$V_{\xi p}^{n+1/2} = V_{\xi p}^n\left(V_\xi^{n+1/2}/V_\xi^n\right). \tag{7}$$

It is easy to show that, when using (5)–(7), the particles' total masses will remain unchanged, and the particles' total internal energies will be equal to the energy calculated for the given material as a whole, i.e. the following relationships hold:

$$\rho_\xi^{n+1/2} V_\xi = \sum_p \rho_{\xi p}^{n+1/2} V_{\xi p},\; e_\xi^{n+1/2} m_\xi = \sum_p e_{\xi p}^{n+1/2} m_{\xi p}. \tag{8}$$

4. Eulerian step

Major difficulties in implementing the particle method are associated with the Eulerian step, and namely, with calculations of mass and internal energy fluxes from cell to cell. In the PIC method, when a particle migrates to a neighbor cell, its mass and energy "migrate" with it. Because of the discrete (and, accordingly, non-monotonic) character of the transfer of mass and all the quantities defined per unit mass, this method delivers highly non-monotonic quantity profiles. Section 4 provides a detailed description of the monotonization algorithm for the PIC method.

In calculations, it is not always efficient to represent all constituent materials by particles because this requires extra calculations and computer memory. Therefore, it is reasonable to use particles for the constituent materials, for which the errors due to the solution of the advection equation are most essential, for example, for thin layers or materials, which require remembering the history of their Lagrangian particle.

As part of the proposed SP method, the following algorithms have been developed:

1. Interaction of materials described by particles and materials calculated by the code's standard scheme;

2. Support of existence, creation and removal of particles only in the vicinity of the interface;

3. Particles merging;

4. Remapping of particles density and energy to the cell as a whole.

These algorithms are listed in the order of their execution in the Eulerian step after the monotonization algorithm.

5. Monotonization algorithm for the particle method

5.1 One-dimensional case

Let us discuss the concept of the algorithm as applied to a one-dimensional flow for a single particle migrating from cell to cell (**Figures 1** and **2**). The figures show two cells containing particles represented by dots and imaginary boundaries of volumes represented by dashed segments. Note that calculations by this technique require only numerical values of the volumes, not their layout.

The flow is directed from left to right as indicated by velocity vector (**Figure 1**). Volume ΔV flowing out of the left cell (in what follows we call it the *volume flux*, darker color) is then equal to the product of the cell's lateral side length L and quantity $S = u \cdot \tau$:

$$\Delta V = L \cdot u \cdot \tau. \tag{9}$$

The non-monotonic behavior of the classical PIC method stems from the discrepancy between the real volume flux (and, accordingly, the mass flux) calculated by (9) and the volume of the particle crossing the cell side. In one case (**Figure 1a**), the volume moving from the left cell is smaller than the particle volume, and in the

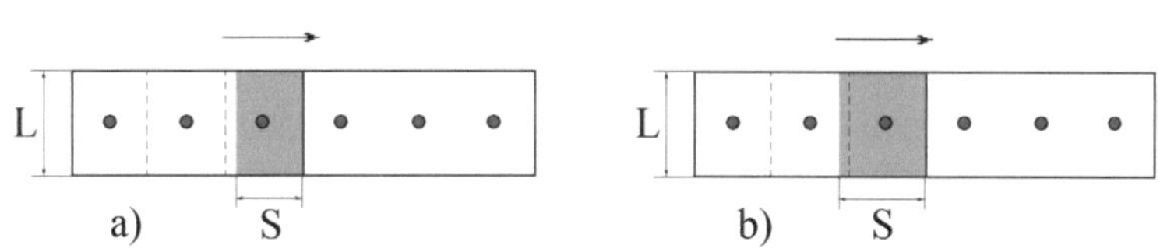

Figure 1.
Illustration of the reason for the non-monotonic behavior: a) volume flux is smaller than the particle volume; b) volume flux is larger than the particle volume.

other (**Figure 1b**), it is larger than the particle volume. In the 2D case and in the case of several migrating particles, the situation remains fundamentally the same.

Let us introduce the following notation:

$$\delta V = \Delta V\text{-}V_p, \tag{10}$$

where ΔV is the volume flux calculated by (4), V_P is the volume of particle p migrating from one cell into another.

Below we explain the monotonization algorithm for both of these cases.

1. *Volume flux ΔV is smaller than the migrating particle volume* ($\delta V < 0$). This case is illustrated in **Figure 2a**. Left is the state at $t^{n+1/2}$; right, at t^{n+1}. In this case, the particle migrating from the donor to the acceptor cell splits into two parts, a mother and a daughter particle. The mother particle migrates into the acceptor cell and now has new coordinates corresponding to its velocity and a new volume equal to the volume flux leaving the donor cell ΔV. The daughter particle, whose volume is equal to the difference between the initial particle volume and volume flux ΔV, is placed in the donor cell and acquires coordinates on the cell side. The link between the mother and the daughter particle is indicated by a broken line.

2. *Volume flux ΔV is larger than the migrating particle volume* ($\delta V > 0$). This case is illustrated in **Figure 2b**. In this case, the missing volume of the migrating particle must be made up by forced transfer of some particles or particle fragments from the donor cell to the acceptor. To be split is the particle lying next to the side of these cells and not yet transferred to the acceptor cell. It produces a daughter particle of volume δV, which migrates into the acceptor cell with donor-acceptor side coordinates.

If more than one cell migrates from cell to cell, formula (10) will take the form of

$$\delta V = \Delta V - \sum_P V_p, \tag{11}$$

where ΔV is the volume flux calculated by (9); V_P is the volume of the particle with index p migrating from cell to cell; summing is performed for all transferred particles.

Here, let us describe the differences from the algorithm described above.

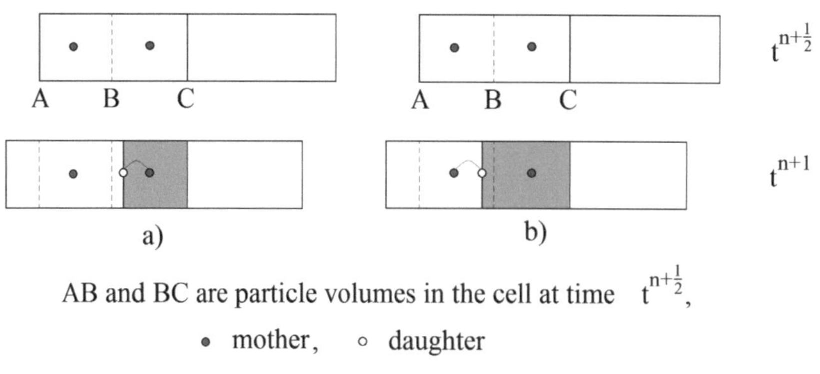

Figure 2.
Illustration of the monotonization algorithm in the 1D case: a) $\delta V < 0$; b) $\delta V > 0$.

1. *The volume flux* ΔV *is smaller than the migrating particles' total volume* ($\delta V < 0$). In this case, to be split are all the migrating particles. The mother particles migrating into the acceptor cell have volumes

$$V_p^M = V_p \left(\Delta V / \sum_p V_p \right). \tag{12}$$

The daughter particles stay in the donor cell and have the following volume each:

$$V_p^D = V_p - V_p^M. \tag{13}$$

2. *Volume flux* ΔV *is larger than the migrating particles' total volume* ($\delta V > 0$). The particle staying next to the interface on the donor side is split to fill the remaining volume δV. Note that if $V_p < \delta V$, the mother particle's volume entirely goes to the daughter particles, and to be split is the next particle from the donor cell. If the donor cell is mixed, and the acceptor cell is pure and filled with the material present in the donor cell, the particles of this material will be split first.

5.2 Two-dimensional extension

Of particular interest in this case is the particle transition to the neighbor cell located diagonally from the donor cell (**Figure 3**). This case is special, because EGAK solves the advection equation using decomposition in directions, whereas no provision is made for diagonal cell-to-cell fluxes.

Consider the particular case depicted in **Figure 3**. Suppose only one node A moves to the new position B in the Eulerian step. The dashed lines in the figure show the locations of the cell sides, for which the grid node is a common vertex, at tn+1/2. C and D denote the points of intersection of straight lines AG and BE, and AF and BH, respectively.

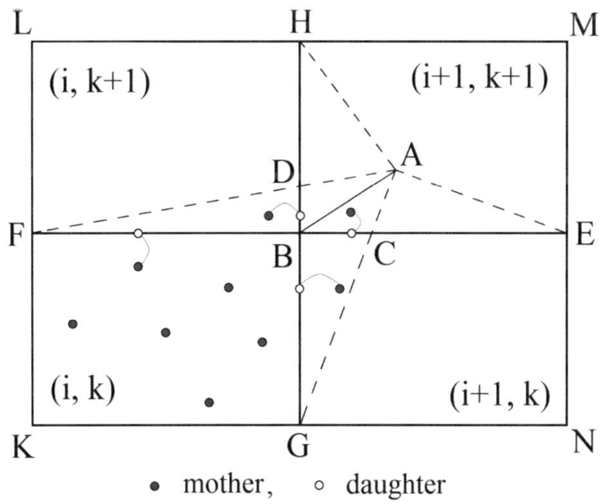

Figure 3.
Illustration of volume flux calculations in 2D case.

In EGAK, the cell-to-cell volume fluxes are defined as follows. The volume flux corresponding to triangle ABG (for short, *volume in triangle ABG*) is transferred from cell (i, k) to cell (i + 1, k), that is:

$$V_{i,k}^{n+1} = V_{i,k}^{n+1/2} - V_{ABG},$$
$$V_{i+1,k}^{n+1} = V_{i+1,k}^{n+1/2} + V_{ABG}. \tag{14}$$

Accordingly, the following relationships apply to all the cells under consideration including all the fluxes:

$$V_{i,k}^{n+1} = V_{i,k}^{n+1/2} - V_{ABG} - V_{ABF} = V_{BFKG},$$
$$V_{i+1,k}^{n+1} = V_{i+1,k}^{n+1/2} + V_{ABG} - V_{ABE} = V_{BGNE},$$
$$V_{i,k+1}^{n+1} = V_{i,k+1}^{n+1/2} - V_{ABH} + V_{ABF} = V_{BFLH}, \tag{15}$$
$$V_{i+1,k+1}^{n+1} = V_{i+1,k+1}^{n+1/2} + V_{ABE} + V_{ABH} = V_{BHME}.$$

Note that in accordance with (15) the volume of triangle ABC is included in the volume of cell (i + 1,k) twice – as part of triangles ABG and ABE – but in one case it is positive, and in the other, negative. Thus, in fact it is not included in the updated volume of this cell; but it will be included in the volume of cell (i + 1,k + 1). The same applies to the volume of triangle ABD, which will be included in the volume of cell (i + 1,k + 1) and not included in the volume of cell (i,k + 1).

In accordance with the above, when considering particle contributions, flux calculations between cell (i,k) and its non-diagonal neighbors assume that the particle lying in triangle ABC migrates into cell (i + 1,k), and the particle lying in triangle ABD, into cell (i,k + 1). Then, in calculations of the flux between cells (i + 1, k), (i + 1,k + 1) and (i,k + 1), (i + 1,k + 1), these particles migrate into cell (i + 1, k + 1). Therefore, when considering this process in terms of flux monotonicity, corresponding daughter particles are introduced as shown in **Figure 4**. Note that the mother particle's position during the flux calculations is nevertheless defined in the true acceptor cell (i + 1,k + 1).

The particle splitting is based on the following principles:

- Both particles produced from the particle being split share its thermodynamic state (to comply with the conservation laws);
- The index assigned to the daughter particle is the same as the index of its mother particle, which also indicates that the particle is a daughter;
- The mother particle "knows" nothing about its daughter particles;

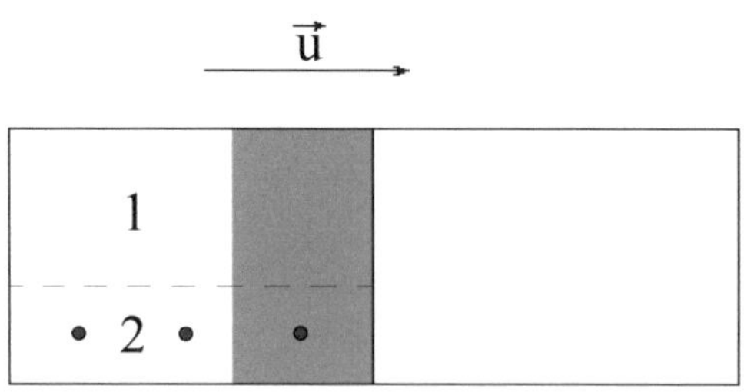

Figure 4.
Volume flux from a mixed cell.

- One mother particle may have several daughter particles;
- One daughter particle may have only one mother particle;
- The daughter particles may also split, but all the subsequent daughter particles will "remember" the index of their initial mother particle;
- The daughter particles are placed at the common donor/acceptor side to ensure their quickest possible combination with their respective mother particles.

6. Algorithm of interaction between particles and particle-free materials

The algorithm for volume flux calculations has a modification to deal with mixed cells containing materials with and without particles.

Consider the case of a cell filled with heterogeneous materials, one described only by grid quantities, and the other, by particles (1 and 2, respectively, in **Figure 4**). Suppose we need to divide the flux moving from left to right (shown with a darker color) between the materials. The volume flux leaving the cell is first filled with the volume of migrating particles. If the migrating particles' total volume exceeds this volume flux, then the above particle splitting algorithm is engaged (see Section 4, case $\delta V < 0$).

Otherwise, the missing part of the outflowing volume flux is filled with the particle-free material. If there are several particle-free materials, the volume is distributed among them based on the VOF algorithm [1]. If the particle-free materials are still not enough, the remaining volume flux is filled with particles using the splitting algorithm for the case of $\delta V > 0$ from Section 4.

7. Near-interface algorithms

As part of the proposed method, we have developed an algorithm involving the particles located only near the interface. The region near the interface includes mixed cells and one layer of adjacent pure cells of each material on each side.

Figure 5 shows possible particle layouts relative to the interface. The dark and light cells (**Figure 5a**) are pure, and the intermediate-color cell (**Figure 5b**) is

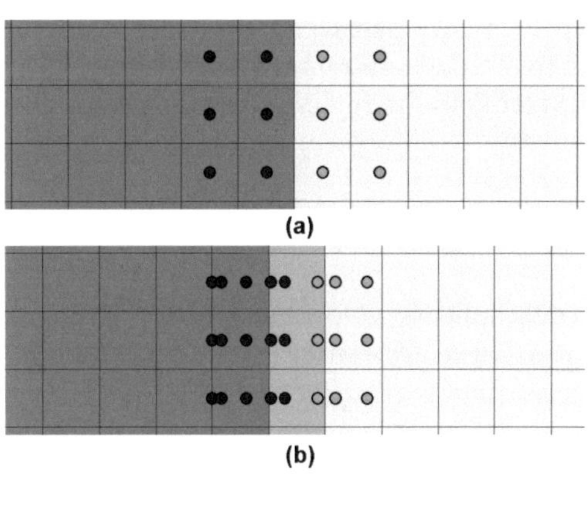

Figure 5.
Particle layout near the interface: a) t = 0, b) t > 0.

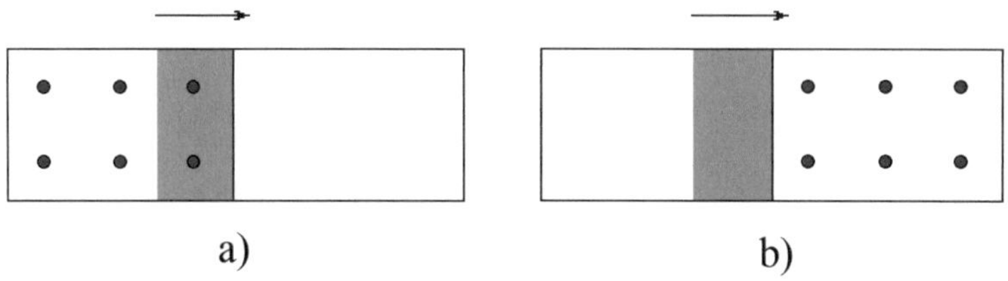

Figure 6.
Conflicting cases of two representations of the same material: a) volume flux from a cell with particles to a particle-free cell; b) volume flux from a particle-free cell to a cell with particles.

mixed. The particles in the cells are marked with a contrasting color. **Figure 5**, a illustrates the case when both materials are represented by particles only near the interface at t = 0, and **Figure 5b** shows the particle layout during the computation.

Thus, if the material is represented by particles only near the interface, then the same material can be represented both by particles and without particles depending on the interface location. To simplify the algorithm, the same material described by particles and without particles cannot occupy the same cell. This condition for each cell is formally given by

$$V_\xi = \sum_{\xi p} V_{\xi p}, \text{ or } \sum_{\xi p} V_{\xi p} = 0. \tag{16}$$

Let us consider two reasons leading to the case when two different representations of the same material are present in the same cell. **Figure 6a, b** shows cells filled with the same material. In each case, however, in one of the cells the material is represented by particles (black dots), and in the other, without particles. The darker color shows the volume flux relative to the cells' total volume; the arrow indicates its direction. Below we describe the unwanted cases and the ways to avoid them.

Figure 6a shows a volume flux from a cell with particles to a particle-free cell. In this case, the particle splitting algorithm described in Section 4 stays unchanged in the donor cell, but the particles that were supposed to migrate into the acceptor cell are removed with an update of the thermodynamic state, and the particles staying in the donor cell become ordinary (they are no daughter cells any more if they were). **Figure 6b** shows a volume flux from a particle-free cell to a cell with particles. In this case, the acceptor cell receives a particle, the volume of which is equal to the volume flux and the state of which is the same as the donor-cell material parameters. The added particle then immediately combines with one of the particles in the acceptor cell. The combination rules are given below.

To preserve the layout, where particles are present only near the interface, the particles lying beyond this region are removed from the cells and new particles are created in the cells appearing in the region near the interface.

8. Particle combination algorithm

To balance the particle splitting algorithm (Section 4), we have developed a particle combination procedure. The latter serves to prevent uncontrolled multiplication of particles as a result of their splitting.

Two particles of the same material within the same cell must be combined if one of the following criteria is met:

- One of the particles is a daughter of the other one;

- Two daughters have the same mother;
- The particles have close (to within a constant) coordinates;
- The particle number exceeds the maximum number specified for the cell;
- One of the particles has a relatively small volume.

Particle combination rules:

- If a daughter is combined with its mother, the resulting particle inherits the mother's coordinates;
- If two daughters of the same mother are combined (p_1 and p_2), the coordinates of the resulting particle (p) are chosen in accordance with their mass ratio:

$$\tilde{x}_p = x_{p1} + \left(x_{p2} - x_{p1}\right)\left(m_{p2}/\left(m_{p1} + m_{p2}\right)\right), \qquad (17)$$
$$\tilde{y}_p = y_{p1} + \left(y_{p2} - y_{p1}\right)\left(m_{p2}/\left(m_{p1} + m_{p2}\right)\right);$$

- The parameters of the resulting particle are calculated subject to the laws of conservation of their mass, specific internal energy and volume:

$$\tilde{e}_p = \frac{e_{p1}\rho_{p1}V_{p1} + e_{p2}\rho_{p2}V_{p2}}{\rho_{p1}V_{p1} + \rho_{p2}V_{p2}},$$
$$\tilde{\rho}_p = \frac{\rho_{p1}V_{p1} + \rho_{p2}V_{p2}}{V_{p1} + V_{p2}}, \qquad (18)$$
$$\tilde{V}_p = V_{p1} + V_{p2}.$$

9. Particle-to-cell density and energy remapping algorithm

For each cell containing particles, the quantities are remapped as follows:

$$\rho_\xi^{n+1} = \sum_p \rho_{\xi p} V_{\xi p} / \sum_p V_{\xi p},$$
$$e_\xi^{n+1} = \sum_p e_{\xi p} \rho_{\xi p} V_{\xi p} / \sum_p \rho_{\xi p} V_{\xi p}, \qquad (19)$$

where summing is performed for the particles of material ξ in the cell.

10. Method testing

10.1 Test problem 1. A moving cruciform density discontinuity

Domain $0 < x < 12$, $0 < y < 12$ is divided into two subdomains (0 and 1). In subdomain 0: $\rho_0 = 1$, $e_0 = 0$, $u_x = 1$, $u_y = 1$, no particles are specified; in subdomain 1: $\rho_0 = 10$, $e_0 = 0$, $u_x = 1$, $u_y = 1$, each cell contains one particle. P = 0 all over the domain, so the problem involves virtually no gas dynamics, only convective flow. The calculations were performed on a fixed grid of 60x60 cells.

Results calculated by the SP method and EGAK (in what follows, SP is the particle method with flux correction described above, PIC is the correction-free particle method similar to the PIC method, and EGAK is used to denote the particle-free code) are shown in the form of density distributions at t = 7.08 (**Figure** 7). The figure shows that the result produced by the SP method is exact.

10.2 Test problem 2. A one-dimensional steady shock wave

We consider the following 1D problem statement. Domain $0 < x < 50$, $0 < y < 4$ is occupied by an ideal gas with $\rho = 1$, $P = 0$, $u = 0$, $\gamma = 3$. A plane shock wave propagates in the material from left to right. Its parameters behind the shock front are $\rho = 2$, $e = 2$, $P = 8$, $u = 2$. The calculations were performed on a fixed grid of 100x4 cells. In the PIC and SP calculations, there were four particles in each cell.

Figure 8 shows the plots of density as a function of coordinate at t = 10 calculated by the SP, PIC and EGAK methods. One can see that the SP method gives nearly the same result as EGAK and a better result compared to PIC. The exact shock position is x = 40.

10.3 Test problem 3. A point explosion

Domain $0 < x < 20$, $0 < y < 20$ contains two materials: a circle of radius 0.1 with its center at the origin is occupied by an ideal gas with $\rho = 1$, $e = 1$, $P = 0$, $\gamma = 1.4$; the remaining part is occupied by an ideal gas with $\rho = 1$, $e = 0$, $P = 0$, $\gamma = 1.4$.

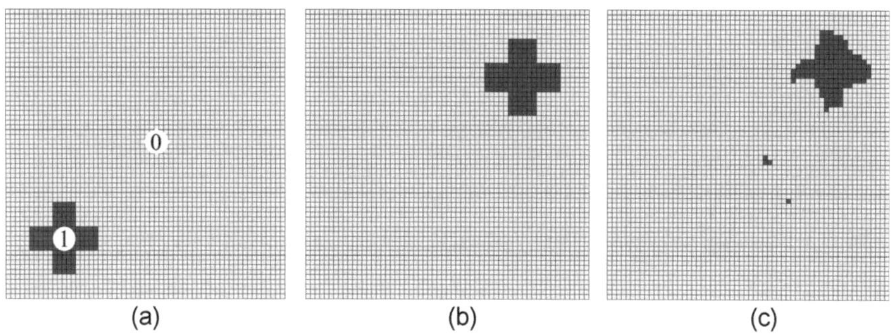

Figure 7.
Test problem 1. Density distributions: a) t = 0; b) t = 7.08, method SP; c) t = 7.08, method EGAK.

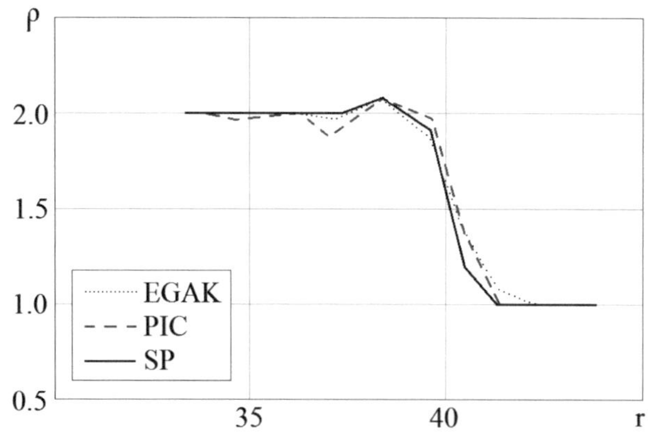

Figure 8.
Test problem 2. Density as a function of shock position, t = 10.

Calculations were performed by the SP, PIC and EGAK methods. In the SP and PIC calculations, each cell initially contained four particles. In the SP problem calculation, the number of particles in the region with gas increased because of its strong expansion. The calculations were done on a fixed regular grid of 100x100 cells.

The SP calculation results are presented in **Figure 9** as a density distribution at t = 100. **Figure 10** shows plots of density as a function of radius for all the cells in the domain occupied by the shock wave at a given instant. They demonstrate how efficient the methods are in preserving the flow's spherical symmetry. **Figure 11** shows plots of density as a function of radius for section x = y.

Figure 9.
Test problem 3. Density distribution at t = 100.

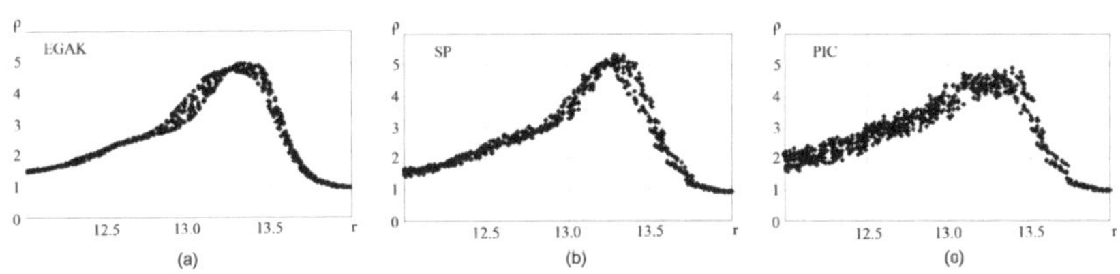

Figure 10.
Test problem 3. Density across the cells as a function of radius, t = 100: a) EGAK; b) SP; c) PIC.

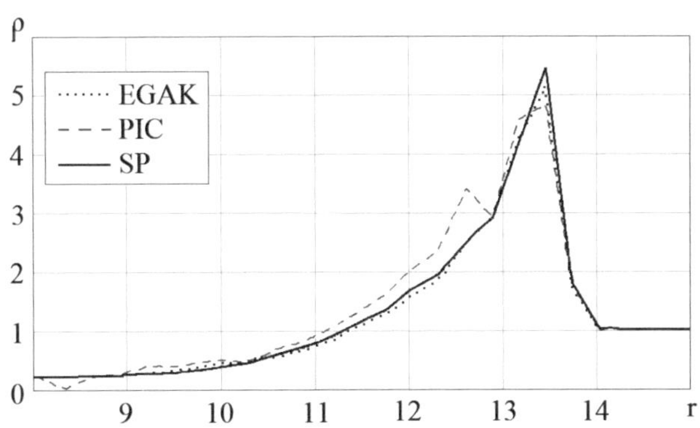

Figure 11.
Test problem 3. Density as a function of radius in section x = y, t = 100.

These figures demonstrate that the SP result is again nearly as good at the EGAK result and noticeably more accurate than the PIC result.

10.4 Test problem 4. A spherically converging shell

The initial problem geometry is borrowed from paper [16] and shown in **Figure 12**. Domain 1: $R_1 = 0.8$, $\rho_0 = 0.01$, $e_0 = 0$, $u_0 = 0$, ideal gas with $\gamma = 5/3$. Domain 2: $R_2 = 1$, $\rho_0 = 10$, $e_0 = 0$, $U_0^R = -1$, equation of state of Mie-Grueneisen type with constants $\rho_0 = 10$, $c_0 = 4$, $n = 5$, $\gamma = 2$. Boundary condition at R2: pressure $P = 0$. Domain 3: vacuum with $P = 0$. Units of measurement: ρ - [g/cm^3], t - [10 μs], L - [cm]. The calculations were done on a fixed regular grid of 110x110 cells. The SP calculations involved particles in both gas and shell domains (one particle per cell, with a limitation of no more than four particles). Their results are compared with the EGAK results. An interesting feature of this problem is that the number of particles in the computation constantly decreases, because both materials are compressed.

The main target result in this problem is maximum gas compression. Note that the reference density obtained in convergence calculations by the 1D method [17] is ≈25. The maximum average gas density and respective time for SP and EGAK are 16.03 at t = 0.368 and 16.49 at t = 0.369, respectively. As an illustration, **Figure 13a** shows a fragment of the domain with particles at t = 0.368. The figure also shows density values across the cells of the compressed-gas domain from the calculations by SP (**Figure 13b**) and EGAK (**Figure 13c**).

The maximum gas compression ratios and their respective time obtained by EGAK and SP are close, but the maximum compression ratios are much lower than the reference solution, which is explained by a small number of cells in these calculations (the solutions converge to the reference solution with mesh size). A smaller compression ratio in the SP calculation compared to EGAK is attributed to the presence of gas spots "split-off" from the main domain in the SP calculation. As for the flow symmetry preservation, SP is nearly as good as EGAK.

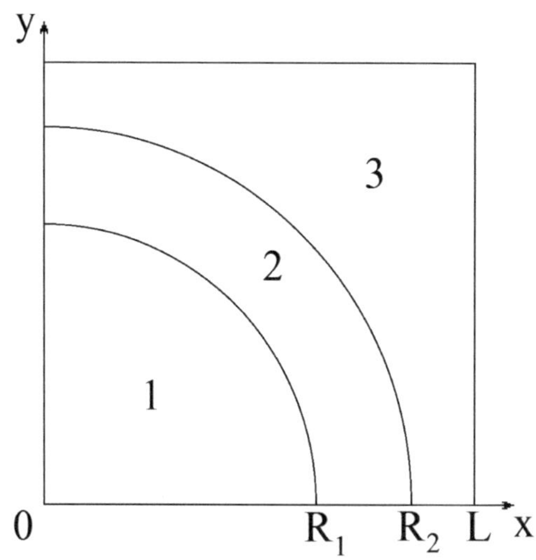

Figure 12.
Geometry of test problem 4.

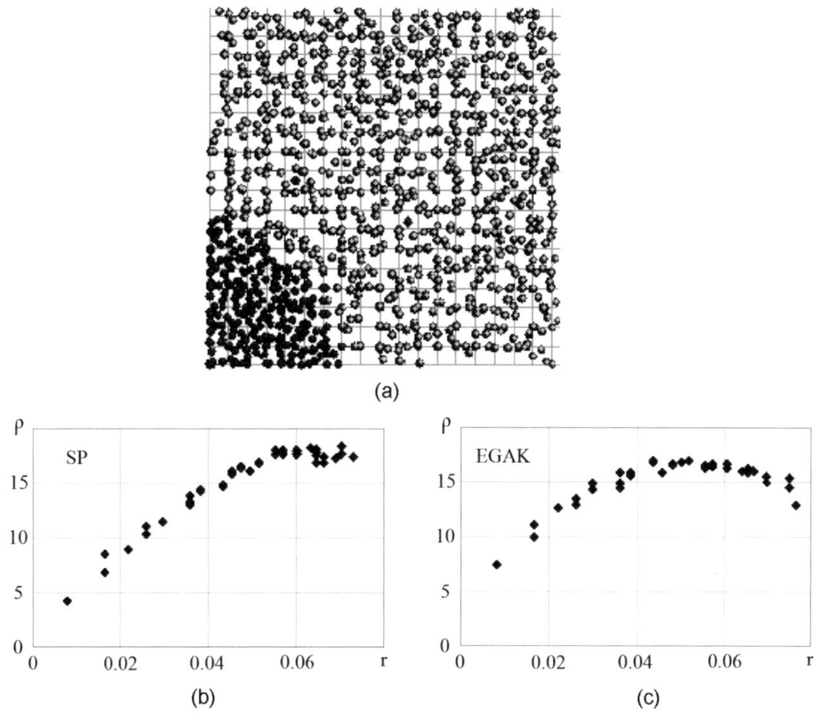

Figure 13.
Test problem 4. a) Particle positions in the central domain; b) Density value across the gas cells calculated by SP, and c) Density value across the gas cells calculated by EGAK.

11. Conclusion

The paper describes a monotonic split-particle method. The method has been developed to simulate multi-material gas dynamic flows using a combination of grid methods implemented in EGAK and the SP method for some layers. The calculations demonstrated that the SP method is close to EGAK in the accuracy of shock-capturing simulations and is much more accurate as applied to convective flow simulations, like the PIC method. At the same time, the SP method is free of the major drawback of the PIC method, namely the severe nonmonotonicity of its solution due to the discrete mass transfer. In addition, this method uses a relatively small number of particles.

Further prospects of the SP method are related to its application to the problems that require "remembering" the process history at Lagrangian points, like detonation and combustion of explosives, elastoplastic behavior and fracture of materials, etc. In particular, implemented to date have been the kinetic model of explosive HE transformation by Morozov and Karpenko [18] and the model of materials fracture by Kanel et al. [19]. This method has also been implemented in the 3D extension of the EGAK code.

Acknowledgements

We would like to thank Tatiana Zezyulina for the translation of this paper.

Author details

Yury Yanilkin*, Vladimir Shmelev and Vadim Kolobyanin
Russian Federal Nuclear Centre, VNIIEF, Sarov, Russia

*Address all correspondence to: n.yanilkina@mail.ru

References

[1] Hirt CW. Nicols BD. Volume of Fluid (VOF) Method for the Dynamics of Free Boundaries. Journal of Computational Physics. 1981; 39: 201–225.

[2] Bakhrakh SM, Glagoleva YuP, Samigulin MS, Frolov VD, Yanenko NN, Yanilkin YuV. Gas dynamic flow simulations by the concentration method. DAS USSR. 1981; **257 (3)**: 566–569. (In Russian).

[3] Harlow FH. The particle-in-cell computing methods for fluid dynamics. Meth. Comput. Phys. 1964; 3: 319–343.

[4] Tskhakaya D, Matyash K, Schneider R, Taccogna F. The Particle-In-Cell Method. Contributions to Plasma Physics. 2007; 47(8–9): 563–594.

[5] Shalaby M, Broderick AE, Chang P, Pfrommer C, Lamberts A, Puchwein E. SHARP: A Spatially Higher-order, Relativistic Particle-in-Cell Code. The Astrophysical Journal. 2017; 841(1): 52.

[6] Jiang C, Schroeder C, Selle A, et al. The affine particle-incell method. ACM Trans. Graph. 2015; 34:4.

[7] Bogomolov SV, Zvenkov DS. Explicit particle method, nonsmoothing gas-dynamic discontinuities. Matematicheskoe modelirovanie. 2007; 19:3: 74–86. (In Russian).

[8] Jiang C, Schroeder C, Teran J. An angular momentum conserving affine-particle-incell method. Journal of Computational Physics. 2017; 338: 137–164.

[9] Fu C, Guo Q, Gast T, et al. A polynomial particle-in-cell method. ACM Trans. Graph. 2017; **36**: 222:1–222:12.

[10] Grigoryev YuN, Vshivkov VA, Fedoruk MP. Numerical "Particle-in-Cell" Methods. Theory and Applications. Utrecht Boston. 2002.

[11] Lapenta G, Brackbill JU. Dynamic and selective control of the number of particles in kinetic plasma simulations. Journal of Computional Physics. 1994; 115: 213–217.

[12] Welch DR, Genoni TC, Clark RE, Rose DV. Adaptive particle managment in a Particle-in-Cell Code. Journal of Computional Physics. 2007; 227: 143–155.

[13] Sapojnikov GA. A combine method of fluid fluxes and particle-in-cell for calculations of the gasdynamical flows. In "Voprocy razrabotki I ekspluatacii paketov prikladnyx program". Novosibirsk: ITPM SO AN SSSR. 1981; 89–97. (In Russian).

[14] Yanilkin YuV, Belyaev SP, Bondarenko YuA, et al. EGAK and TREK: Eulerian codes for multidimensional multi-material flow simulations. RFNC-VNIIEF Transactions. Research Publication, Sarov: RFNC-VNIIEF. 2008; **12**: 54–65. (In Russian).

[15] Yanilkin YuV, Goncharov EA, Kolobyanin VYu, Sadchikov VV, Kamm JR, Shashkov MJ, Rider WJ. Multi-material pressure relaxation methods for Lagrangian hydrodynamics. Computers & Fluids. 2013; 83: 137–143.

[16] Yanilkin YuV, Toporova OO. Two-dimensional scalar artificial viscosity of the EGAK code in spherical systems. VANT, Series MMFP. 2010; 3: 46–54. (In Russian).

[17] Bondarenko YuA. The order of approximation, the order of numerical convergence and cost efficiency of Eulerian multi-dimensional gas dynamics computations illustrated by "blast waves" problem simulations. VANT, Series MMFP. 2004; 4, 51–61. (In Russian).

[18] Morozov VG, Karpenko II, Kuratov SE, Sokolov SS, Shamraev BN, Dmitrieva LV. Theoretical Substantiation of the Phenomenological Model of HE Sensitivity to Shock Waves Basing on TATB. Chemical Physics. 1995; **14 (2–3)**: 32–37. (In Russian).

[19] Kanel GI, Pazorenov SV, Utkin AV, Fortov VE. Studies of mechanical response of materials to their shock-wave loading. Izvestiya RAN. MTT. 1999; 5:173–188. (In Russian).

Chapter 4

Numerical Simulation Modelling of Building-Integrated Photovoltaic Double-Skin Facades

Siliang Yang, Francesco Fiorito, Deo Prasad and Alistair Sproul

Abstract

Building-integrated photovoltaic (BIPV) replaces building envelope materials and provides electric power generator, which has aroused great interest for those in the fields of energy conservation and building design. Double-skin façade (DSF) has attracted significant attention over the last three decades due to its bi-layer structure, which improves thermal and acoustic insulation and therefore increases the energy efficiency and thermal comfort of buildings. It is hypothesised that the integration of BIPV and DSF (BIPV-DSF) would help buildings in reducing energy consumption and improving indoor thermal comfort concurrently. However, the prototype of the BIPV-DSF has not been well explored. Thus, the investigations of the BIPV-DSF are worthwhile. Numerical simulation is a cost and time effective measure for the design and analysis of buildings. This chapter spells out a comprehensive method of numerical simulation modelling of the novel BIPV-DSF system in buildings, which is carried out by using a graphically based design tool – TRNSYS and its plugins. TRNSYS has been validated and widely used in both the BIPV and building related research activities, which are capable in analysing the effects of BIPV-DSF on building performance such as energy consumption and indoor thermal condition.

Keywords: numerical simulation modelling, TRNSYS, TRNFlow, double-skin facades (DSF), building-integrated photovoltaic (BIPV)

1. Introduction

Either building-integrated photovoltaic (BIPV) or double-skin façades (DSF) is widely adopted in buildings; however, few studies or real applications of the hybrid mechanism of BIPV and DSF have been implemented in either academic or industrial settings [1]. Therefore, it is worth understanding the behaviour of this novel integrated system – BIPV-DSF. As shown in **Figure 1**, the BIPV-DSF is analogous to a typical DSF, which consists of an outer façade, an inner façade and an air cavity between the two facades; and there are ventilation openings within the BIPV-DSF [2]. However, a semi-transparent PV panel serves as the external window glazing on the outer facade of the BIPV-DSF, while the window of the inner façade is composed of a normal glazing unit. The outer façade of the BIPV-DSF provides protection against the outdoor environments and the ventilation in the air cavity is driven by the stack effect through the ventilation openings to cool the indoor area down in

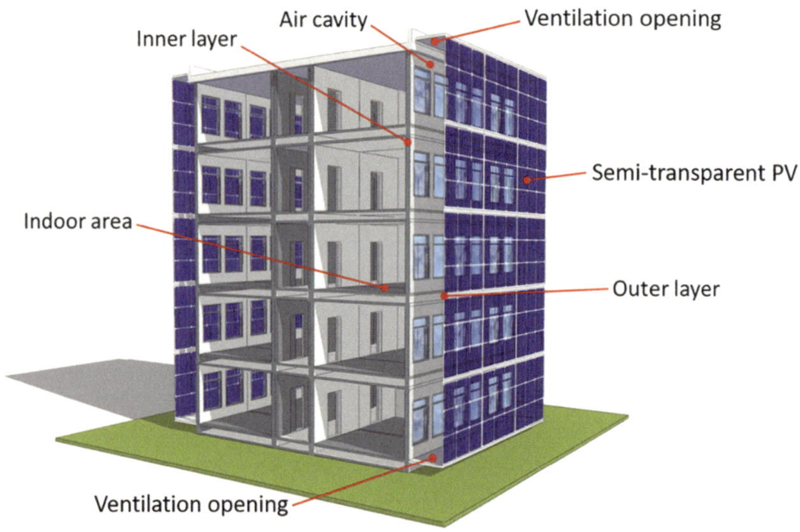

Figure 1.
Schematic diagram of the BIPV-DSF.

summer, while the openings are closed to reduce heat loss in winter, hence delivers a comfortable indoor thermal condition [2]. Moreover, the energy production (electric power generated by PV panel and thermal energy from the heated air in the cavity) can be obtained from the BIPV-DSF, which contributes to the reduction of the energy consumption of buildings.

Numerical simulation modelling of buildings and building systems has been developing and carrying out for over 30 years, during which time the accuracy, depth and speed of the simulation have been widely verified and significantly improved for the design and analysis of new buildings [3]. It has been proved that numerical simulation modelling is much more cost-effective and less time consuming than experimental study, especially for those inaccessibly experimental conditions in real life [4]. In this context, the proposed chapter presents a comprehensive method of numerical simulation modelling of the novel BIPV-DSF system in buildings.

The proposed numerical BIPV-DSF model can be used to investigate the advantages of double-skin façades and building-integrated photovoltaic technology in terms of thermal and electrical performances of the entire BIPV-DSF integrated onto buildings, which are related to indoor thermal condition and energy consumption of buildings.

2. Overview of the proposed numerical simulation modelling

TRNSYS and TRNFlow programmes are selected for carrying out the proposed numerical modelling, which are able to predict the effects of the implementation of the BIPV-DSF on building performance such as energy consumption and indoor thermal condition, based on the capabilities of the software. TRNSYS is a graphically based software, which has been validated and widely used in the BIPV and building related research activities [5–8]. In addition, TRNSYS output data files are in a (human) readable and editable plain text format, which allows users to convert dataset simulation results to Excel spreadsheet; hence, they precisely present the result plots and effectively observe the errors of simulations. Energy consumption of thermal building models and electricity productions of the BIPV systems can be directly predicted in TRNSYS. TRNFlow is an external engine for TRNSYS for assisting in the calculation of ventilation [9], which can be integrated into the TRNSYS thermal building model (Type56) to analyse the performance of ventilation within the DSF and therefore the indoor thermal comfort due to the ventilated DSF.

3. Building simulation modelling in TRNSYS

TRNSYS was initially developed by the University of Wisconsin and the latest programme and external plugins of the software are the outcomes of international collaboration of the US, France and Germany [10]. TRNSYS and its plugins can be used for functions, such as electric power simulation, solar design, building thermal performance analysis, HVAC system sizing as well as airflow analysis [10]. TRNSYS provides a modular structure for simulating the transient systems, especially energy systems, such as solar systems (PV systems) and HVAC systems [11]. The simulation activities in this chapter are demonstrated by using the version of TRNSYS 17. The main programmes of TRNSYS 17 [11] include:

- TRNSYS Simulation Studio (**Figure 2**) – a visual interface used for model design, parametrisation and system simulation. Type 56 (the "Building" icon in **Figure 2**) includes the details of a multi-zone building model such as building geometry, load profiles, construction and window glazing properties, while the desired weather file can be uploaded in the "Weather data" component (**Figure 2**).

- TRNDII – a simulation engine of TRNSYS which is called by TRNEXE (an executable program).

- TRNBuild (as shown in **Figure 3**) – a visual interface used for creating and editing the simulation input data for non-geometry information of the multi-zone buildings.

- TRNEdit or TRNSED – an editor programme that to create stand-alone applications for proposed models.

In general, TRNSYS Simulation Studio in association with TRNBuild are used to assess energy performance of the proposed BIPV-DSF building model. In addition, as mentioned earlier, an external plugin, namely TRNFlow (as shown in **Figure 4**), is chosen for the integration with the TRNSYS thermal building model, in order to assess the performance of ventilation (for example, natural ventilation) within the DSF.

In the Simulation Studio interface (as shown in **Figure 5**), Type 56-TRNFlow (the green "Building" icon) contains the specific numerical information of the multi-zone building model, which is not only including building geometry, load profiles, construction and window glazing properties, but also including the airflow input and output values for the purpose of the simulation of ventilation.

The functionalities of those programmes of TRNSYS are implemented by using the common computer programming languages for instance "Fortran", which establishes mathematical models for the components and types (in TRNSYS) of the proposed systems in terms of their ordinary differential or algebraic Equations [12]. The general procedures of the simulation modelling in TRNSYS are as follows, in sequence:

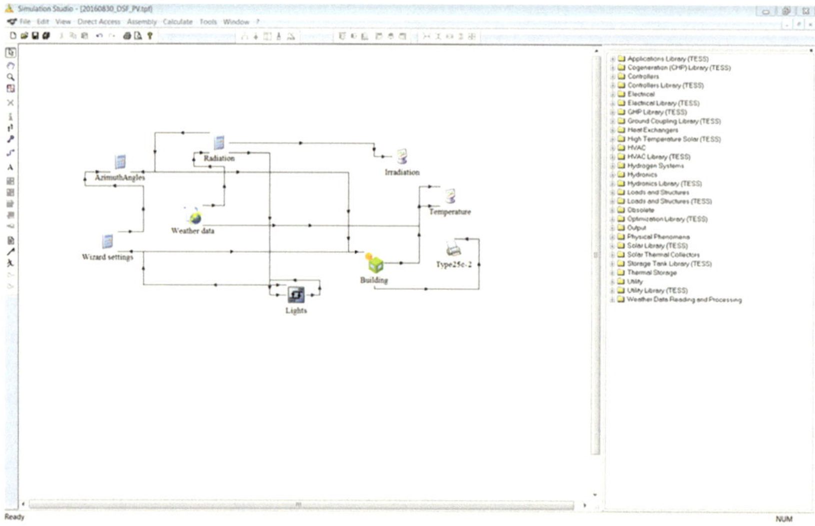

Figure 2.
Simulation studio user interface in TRNSYS 17.

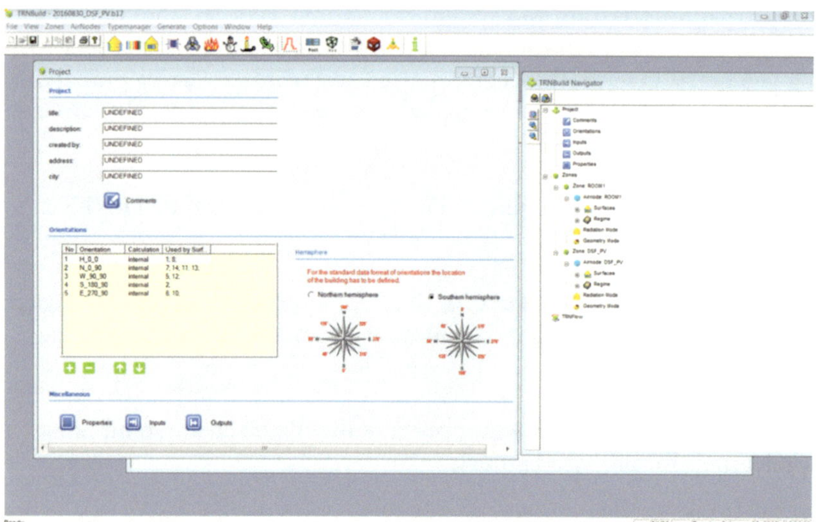

Figure 3.
TRNBuild user interface in TRNSYS 17.

First, the user needs to identify if the available system component and the type of models of the proposed system in TRNSYS will satisfy the specific needs of the real system [11].

The simulation modelling can proceed in TRNSYS if the proposed real system can be designed by the available components and types; otherwise, the missing components and types must be created by writing the corresponding programming language such as Fortran and C++. The "type" normally consists of inputs (variables such as input temperature or airflow rate), parameters (the fixed values such as the area of solar PV panel, dimension of buildings, and so on) and outputs (the resultant values such as output temperature or airflow rate). Moreover, the "type" in TRNSYS can also include external files (for example, weather file) and derivatives which specify the initial values (for example, initial room air temperature in a building) for the "type" [11]. For the proposed BIPV-DSF building model, all the

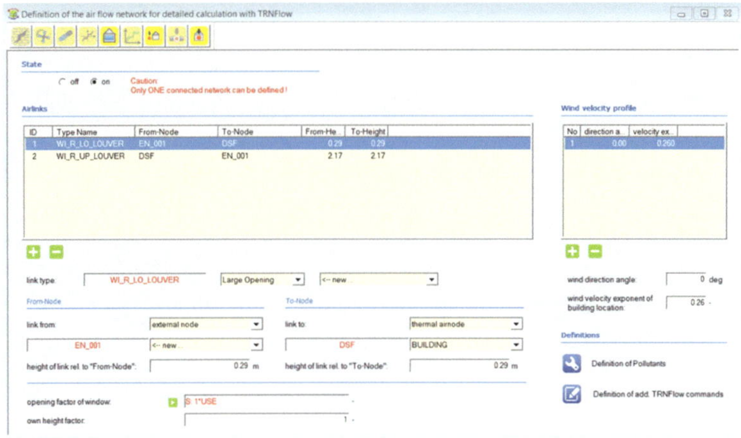

Figure 4.
TRNFlow user interface.

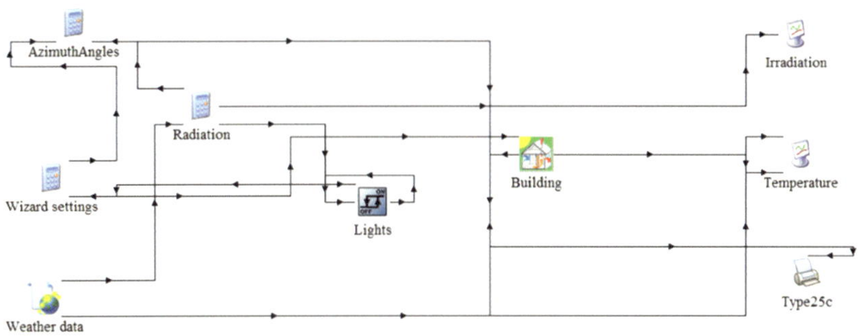

Figure 5.
Type56-TRNFlow in TRNSYS model.

required system components and types are available in TRNSYS 17, so the additional programming procedures are left out.

Once the components and types that represent the proposed real system are fully prepared, the system modelling can proceed in the visual interface – TRNSYS Simulation Studio. All the selected system components and types need to be connected to each other in the Simulation Studio, then the simulation parameters are defined accordingly by the user [11]. For the proposed simulation modelling (the BIPV-DSF), the building geometry information are created in TRNSYS 3D (**Figure 6**). TRNSYS 3D is a plugin for SketchUp (a 3D modelling software developed by Google) to draw 3D multi-zone buildings and export the building geometry information directly from the SketchUp interface into the visual interface of TRNBuild [13].

All the non-geometry information of the TRNSYS building model are then created and edited in TRNBuild (see **Figure 3**). The user can flexibly edit the material properties of building envelope, create ventilation and infiltration profiles, add internal gain and position occupants for the indoor comfort calculations [13].

As mentioned earlier, TRNFlow is used for the calculation of ventilation for the proposed BIPV-DSF building model. TRNFlow will work as a plug-in of TRNSYS, which can be accessed through the user interface in TRNBuild (see **Figure 7**). Basically, an airflow network of the ventilated building model needs to be created in terms of the selected airlinks (for example, the DSF and outdoor air) in TRNFlow. In detail, the model of airflow between the selected airlinks must be

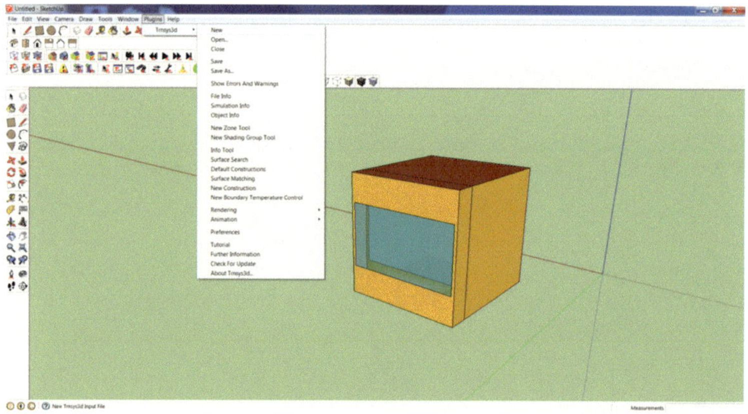

Figure 6.
TRNSYS 3D user interface – A 3D BIPV-DSF model sample.

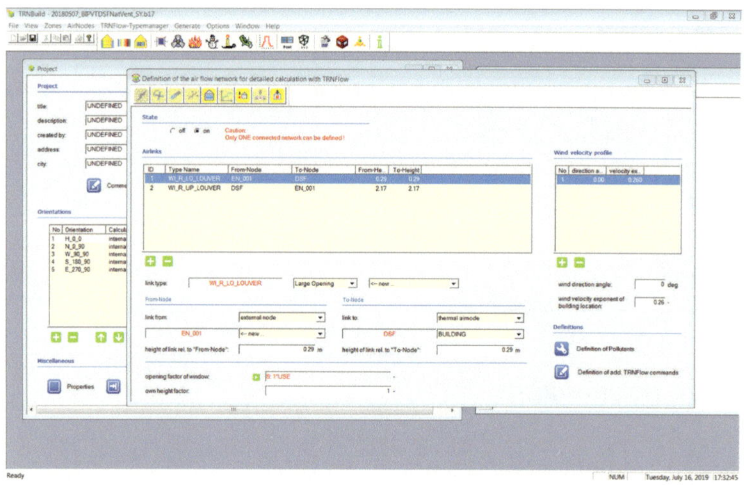

Figure 7.
TRNFlow interface in TRNBuild.

modified accordingly. For the modelling of the BIPV-DSF, both the external air and DSF are linked together, in which the ventilation openings (for example, ventilation louvres) are modelled as large opening through the Link Type Manager (see **Figure 8**). There are 6-category of "links" in the selected version of TRNFlow (version 1.4) as follows:

a. Crack

b. Fan

c. Straight duct

d. Flow controller

e. Large opening

f. Test data

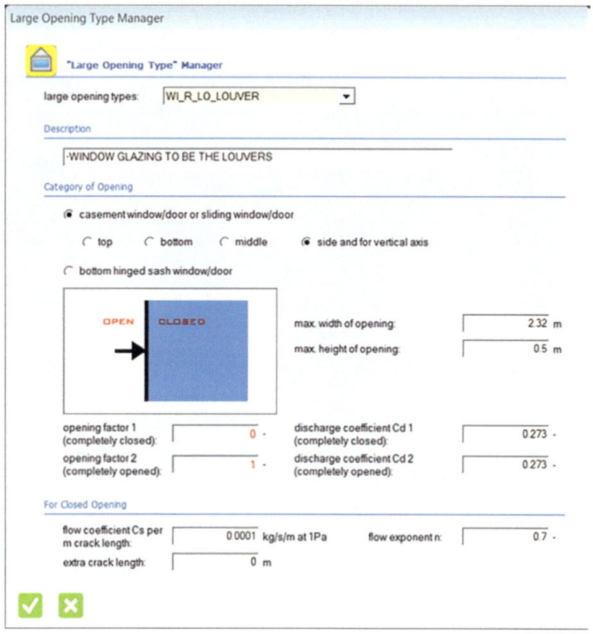

Figure 8.
The link type manager of "large opening" in TRNFlow.

According to the TRNFlow user manual [14], "large opening" is the most fit type of link for modelling the ventilation openings.

The simulation can then be performed when all the inputs, parameters, and outputs are completely edited. The TRNSYS Simulation Studio will report errors if the connections among different components and types are not correct or logically functionable. It will also report errors if any non-geometry information is not correctly created and edited in TRNBuild. The simulation results will be printed in external files through a printer component (see **Figure 9**) or visualised in an online plotter (implemented by TRNEXE, as shown in **Figure 10**, reporting the value of solar radiation as an example) in the Simulation Studio interface [11]. In the proposed simulations of the BIPV-DSF, the online plotter (**Figure 11**) is used to view the output variables during and after the simulations for ensuring the simulations run properly, while a .OUT formatted file (**Figure 12**) as an external text file is used to print the numerical simulation results. The various numerical results can be directly copied from the .OUT file and pasted in an Excel sheet, which are then plotted as viewable graphs accordingly to simplify the analysis of the results.

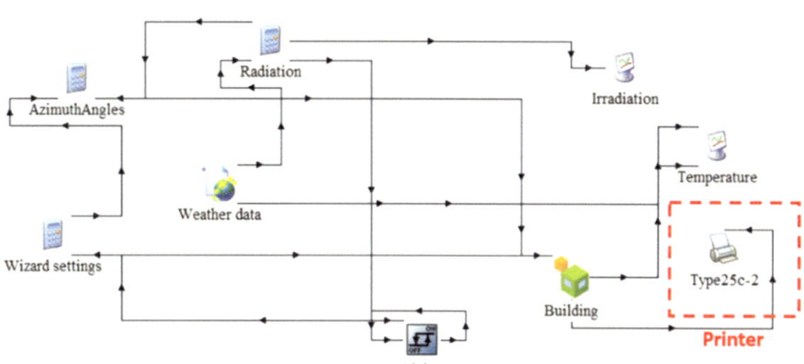

Figure 9.
The printer that loads the external files.

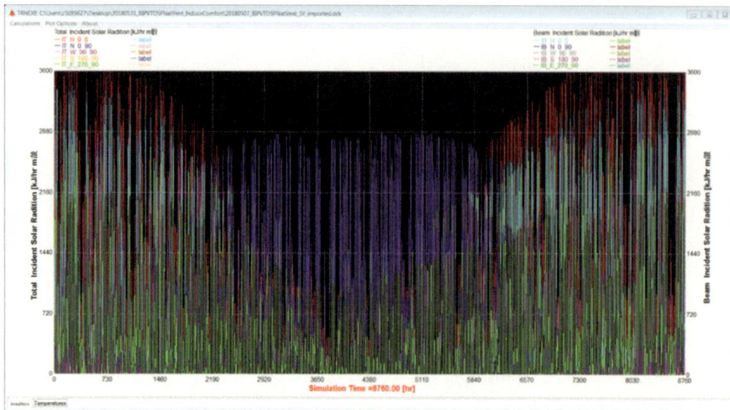

Figure 10.
The online plotter is implemented by TRNEXE.

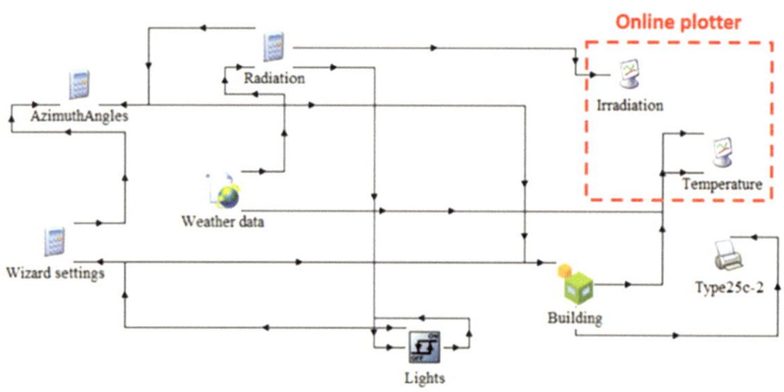

Figure 11.
The online plotter as part of the simulation model.

4. The BIPV-DSF simulation modelling in TRNSYS

As specified in Section 3, the "Type 56-TRNFlow" component is adopted in the TRNSYS model, which contains the numerical information of the multi-zone building model including building geometry, load profiles, construction, window glazing properties as well as the airflow input and output values for the proposed BIPV-DSF. The numerical building information is read by the Type 56 or Type 56-TRNFlow from a set of external files having the extensions *.bui, *.bld and *.tm. These files can be generated by running TRNBuild [15]. An available weather data processor Type 15–3 in TRNSYS, which can read data in the IWEC (International Weather for Energy Calculations) format, is used to read the weather data at regular time intervals (for example, one hour for the simulations) from the external IWEC weather file [15]. As shown in **Figure 13**, both the "Irradiation" and "Temperature" (they both are Type 65d) are online graphical plotters that display the selected system variables (that is, solar irradiation and air temperature) while the simulation is progressing, which allows the users to immediately view the output variables; hence, they can know if the simulations run properly [15].

Type 25c is a printer component that prints the variables (for example, zone air temperature, ambient temperature and heating/cooling demand) of the proposed multi-zone building model (for example, the BIPV-DSF) at specified intervals of

20180507_BIPVTDSFNatVent_SY_imported.out - Notepad

File Edit Format View Help

TIME	T_amb	T_zone	Q_heating	Q_cooling
+0.0000000000000000E+00	+2.1500000000000000E+01	+2.0000000000000000E+01	+0.0000000000000000E+00	+0.000000
+1.0000000000000000E+00	+2.1399999999999999E+01	+2.0716676558455564E+01	+2.4904337697249741E+02	+0.000000
+2.0000000000000000E+00	+2.1149999999999999E+01	+2.1509241987925996E+01	+3.7367662295946798E+02	+0.000000
+3.0000000000000000E+00	+2.0899999999999999E+01	+2.1623349136824334E+01	+2.8669155214414411E+02	+0.000000
+4.0000000000000000E+00	+2.0700000000000003E+01	+2.1673069037546490E+01	+2.4876250213633671E+02	+0.000000
+5.0000000000000000E+00	+2.0500000000000000E+01	+2.1693006968376238E+01	+2.3350210605921018E+02	+0.000000
+6.0000000000000000E+00	+2.0350000000000001E+01	+2.1735078460737881E+01	+2.0154441515450185E+02	+0.000000
+7.0000000000000000E+00	+2.0149999999999999E+01	+2.1870575819052199E+01	+9.8799242450595671E+01	+0.000000
+8.0000000000000000E+00	+2.0500000000000000E+01	+2.2605363261167760E+01	+0.0000000000000000E+00	+0.000000
+9.0000000000000000E+00	+2.0750000000000000E+01	+2.3807816883146600E+01	+0.0000000000000000E+00	+2.062580
+1.0000000000000000E+01	+2.1100000000000001E+01	+2.5375418881523125E+01	+0.0000000000000000E+00	+1.247060
+1.1000000000000000E+01	+2.2350000000000001E+01	+2.5811077057538732E+01	+0.0000000000000000E+00	+4.560541
+1.2000000000000000E+01	+2.2949999999999999E+01	+2.6048051792947795E+01	+0.0000000000000000E+00	+6.362967
+1.3000000000000000E+01	+2.3449999999999999E+01	+2.6181359087380354E+01	+0.0000000000000000E+00	+7.377734
+1.4000000000000000E+01	+2.4000000000000000E+01	+2.6298100424719685E+01	+0.0000000000000000E+00	+8.267579
+1.5000000000000000E+01	+2.4199999999999999E+01	+2.6359300495678035E+01	+0.0000000000000000E+00	+8.737101
+1.6000000000000000E+01	+2.4199999999999999E+01	+2.6382096058710022E+01	+0.0000000000000000E+00	+8.913899
+1.7000000000000000E+01	+2.4000000000000000E+01	+2.6374206151639207E+01	+0.0000000000000000E+00	+8.856084
+1.8000000000000000E+01	+2.3500000000000000E+01	+2.6276989832562773E+01	+0.0000000000000000E+00	+8.118796
+1.9000000000000000E+01	+2.3000000000000000E+01	+2.5977011667777411E+01	+0.0000000000000000E+00	+0.000000
+2.0000000000000000E+01	+2.2500000000000000E+01	+2.5290992029156097E+01	+0.0000000000000000E+00	+0.000000
+2.1000000000000000E+01	+2.2000000000000000E+01	+2.4654940045394628E+01	+0.0000000000000000E+00	+0.000000
+2.2000000000000000E+01	+2.2000000000000000E+01	+2.4244527289095615E+01	+0.0000000000000000E+00	+0.000000
+2.3000000000000000E+01	+2.2000000000000000E+01	+2.3889213691615041E+01	+0.0000000000000000E+00	+0.000000
+2.4000000000000000E+01	+2.1500000000000000E+01	+2.3561317657952550E+01	+0.0000000000000000E+00	+0.000000
+2.5000000000000000E+01	+2.0750000000000000E+01	+2.3150971042908957E+01	+0.0000000000000000E+00	+0.000000
+2.6000000000000000E+01	+2.0250000000000000E+01	+2.2760644989142502E+01	+0.0000000000000000E+00	+0.000000

Ln 1, Col 1 100% Windows (CRLF) UTF-8

Figure 12.
The .OUT formatted external file.

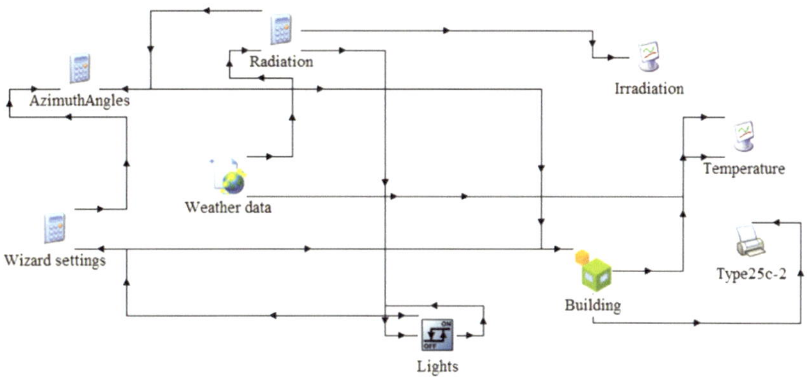

Figure 13.
Basic model of a multi-zone building in TRNSYS simulation studio.

time [15]. In addition, as noted in Section 3, Type 25c is the available component to print simulation results to an output file (an external file), of which the numerical results can be more easily edited by exporting the values from the output file into the Excel spread sheet. It should be noted that daylight or artificial light condition analysis for the multi-zone building cannot be modelled in TRNSYS 17, so the lighting will be only considered as a part of the heat gains for the energy aspect in this case.

The "Weather data" component (Type 15–3) will send the value of outdoor luminous intensity which is directly tied to its calculation of the controllers (AzimuthAngles and Radiation) of solar radiation as shown in **Figure 13** [16]. The "Lights on–off" is an on/off differential controller that generates a control function having a value of 1 or 0 [15]. If the Lights are on, they will stay "on" until the amount of horizontal solar radiation exceeds 200 W/m^2; while if the Lights are off, then they will stay "off" until the amount of horizontal solar radiation drops below 120 W/m^2 [16]. For the proposed simulations, the "Building Wizard" type in TRNSYS Simulation Studio is selected to make faster and easier simulation processes for the multi-zone building, in which the "Wizard setting" as a calculation controller is directly created under the Building Wizard and used for adjusting the orientation of the multi-zone building in terms of the pre-defined degrees of the Z-axis [17].

The thermal and optical properties of the proposed semi-transparent PV panels can be created in WINDOW program (that is, an open access computer program for calculating total window thermal performance indices) [18] if the components of the PV panels are not available in the current TRNSYS database. Electrical productions of the PV panels are calculated separately using Type 567 which is an existing BIPV component in the Thermal Energy System Specialists (TESS) libraries. The TESS libraries (developed by TESS of Madison, Wisconsin) are the new component libraries that contain more than 250 components used in the TRNSYS simulation, in which the components are used for the simulation of PV systems, solar thermal systems, geothermal systems, HVAC systems, and so on. Each of the components has been extensively tested and has the compatible format with TRNSYS Simulation Studio environment [19]. Type 567–5 (**Figure 14**) is used to model a glazed solar collector for the numerical results of electric power and also take the collected thermal energy into consideration, which is the most suitable PV model for the simulation of the electrical performance of the proposed semi-transparent PV glazing in TRNSYS. This model type can be combined with the detailed multi-zone building model (Type 56) that provides the temperature of the back surface of the PV glazing by giving the mean surface temperature [20].

As mentioned earlier, TRNSYS has already been validated and widely used in the BIPV related research activities. Egrican and Akguc [5] calculated cooling and heating demands of an office building with BIPV building façade through TRNSYS simulation, in which the PV panel model (Type 567) was well coupled with the building façade model (Type 56). Kim and Kim [6] evaluated both the electrical and thermal performances of a building with BIPV façade by means of the simulation modelling in TRNSYS, which modelled different cases of the building façade (that is, single-skin and ventilated double-skin façades) using Type 567 as the PV panel model and Type 56 as the building model.

Kamel and Fung [7] developed a roof system model that integrated a BIPV collector with an air source heat pump (ASHP) for a residential house in TRNSYS, in which the Type 567 was selected to model the glazed PV collector and connected to the multi-zone building model (Type 56). The simulation successfully predicted the seasonal performance of the heat pump and the electricity cost savings from the PV production. Elarga et al. [8] created a similar model typology as the proposed BIPV-DSF in this chapter, and investigated the energy performance of a DSF coupled with BIPV system for an office building, but the semi-transparent PV panel

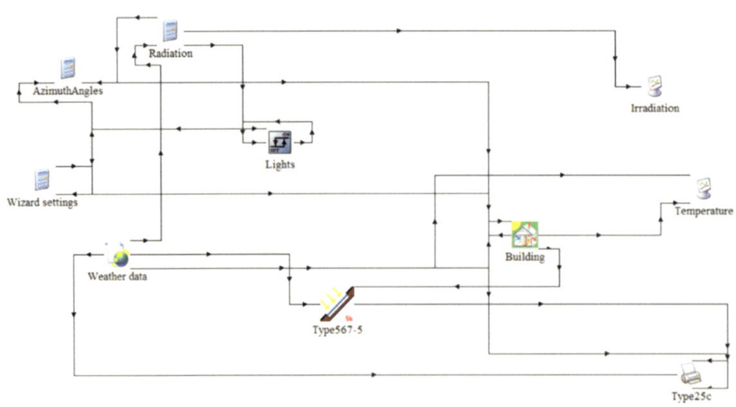

Figure 14.
Connection of Type567 and multi-zone building model in TRNSYS.

was mounted in-between the two skins of the DSF and Type 562 (it modelled the electric power of the PV only) was selected as the model of the PV panel. Also, in this case, the TRNSYS numerical model was validated against the experimental data and used for further analysis for different cases. Generally, the previous simulation studies have well proved the ability and capability of TRNSYS simulation in modelling the buildings adopt double-skin façade coupled with BIPV system, and the Type 567 is considered as the most appropriate component to model the electric and thermal characteristics of the proposed PV window glazing by coupling with the multi-zone building model (that is, Type 56 or Type 56-TRNFlow).

5. Summary of the BIPV-DSF model in TRNSYS

According to the descriptions in Section 4, the proposed BIPV-DSF model can be created in TRNSYS by employing the determined essential components and types. **Figure 15** shows a complete BIPV-DSF model with all the linked components and types in TRNSYS (Simulation Studio). The system components and types of the TRNSYS model presented in **Figure 15** are further described in **Table 1**.

"Type56-TRNFlow" is used to carry out the numerical building information of the BIPV-DSF model, which is able to model the ventilation of the façade/building model in TRNSYS Simulation Studio based on the characteristics of building geometry, while the non-geometry information (for example, building envelope properties, internal heat gains and occupancy) of the façade/building model is edited in TRNBuild.

As stated in above, the selected system components and types for the proposed TRNSYS models have already been validated and widely used in the BIPV related research activities, which can be confidently employed to perform the proposed simulation analysis.

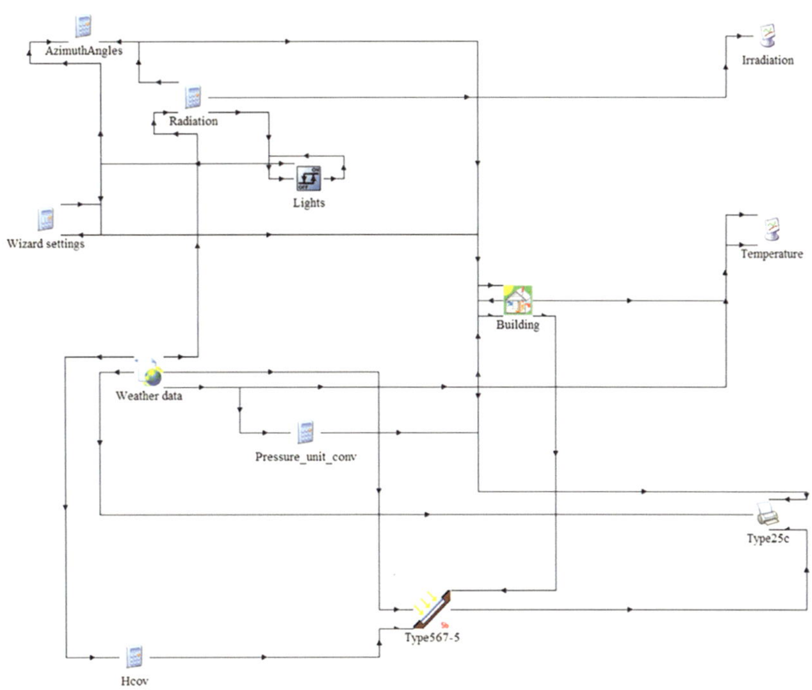

Figure 15.
Schematic diagram of the BIPV-DSF model in TRNSYS.

System components/types	Functions
Azimuth Angles	Determining solar azimuth angle
Radiation	Calculating solar radiation
Lights (Type2d)	Generating ON/OFF control functions
Wizard settings	Adjusting building orientation
Weather data (Type15–3)	Reading weather data at regular time intervals from the external weather file
H_{cov}	The convective heat transfer coefficient from the PV panel surface to ambient air
Pressure_unit_conv	Converting the default pressure unit in the Type15–3 from "atm" into "pa"
Building (Type56-TRNFlow)	Containing numerical information of building geometry, load profiles, construction, window glazing properties, airflow input and output and thermal behaviour of the building
Type567–5	The model of the PV panel
Irradiation (Type65d)	Displaying irradiation related variables while the simulation is progressing
Temperature (Type65d)	Displaying temperature related variables while the simulation is progressing
Type25c	Outputting the selected system variables at specified time intervals

Table 1.
Description of system components and types of the BIPV-DSF model.

6. Conclusions

This chapter presents a comprehensive method of numerical simulation modelling of a novel façade system – BIPV-DSF – in buildings. The proposed simulation modelling is carried out by using a graphically based design tool – TRNSYS and its relevant plugins (TRNFlow and TRNSYS 3D), and the performance of the BIPV-DSF is predicted accordingly.

It is demonstrated the applicability of the TRNSYS programme and its plugins in predicting the performance (for example, indoor thermal comfort and energy consumption) of the buildings incorporate the ventilated BIPV-DSF. The chapter also shows the applicability of TRNSYS in predicting the electric power produced by the semi-transparent PV panels, which serve as the external window glazing in the BIPV-DSF.

However, the BIPV-DSF model in TRNSYS should be further validated in order to reduce the discrepancies in-between the simulated and the actual building/façade behaviours, and therefore ensure the proposed BIPV-DSF model is created accurately and reliably. At this point, future work is supposed to calibrate the BIPV-DSF model against the real building/façade settings, hence it is possible to check validity and accuracy of the simulation results as well as to eliminate the random errors.

Author details

Siliang Yang[1,2]*, Francesco Fiorito[2,3], Deo Prasad[2] and Alistair Sproul[4]

1 School of Built Environment, Engineering and Computing, Leeds Beckett University, Leeds, United Kingdom

2 School of Built Environment, University of New South Wales, Sydney, Australia

3 Department of Civil, Environmental, Land, Building Engineering and Chemistry, Polytechnic University of Bari, Bari, Italy

4 School of Photovoltaic and Renewable Energy Engineering, University of New South Wales, Sydney, Australia

*Address all correspondence to: s.yang@leedsbeckett.ac.uk

References

[1] Peng, J., L. Lu, and H. Yang, *An experimental study of the thermal performance of a novel photovoltaic double-skin facade in Hong Kong.* Solar Energy, 2013. **97**: p. 293-304.

[2] Ding, W., Y. Hasemi, and T. Yamada, *Natural ventilation performance of a double-skin façade with a solar chimney.* Energy and Buildings, 2005. **37**(4): p. 411-418.

[3] Parker, N., et al., *Optimising environmental performance using building performance simulation.* Environment Design Guide, 2017: p. 1-16.

[4] Nimlyat, P.S., E. Dassah, and E.L.A. Allu, *Computer Simulations In Buildings: Implications For Building Energy Performance.* IOSR Journal of Engineering, 2014. **4**(3): p. 56-62.

[5] Egrican, N. and A. Akguc, *Thermal performance estimation of the office building with the building integrated photovoltaic system*, in *ASME 2011 5th International Conference on Energy Sustainability*. 2011, American Society of Mechanical Engineers. p. 91-101.

[6] Kim, J.-H. and J.-T. Kim, *A Simulation Study of Air-Type Building-Integrated Photovoltaic-Thermal System.* Energy Procedia, 2012. **30**: p. 1016-1024.

[7] Kamel, R.S. and A.S. Fung, *Modeling, simulation and feasibility analysis of residential BIPV/T+ASHP system in cold climate—Canada.* Energy and Buildings, 2014. **82**: p. 758-770.

[8] Elarga, H., A. Zarrella, and M. De Carli, *Dynamic energy evaluation and glazing layers optimization of façade building with innovative integration of PV modules.* Energy and Buildings, 2016. **111**: p. 468-478.

[9] Weber, A., et al. *TRNFLOW: Integration of COMIS into TRNSYS TYPE 56*. in *Proceedings of the 3rd european conference on energy performance and indoor climate, EPIC 2002*. 2002.

[10] Lalwani, M., D.P. Kothari, and M. Singh, *Investigation of solar photovoltaic simulation softwares.* International journal of applied engineering research, 2010. **1**(3): p. 585-601.

[11] Jonas, D., et al., *A TRNSYS-based simulation framework for the analysis of solar thermal and heat pump systems.* Applied Solar Energy, 2017. **53**(2): p. 126-137.

[12] Kalogirou, S.A., *Use of TRNSYS for modelling and simulation of a hybrid pv–thermal solar system for Cyprus.* Renewable Energy, 2001. **23**(2): p. 247-260.

[13] TRNSYS. *Suite of Tools*. 2018 [cited 2021 20 January]; Available from: http://www.trnsys.com/features/suite-of-tools.php.

[14] Transsolar, *A module of an air flow network for coupled simulation with Type56 (multi-zone building of TRNSYS)*. 2009.

[15] Solar Energy Laboratory, *TRNSYS 17 a TRaNsient SYstem Simulation Program*, in *Volume 3 Standard Component Library Overview*. 2014, Solar Energy Laboratory, University of Wisconsin-Madison: Madison, Wisconsin.

[16] Bradley, D. *[TRNSYS-users] Artificial lighting*. 2014 [cited 2021 10 January]; Available from: http://lists.onebuilding.org/pipermail/trnsys-users-onebuilding.org/2014-March/026133.html.

[17] Claudino, P.A., *Experiment and modelling study of a geodesic dome solar greenhouse system in Ottawa*. 2016, Carleton University.

[18] Lawrence Berkeley National Laboratory. *WINDOW: A computer program for calculating total window thermal performance indices*. 2019 [cited 2021 15 February]; Available from: https://windows.lbl.gov/software/window.

[19] Aiguasol. *TESS 17 – Component Libraries (Trnsys 18)*. 2021 [cited 2021 10 January]; Available from: https://aiguasol.coop/energy-software/tess-17-component-libraries-trnsys-18/.

[20] Thermal Energy System Specialists, *TESSLibs 17 Component Libraries for the TRNSYS Simulation Environment*, V.E.L.M. Reference, Editor. 2013, Thermal Energy System Specialists, LLC: Madison, Wisconsin.

Chapter 5

Parameter Dependencies of a Biomechanical Cervical Spine FSU - The Process of Finding Optimal Model Parameters by Sensitivity Analysis

Sabine Bauer and Ivanna Kramer

Abstract

The knowledge about the impact of structure-specific parameters on the biomechanical behavior of a computer model has an essential meaning for the realistic modeling and system improving. Especially the biomechanical parameters of the intervertebral discs, the ligamentous structures and the facet joints are seen in the literature as significant components of a spine model, which define the quality of the model. Therefore, it is important to understand how the variations of input parameters for these components affect the entire model and its individual structures. Sensitivity analysis can be used to gain the required knowledge about the correlation of the input and output variables in a complex spinal model. The present study analyses the influence of the biomechanical parameters of the intervertebral disc using different sensitivity analysis methods to optimize the spine model parameters. The analysis is performed with a multi-body simulation model of the cervical functional spinal unit C6-C7.

Keywords: multi-body simulation, sensitivity analysis, cervical spine FSU model, intervertebral disc pressure, stiffness and damping coefficients

1. Introduction

Biomechanical modeling offers a non-invasive possibility to analyze and answer kinematic and kinetic questions. A distinction is made between finite element (FE) simulation and multi-body simulation (MBS). The difference between FE and MBS modeling lies in the basic model structure and thus in the field of application. Further information on FE and MBS are described in [1]. Due to their complexity, FE models make an important contribution to understand the biomechanical function of the spine and the behavior of spinal structures in the state of health, illness or damage [2–4] as well as the influence of the material parameters of various implants and fusion techniques [5–8]. However, the complexity of the FE models usually requires high computing times for each simulation case. If the aspect of predicting kinematic and dynamic reactions of the whole or a larger part of the spinal column during complex movement sequences is the focus of interest, MBS is a suitable simulation method due to the highly efficient short computing times [1]. The existing FE models are mostly based only on a specific or an idealized average model with unique mechanical and geometrical characteristics. According to [9], a better insight into the influence of the biomaterial and the geometrical diversity on the biomechanical behavior of the spine is essential for a better understanding of the spine mechanics and the patient care. Because a model contains numerous parameters that are often only vaguely known and too complex to implement, their effect on the responses is a priori unknown and full validation is largely impossible. Therefore there is a need for sensitivity analysis [10]. Sensitivity analysis can be used to gain the required knowledge about the correlation of the input and output variables in a complex spinal model, which has an essential meaning for the realistic modeling and system optimization. Especially the biomechanical parameters of the intervertebral discs, the ligamentous and muscular structures and the facet joints are significant components of a spine model, which define the quality of the model. Hence, it is important to understand, how the variations of input parameters for these components affect the entire model and its individual structures. The present study analyses the influence of the biomechanical parameters of the intervertebral disc using different sensitivity analysis methods, which enables the direct optimization of the spine model parameters. The analysis is performed with a multi-body simulation model of the cervical functional spinal unit C6-C7.

2. Model configuration

The MBS model of a functional spinal unit (FSU) consists of the vertebrae C7 and C6 represented by rigid bodies. Furthermore, an intervertebral disc, ligamentous structures and facet joints are implemented with specific biomechanical characteristics. When configuring the model, the focus is on the creation of the simplest possible model so that all biomechanical parameters could be adequately defined. In the case of models with many parameters, there is a risk that the parameters cannot be determined sufficiently. Therefore, a model with a high parameter dependency does not necessarily lead to better results.

2.1 FSU setup

The 3D surface of the vertebrae based on artificial vertebrae (Sawbones) and were implemented as triangular meshes. The 3D geometry data of the C7 vertebra can be taken out of **Figure 1** and of C6 out of **Figure 2**. The abbreviations for different vertebral parts are made up of three capitalized letters and adapted from [11]. The first two describe the corresponding vertebral part and the third represents the dimension to be measured. These three letter combinations can be supplemented by a lower case letter that indicates a direction, such as right (r), upper (u) and depending on the content, lower or left (l).

The body reference frame of two vertebrae C7 and C6 are located at the same position. The location of the center of gravity (CG) of both vertebrae is visualized in **Figure 3** and the data can be taken out of **Table 1**. The coordinates of the CG are given relative to the reference frame of the corresponding vertebra. The center of gravity is defined as a point at which the entire mass of the vertebra is united and where the earth gravitational acceleration with $g = -9.81\ m/s^2$ along the z-axis is applied. The mass properties of the rigid vertebrae are automatically calculate from their 3D geometry. For each single vertebra, the tessellated volume of its 3D data is multiplied with the specific density. The density is specified by [12] with 473 mg/cm^3 for C6 and 414 mg/cm^3 for C7.

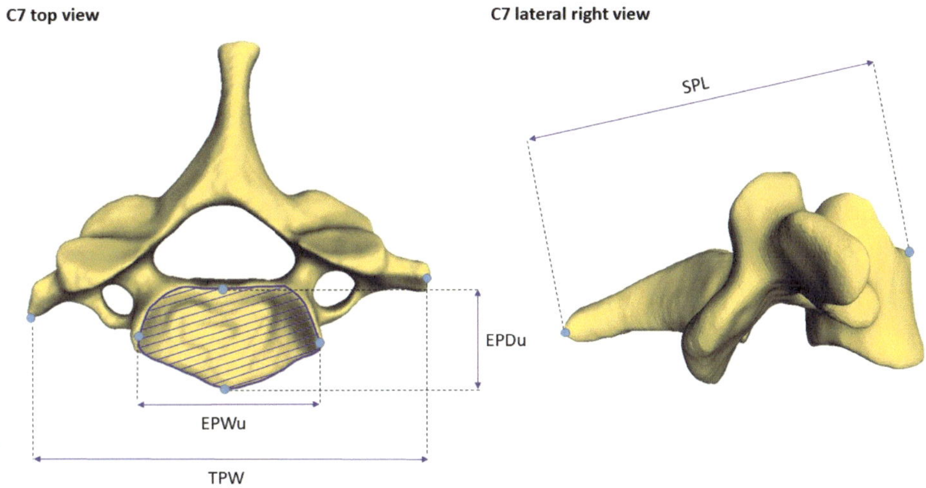

Figure 1.
Geometry data of the vertebra C7.

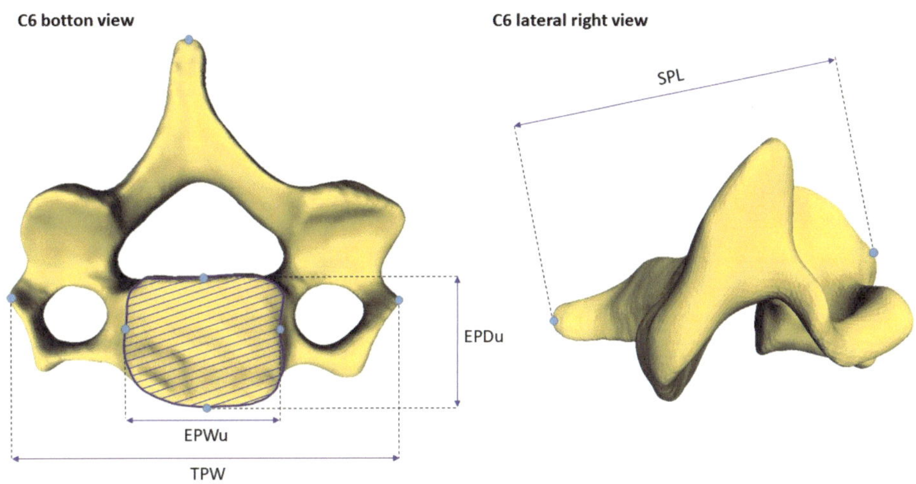

Figure 2.
Geometry data of the vertebra C6.

The information of the inertia moment (**Table 2**) relates to the body's center of gravity. The moment of inertia (I) is defined as a symmetric matrix whose entries are mirror-symmetric with respect to the main diagonal and relative to center of mass of the corresponding vertebra.

2.2 Intervertebral disc modeling

The biomechanical characteristic of the intervertebral disc between C7 and C6 is represented by a simple stiffness-deformation relation and a velocity-dependent damping term. If a load is applied to the model, the disc is deformed and develops reaction forces that depend on the deformation value and the deformation velocity. The forces F_x, F_y, F_z are interacting between two defined markers, one refers to C7 and another to C6 in three translation directions x, y, z. The corresponding force equation is determined by four main components: stiffness constant c, damping constant d, disc deformation and deformation velocity. The stiffness term c as well as

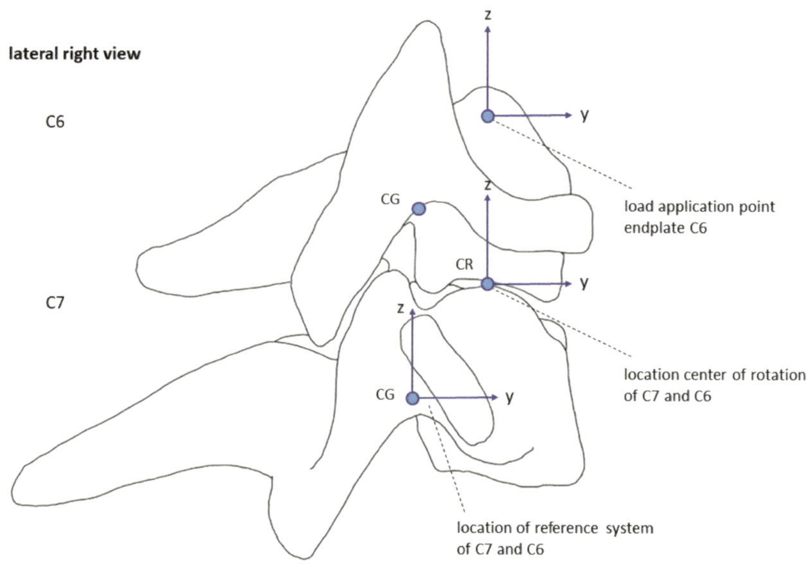

Figure 3.
Center of gravity (CG), reference frame of the vertebra C7 and C6, location of center of rotation (CR) and load application point.

Vertebra	Mass [kg]	CG_x [m]	CG_y [m]	CG_z [m]
C7	0.0070	-8.23×10^{-9}	-9.13×10^{-9}	-8.37×10^{-9}
C6	0.0057	6.2×10^{-4}	4.1×10^{-3}	2.0×10^{-2}

Position of gravity center CG is given with respect to local body coordinates system.

Table 1.
Mass and Center of Gravity of vertebrae C7 and C6.

Vertebra	I_{xx} [kg m²]	I_{yy} [kg m²]	I_{zz} [kg m²]	I_{xy} [kg m²]	I_{xz} [kg m²]	I_{yz} [kg m²]
C7	1.41×10^{-6}	1.48×10^{-6}	2.50×10^{-6}	-3.87×10^{-8}	6.89×10^{-8}	-2.44×10^{-8}
C6	6.77×10^{-7}	1.24×10^{-6}	1.65×10^{-6} 1	-1.47×10^{-8}	6.25×10^{-8}	1.36×10^{-9}

Moments of inertia I are given with respect to the local body center of gravity.

Table 2.
Moments of inertia with respect to local body center of gravity for the vertebrae of FSU C7-C6.

the damping constant d are represented separately for each translation direction c_x, c_y, c_z and d_x, d_y, d_z respectively. The disc deformation value is calculated as a distance between two points and is represented by the variables x_F, y_F, and z_F, where the axis-wise velocities of the markers are $\dot{x}$, $\dot{y}$ and $\dot{z}$. The disc force is defined in Eq. (1).

$$\begin{pmatrix} F_x \\ F_y \\ F_z \end{pmatrix} = \begin{pmatrix} c_x x_F + d_x \dot{x} \\ c_y y_F + d_y \dot{y} \\ c_z z_F + d_z \dot{z} \end{pmatrix} \tag{1}$$

The disc force is implemented in such a way, that its responds depend on specific movement scenarios: if the disc is deformed by an external load and the deformation velocity vector is negative, then the disc force is determined by both the stiffness and the damping terms. If the intervertebral disc is still deformed but begins to relax, then the deformation velocity vector changes into a positive direction. In this state, the disc force is only determined by the stiffness term. If the intervertebral disc is stretched, both terms are set to zero (**Figure 4**).

The initial value for the stiffness bases on [13] and the damping value is set to 10% of the stiffness value, because no actual cervical spine disc damping coefficients have been reported in the literature [14].

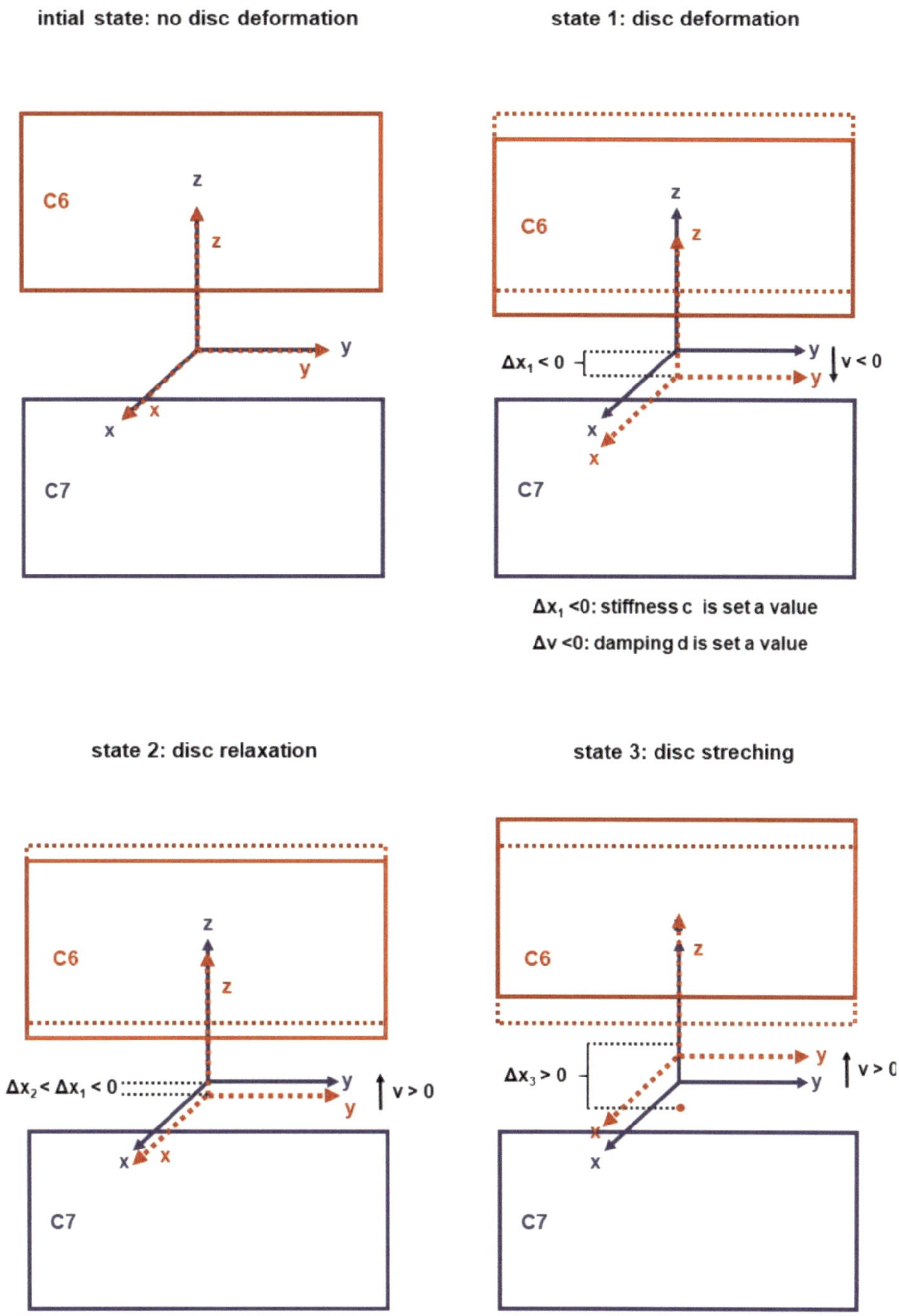

Figure 4.
Schematic representation of the intervertebral disc characteristics under different stress scenarios.

In reality the intervertebral disc is not only deformed by loads, but also bent by external torques. Depending on the action direction of the external torque the intervertebral disc performs a flexion and extension movement, an axial rotation or a lateral flexion. To counteract this rotations, the intervertebral disc develops a counter-torque. This non-linear disc torque is defined by two-dimensional functions that describe the relationship between the disc torque and the relative angle. A specific input function is assigned to the torques acting around three axes of rotation x, y and z. The applied input function bases on [15].

2.3 Facet joint modeling

Through the facets, adjacent vertebrae are connected via a thin layer of cartilage. In the model the facet cartilage layers are approximated by an unilateral contact spring-damper element, whose contact area is determined by the facet geometry. The contact area is a rectangular region, which represents the facet width and height. With an additional dimension the cartilage layer of the facet joint is simulated. The cartilage layer thickness of the lower cervical spine bases on [16] and is determined to be 0.00045 m for the superior layer and 0.00049 m for the inferior. The parameterization of the geometry, positioning and orientation of the 3D facet contact area is determined with respect to the C7 upper facet surface. The modeled facet contact surface is assumed to be an average facet width and facet height of the superior facet surface C7 and the inferior facet surface C6. The average model geometry results in a facet width (FW) of 0.0094 m and a facet height (FH) of 0.009 m. Comparison of the approximate facet area (FCA) of the current model with FCA = 0.000085 m^2 with the average facet area superior C7 and inferior C6 reported in [17], a discrepancy of FCA = 0.000089 m^2 (**Figure 5**) can be observed. This information is given at this point in order to show the extent to which the model assumption differs from the experimental measurements with regard to the geometry.

The stiffness coefficients are taken from [18] and the damping values is defined as 10% of the stiffness term. The damping coefficient is used to obtain a better attenuation of the maximum linear and angular accelerations of the head [19].

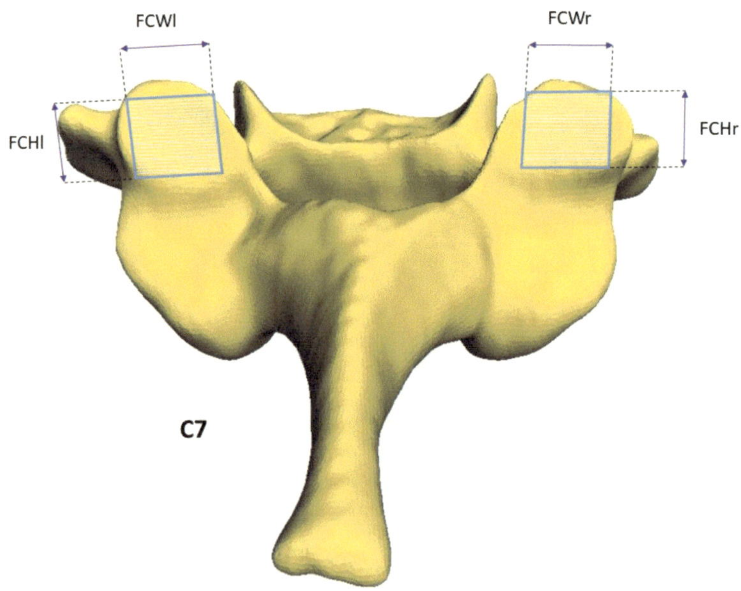

Figure 5.
Representation of the facet width and height, which builds the basis area for the facet contact simulation.

2.4 Ligament modeling

The spinal ligaments provide stability to the motion segments allowing motion within physiological limits. Ligaments are uniaxial structures that resist only tensile or distractive forces becoming slack in compression [14, 20].

In the FSU model the following ligaments are incorporated: anterior and posterior longitudinal ligament (ALL and PLL), flava ligament (FL), interspinous ligament (ISL), nuchal ligament (NL) and the left and right capsular ligaments (CL) (**Figure 6**). Ligaments, which have a broad structure, are represented by several fiber bundles. For instant, ALL and PLL are composed of a right, left and middle ligament structure. CL is approximated by four individual ligament structures that attach to the top, bottom, left and right surfaces of the articular processes. The ISL

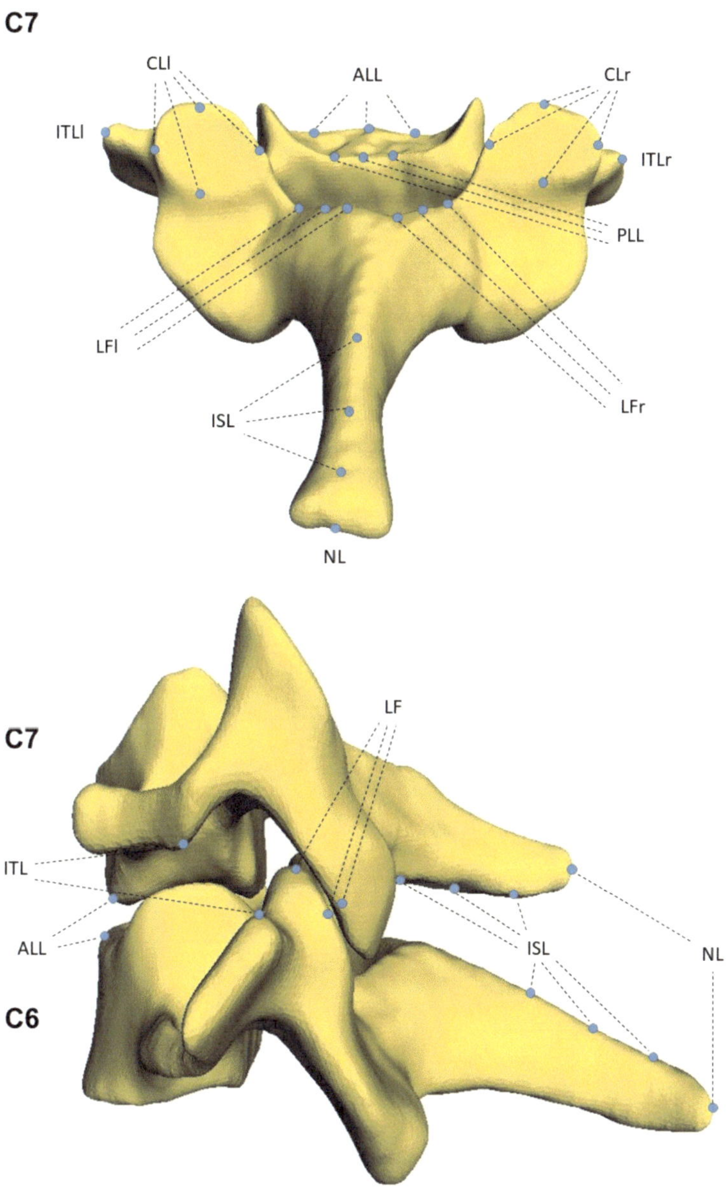

Figure 6.
Representation of the ligament attachment points.

extends over the entire edge of the spinous process and is therefore modeled using three bundles of ligaments. The LF attaches to the proximal edge of the lamina and is represented by six ligament fibers. The NL is an extension of the SSL which extends from the external occipital protuberance to the spinous process of C7 and attaches all the posterior tips of the spinous processes in between [21].

The determination of the ligament attachment points is carried out on the basis of the vertebral geometry and is checked by an expert.

The ligament's characteristic is modeled by the load displacement curves [13, 22, 23]. When a ligament is stretched, it develops a force that is specific to the ligament in question. It acts against the direction of the stretch with no resistance in compression.

2.5 Load case configuration

In order to analyze the reaction of the spinal structures to a load, an external force of 80 N is applied to the endplate of the vertebra C6. This loading case is chosen because the cervical spine is permanently loaded by the weight of the head [24]. To prevent additional torques, the y-coordinates of the external load markers have the same position as the y-coordinates of the disc joint, so that there is no initial lever arm that could lead to unintentional torques.

2.6 Model validation

An important step in the simulation process is the model validation, with which the simulation results are checked for correctness. The correctness of the FSU is proven by comparing the intervertebral disc pressure and disc deformation to existing published data. After researching the literature, it turned out that there is only a limited possibility of validation data that exactly depicts the simulation scenario we have modeled at the moment. In general, there is the difficulty that the own model configuration does not necessarily exactly match to that of other researchers, since different specific research questions have to be answered. In order to get the response of the FSU model to different loads, the FSU is exposed to small and large external loads. The disc pressure and deformation are compared (**Table 3**).

Model	C7 EPWu [mm]	C7 EPDu [mm]	C6 EPWl [mm]	C6 EPDl [mm]	C7 EPAu [mm^2]	C6 EPAl [mm^2]	C6-C7 DW [mm]	C6-C7 DD [mm]	C6-C7 DH [mm]	C6-C7 DA [mm^2]
Current FSU model	21.7	16.9	19.2	15.4	288.0	232.2	20.45	16.15	0.0068	260.1
Hueston et al. [23]	19.0	15.1	19.5	15.7	220.8	316.3				268.6
Tan et al. [25]	19.0	15.1	19.5	15.7	220.8	316.3				
Yoganandan et al. [20]									0.005–0.0075	168–502
Pooni et al. [26]										200–502

The width (W), depth (D) and area (A) of upper (u) and lower (l) endplates (EP) are presented of different studies. Further, the disc width (DW), the disc depth (DD), the disc area (DA) and the disc height (DH) is presented. The idea is to present the various possible measures to be able to assess the model parameters of the current model.

Table 3.
Comparison of the vertebra C7 and C6 anthropometry.

2.7 Motion segment response to small loads

A validated intact FE model of the C4-C5-C6 cervical spine to simulate progressive disc degeneration at the C5-C6 level is presented by [24]. The intact and three degenerated cervical spine models are exercised under the compression load of 80 N. The results of the intact spine model are used to compare the intervertebral disc pressure between vertebrae C5-C6 in the current FSU model. The motion segments were subjected to a small static compression load of 80 N in z-direction. While in the current model the resulting displacement of the intervertebral disc is measured, in the FSU model the overall force displacement response of C4 with respect to C6 is determined. Therefore, the comparison can only be taken as a rough evaluation of the models deformation.

2.8 Motion segment response to large loads

In the second stage of validation, the FSU model is subjected to larger loads of 200 N, 500 N and 673 N to determine its intervertebral disc pressure and disc deformation. The load of 200 N is chosen to represent the combined effects of head weight and muscle tension [27]. The human cervical disc pressure using a pressure transducer, side-mounted in a 0.9 mm diameter needle is investigated by [27]. Forty-six cadaverous cervical motion segments aged 48–90 years are subjected to a compressing load of 200 N for 2 s. Due to the lack of data available for high load cases, these data are used to analyze the characteristics of the intervertebral discs. The deformation value under a certain load is only provided for the specific healthy disc segment C7-T1. These results are used to compare the characteristics of the intervertebral discs in the current model.

A MBS model of human head and neck C7-T1 is presented by [14]. The MBS model comprise soft tissues, i.e. muscles, ligaments, intervertebral discs and supported through facet joints. Also eighteen muscle groups and 69 individual muscle segments of the head and neck are included in the model. For load–displacement testing, each motion segment is mounted so that the inferior vertebra is rigidly fixed whereas the superior vertebra is free to move in response to the

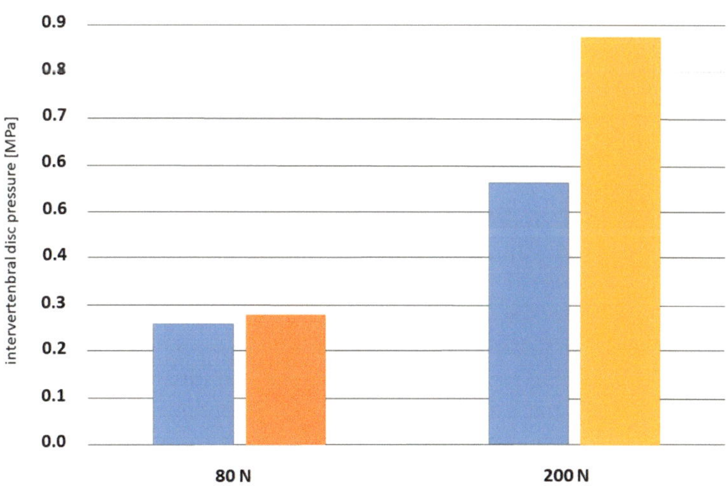

Figure 7.
Response of model motion segments to applied compressive loads of 80 N and 200 N. the orange bar shows the intervertebral disc pressure for the corresponding motion segment as reported by [24] and the yellow one as reported by [27]. The results of the current FSU model is highlighted in blue.

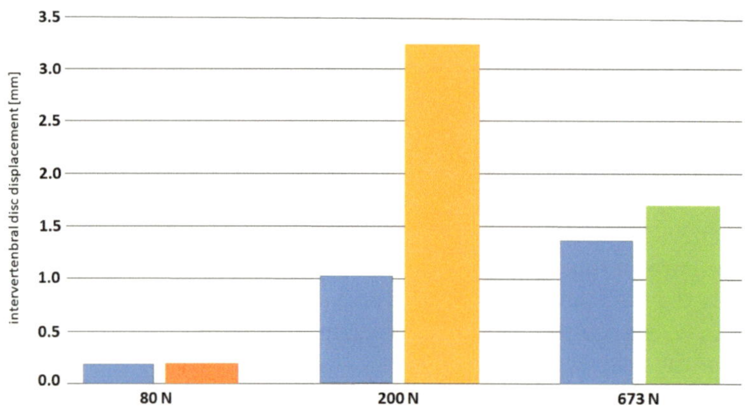

Figure 8.
Comparison of the intervertebral disc displacement with experimental results [20], FE model [24] and MBS model [14]. The results of the currant FSU model are presented in blue bars, die orange bar represent the results of [24], the yellow one of [14] and the green one of [20].

applied loads. The response of model motion segments C5–C6 to the applied translational load of 500 N is shown.

A review with the focus on soft tissue structural responses with an emphasis on finite element mathematical models is done by [20]. Biomechanical data of intervertebral disc under compression test are provided for the FSU C6-C7. Under a load of 673 N the intervertebral disc between vertebrae C6 and C7 is deformed with 1.7 mm.

The comparison of the intervertebral disc pressure under the loads of 80 N and 200 N are shown in **Figure 7**.

Figure 8 compares the effect on the intervertebral disc deformation of the loads 80 N, 500 N and 673 N. The intradiscal pressure almost agrees with the published pressure at the loads of 80 N and 673 N. The current model has about one-third lower disc displacement than the comparison model [14]. One reason for this can be, that under certain circumstances the muscles, that are not taken into account in the current FSU model, are accompanied by a lack of muscle tension, which leads to the less compression of the intervertebral discs.

3. Effects of stiffness and damping variations

3.1 Method

In the biomechanical modeling the quality of the simulation can be considered valid only when the model and the input parameters are accurate and robust [28]. To examine the robustness of the modeled system a method called sensitivity analysis is the first choice. Generally speaking, *sensitivity analysis is collection of approaches, that determine, quantify and analyze the impact of the input parameters on the model outcome* [29]. The sensitivity analysis can also identify those components of the model that might need additional studies to be performed. Further, in the model optimization the sensitivity analysis can be used to refine the values of the critical parameters as well as to simplify or ignore those factors, which do not show any impact on the model response [30].

One of the simplest and effective techniques used to determine the level of the sensitivity or insensitivity of the model outputs to the plausible variation of one particular parameter is *one-way sensitivity analysis* [31, 32].

In order to identify the effect of both stiffness and damping variations on the current FSU model, one-way sensitivity analysis is performed. After every simulation run the corresponding changes in the intervertebral disc pressure in the C6-C7 segment are reported as a difference between initial and current disc pressure $\delta P_i = P_i - P_{init}$, where the initial pressure value is $0.301315\,MPa$.

3.2 Simulation results of stiffness term alternation

The first series of simulations consider the variation of the stiffness term c, which is expressed in Eq. (1), however the damping value d is held constant by $50000\,Ns/m$. For the sake of simplicity, the same value c_i is assigned to c_x, c_y, c_z in every i-th simulation. Starting from the initial value of $500000\,N/m$ in each experiment repetition the stiffness parameter is increased and decreased by a fraction $f \in \{0.1, 0.2, 0.3, 0.4, 0.5\}$.

It can be seen in **Figure 9**, that the variation in the stiffness term results in the linear change of the disc pressure, which points out to the linear relationship between these two variables. However, the change plot indicates the opposite linear relationship between the stiffness and the disc pressure, where increasing of the stiffness causes decreasing of the pressure. Note, that the course of the disc pressure changes is symmetrical, i.e. the minimum and maximum changes in the pressure value are of the same magnitude. According to the revealed results the maximal absolute pressure change is reported to be $7.0935249 \times 10^{-5}\,MPa$ or 0.02% of the initial value.

3.3 Simulation results of damping term alternation

The second part of the experiments aims the analysis of the system sensitivity with respect to alternations of the damping constant d (see Eq. (1)). In order to simplify the experiment execution, the same value d_i is assigned to d_x, d_y, d_z in each i-th simulation run. The stiffness term is set to be constant, i.e. $c_i = 500000\,N/m$ for all N trials. The damping parameter d_i is increased and decreased by a fraction $f \in \{0.1, 0.2, 0.3, 0.4, 0.5\}$ of its initial value $d_{init} = 50000\,Ns/m$.

The impact of the damping parameter changes on the disc pressure is presented in **Figure 10**. Similarly to the results obtained in Section 3.2, an obvious influence of the damping term on the disc pressure can be observed, where the linear changes of d_i are reflected in the linear changes of the disc pressure p_i. However, the magnitude of the change is not symmetrical for decreasing and increasing values of the

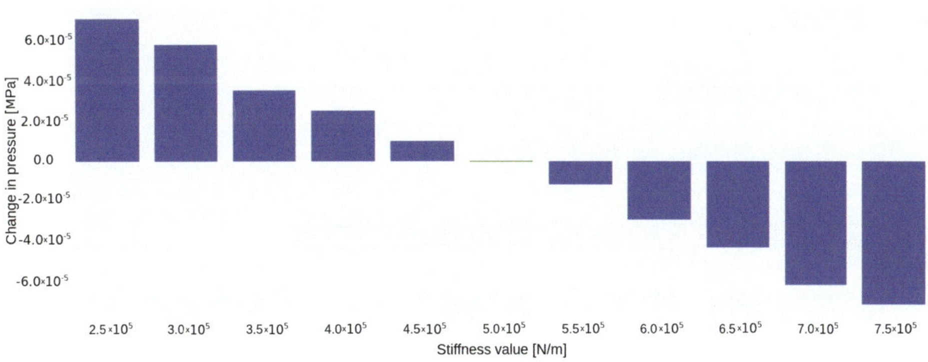

Figure 9.
Decreasing (left side) and increasing (right side) the stiffness term c by factor f impact the intradiscal pressure. The pressure change at the initial point is 0 and is marked green.

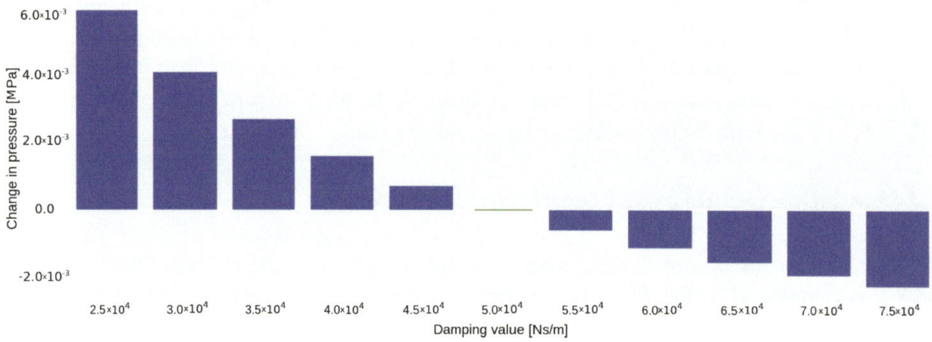

Figure 10.
Decreasing (left side) and increasing (right side) the damping value by f effects the disc pressure in a linear manner. The disc pressure at the initial point is 0.301315MPa, the change of pressure at initial point is 0 (marked green).

damping factor. Moreover, the smaller damping values result in higher changes of the intradiscal pressure. The maximal pressure change is found to be $0.00593\ MPa$ for $d = 25000\ Ns/m$, which is approximately 1.991% of the initial pressure value. In comparison, for $d = 75000\ Ns/m$ the change is $-0.00226\ MPa$ and 0.6% respectively.

In **Figure 11** the disc pressure changes affected by percentage decreasing of both model factors d and c following the one-way sensitivity analysis approach are depicted. It can be seen, that the same alternation of the damping term causes approximately two orders higher magnitude of the disc pressure. To examine the hypothesis, that the damping parameter has much stronger influence on the system, the calculation of a further sensitivity metric called **sensitivity coefficient** is elaborated [33]. In our particular setting the sensitivity coefficient s_v is defined to be an average quotient of the disc pressure change p_i to the i-th change in the parameter value v_i:

$$s_v = \frac{1}{N} \sum \frac{\delta p_i}{\delta v_i}, \tag{2}$$

where N is a total number of trials and δv_i is the i-th change in the observing parameter, $v_i \in \{c_i, d_i\}$.

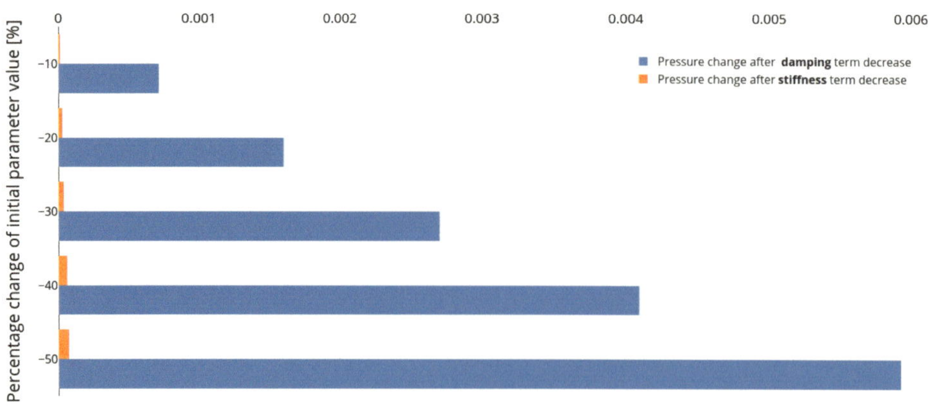

Figure 11.
Comparison of the maximal disc pressure changes [MPa] given with respect to the variability of the input parameters c and d. The stiffness (marked red) and damping (marked blue) terms are decreased by the factor of 10–50%.

	s_c	s_d
Sensitivity coefficient value	-1.145×10^{-5}	$-9.093^{-5}10^{-7}$

Table 4.
Sensitivity coefficient determined for parameters c and d using Eq. (2).

Determined coefficients for the stiffness s_c and damping term s_d are shown in **Table 4**. The obtained results support the above statement, that the behavior of the current model is approximately 12.59 times more sensitive to the damping term than to the stiffness parameter.

4. Impact of different load cases and intervertebral disc areas on intervertebral disc pressure

4.1 Impact of different loads on intervertebral disc pressure

One of the main functional task of the intervertebral disc is transmitting the compressive loads through the spine [34]. Therefore, it is important to study the sensitivity of the input parameters as well as mechanical responses of the model considering multiple loading cases. For this experiment, the acting of various external loads $l \in L$, where $L = \{100N, 200N, \ldots, 800N\}$ on the upper endplate of the C6 vertebra (see **Figure 3**) is simulated. Such high forces are selected in order to investigate the model behavior under different boundary conditions.

The disc pressure responded by the current model is reported in **Figure 12**. It can be seen, that the stiffness alternations among the load cases do not lead to significant change in the disc pressure. An unusual pattern is observed in each particular load situation, where the stiffness variation causes the linear growth in the disc pressure followed by piece-wise non-linear regions. Please note, that this disc

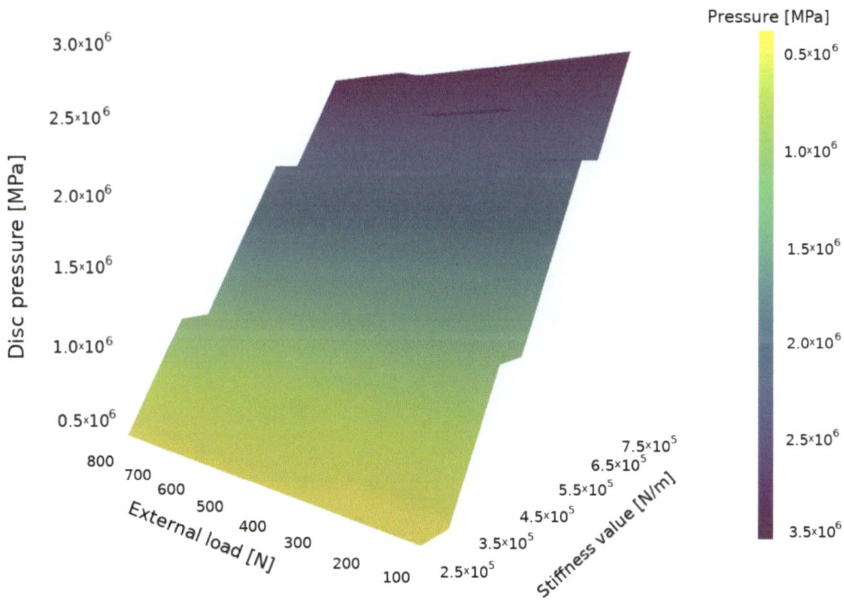

Figure 12.
Maximal intradiscal pressure for C6-C7 segment calculated for multiple compressive loads and different stiffness value. The initial stiffness term is decreased and increased by a factor up to 50% of its initial value $c = 500000\ N/s$. The damping term is help constant $d = 50000\ Nm/s$.

behavior is not detected in the simulations applying the load of 80 N (see Sections 3.2 and 3.3). In details, the non-linear pressure change is detected withing the following ranges of the stiffness term: $c \in [2.5 \times 10^6, 3.0 \times 10^6]$ as well as for $c \in [4.0 \times 10^6, 4.5 \times 10^6]$ and $[6.5 \times 10^6, 7.0 \times 10^6]$.

Figure 13 illustrates the results of the simulations where the applied loads $l \in L$ and the damping factor d are varied simultaneously. The perspective view of this diagram is slightly different in order to emphasize the regions where the non-linearity in the disc pressure occurs. The dark spots in the plot indicate the jumps in the disc pressure value over the loads, where the step-wise patterns show the non-linear responses of the current model for the following cases: 3.0×10^4 Ns/m for the applied force of 200 N, 3.5×10^4 Ns/m for 100 N load, another peaks are observed for the exerting force of 500 N at damping value of 5.0×10^4 Ns/m.

4.2 Impact of different intervertebral disc areas on disc pressure

The size of the disc area is presented in the literature with different values. This leads to the question how different disc areas influence the disc pressure. In order to investigate this effect, approximated intervertebral disc areas from the literature [11, 20, 25, 26] are used as examples. Values of the FSU C6-C7 disc area with minimum of 168 mm^2 and maximum of 502 mm^2 are published in [20]. Estimated disc areas of 180 mm^2, 230 mm^2 and 295 mm^2 are published by [26] and represented as mean values from 3 specimens of their cervical spine. The EPAu of C7 and the EPAl of C6 is specified in [11, 25]. The mean values of EPAu of C7 and EPAl of C6 with 284 mm^2 and 269 mm^2 respectively are taken to approximate the area of the corresponding intervertebral disc.

All disc area values listed above serve as input for the analysis of the relationship between intervertebral disc pressure and disc area. The effects of different

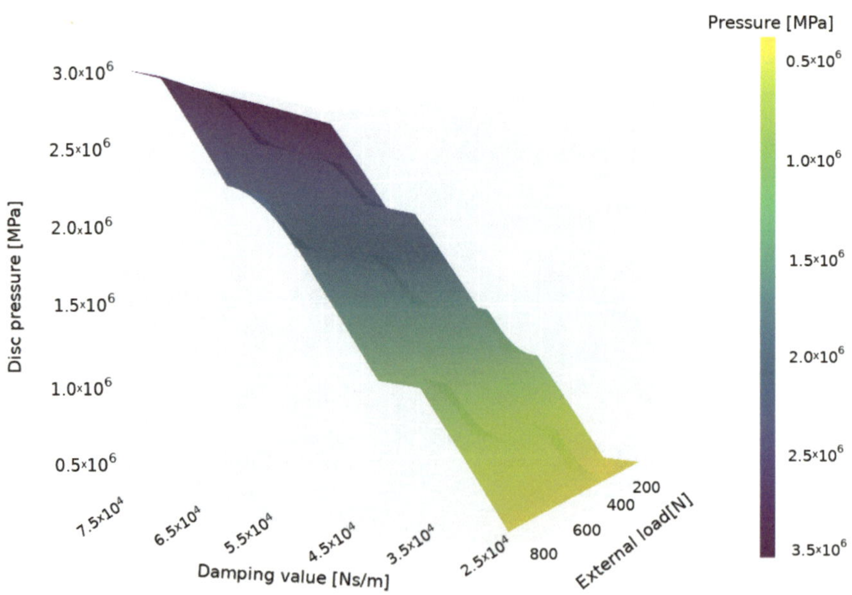

Figure 13.
Maximal intradiscal pressure calculated for C6-C7 segment. Different compressive loads and values of the damping term are simultaneously changed, however the stiffness factor is set constant to $c = 500000$ N/s. The simulation results reveal the areas (dark spot regions in the plot) with a non-linear changes of the maximal disc pressure.

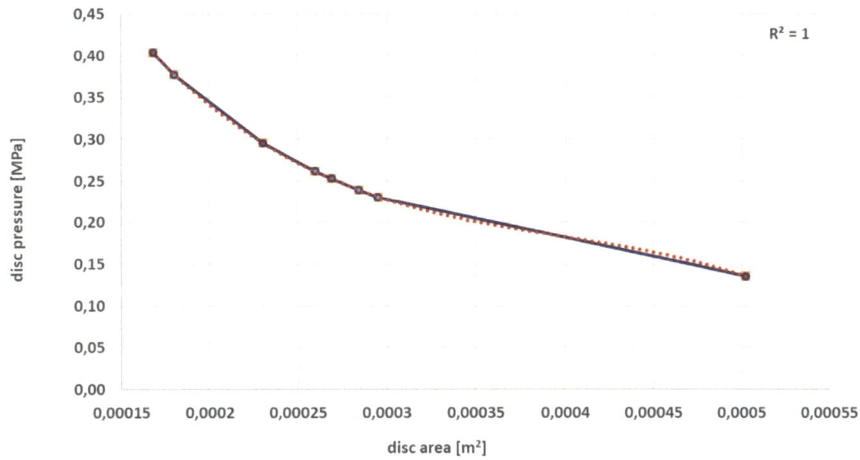

Figure 14.
Representation of the relationship between disc area size and intradiscal pressure. The pressure is determined for eight disc areas of different sizes and an external load of 80 N. the blue points in the plot are the data points connected by a best-fit straight line. The disc areas are based on literature data. The assignment of the data points with their specific intervertebral disc areas to the corresponding literature is as follows (starting from the top left side): Data point 1 [20], data point 2 [26], data point 3 [26], data point 4 (current FSU model), data point 5 [25], data point 6 [11], data point 7 [26], data point 8 [20].

intervertebral disc area on the intradiscal pressure are considered under the load case of 80 N.

In **Figure 14** it can be clearly seen that the size of the disc area has a direct effect on the disc pressure. The course can be approximated by a 3-degree polynomial. In order to assess the goodness of the polynomial fit, the coefficient of determination R^2 is calculated. R^2 is defined to be the square of Pearson correlation coefficient $r_{x,y}$, i.e. $R^2 = r_{x,y}^2$. Pearson correlation coefficient is determined for data pairs $\{(x_1, y_1), \dots, (x_n, y_n)\}$ as follows:

$$r_{x,y} = \frac{\sum_{i=1}^{N}(x_i - \overline{x})(y_i - \overline{y})}{\sqrt{\sum_{i=1}^{N}(x_i - \overline{x})^2 \sum_{i=1}^{N}(y_i - \overline{y})^2}}, \tag{3}$$

where N is sample size, x_i, y_i are the individual sample points, $\overline{x} = \frac{1}{N}\sum_{i=1}^{N} x_i$ the sample mean for x, which is calculated for $\overline{y}$ analogically and $i \in N$.

The calculated R^2 value of 1 (see **Figure 14**) shows a perfect correlation of the variables. Based on this correlation, the disc pressure can be determined by means of the third degree polynomial for given disc surface areas. The resulted polynomial is:

$$r_{x,y} = -1.16 \times 10^{10} x^3 + 1.39 \times 10^7 x^2 + 5.89 \times 10^3 + 1.05. \tag{4}$$

This method offers the possibility of comparing and checking the disc pressure calculated in the simulation model with the one determined by the polynomial.

5. Conclusions

This study should be seen as a first approach to analyze the cervical spine's sensibility to different influencing factors. The focus is on the analysis of the effects of various stiffness and damping parameters and disc area on the intradiscal pressure of the FSU C6-C7 in order to indicate the model weaknesses and optimize the model design.

In the first part of this study an *one-way* sensitivity analysis is performed in order to indicate, whether one of the given input parameter, namely stiffness or damping term, has a dominant influence on the model behavior. The experimental results show, that both parameters exhibit an identical impact on the disc pressure. However the variations of the damping term indicate a slightly stronger effect on the intradiscal pressure measurements, which is reflected in relatively higher value of the calculated sensitivity coefficient. When applying compressive loads from 100 N up to 800 N on the FSU model and varying the analyzing parameters a not foreseeable response pattern in the disc pressure is explored. Simultaneous change of the load and the corresponding parameter values results in a non-linear outcome regarding the intradiscal pressure, which is not detected in the simulations that consider the exerting external force of 80 N.

Further, it could be shown that the correlation between disc area and disc pressure can be approximated by a third-degree polynomial. This allows a further possibility for model validation of the simulated intervertebral disc pressure. For this purpose, the simulation result can be compared with the intervertebral disc pressure calculated by the polynomial with a known disc surface area.

An essential point to be considered in the next step is the implementation of the musculature. This is not taken into account in this model. It is still unclear what influence other cervical parameters, e.g. the facet joints, ligaments or muscles have and how these affect the overall mechanic when changed. Therefore, following this investigation, the effect of model parameters of others spinal structures, such as facet alignment and size, on the load on the intervertebral discs will be evaluated. Further, it must be questioned critically whether these results can be transferred to a model with a larger spinal column section. In order to discuss this question, in a further step not only an FSU should be considered, but the sensitivity of model parameters in a model that contains an entire spinal column section should be analyzed.

In case when additional elements are integrated into the model and the number of input factors grows, another broadly used method called multivariate sensitivity analysis can be applied in order to investigate the model response affected by the simultaneous variations of the underlying parameters. This procedure can help to optimize the model structure by finding the variables, that primarily impact the model outcomes. Moreover, using the sensitivity analysis methods the values of the principal parameters can be determined so that realistic simulation of model behavior is possible.

The experimental design of the presented sensitivity analysis follows the recommendations found in the literature. In the future work, the boundary conditions of the experiments should be extended. For instance, the range of the stiffness value might be increased up to 8.3×10^6 as it was used in the model proposed in [14]. Then the response of the current FSU model can be compared with the outputs of the referenced model.

Acknowledgements

We like to thank Prof. Dietrich Paulus, Institute of Computer Visualistics, University Koblenz-Landau for the fruitful discussion and Dr. Francis Kilian, Head of the Clinic for Spinal Surgery, Head of the Spinal Center Catholic Clinic Koblenz-Montabaur and PD Dr. Roland Jacob, specialist in ear, nose and throat medicine for guidance on medical and anatomical questions.

Conflict of interest

The authors declare no conflict of interest.

Abbreviations

MBS	Multi-Body Simulation
FSU	Functional Spinal Unit
CoG	Center of Gravity
CR	Center of Rotation
MI	Moment of Inertia
ALL	Anterior Longitudinal Ligament
PLL	Posterior Longitudinal Ligament
FL	Flava Ligament
ISL	Interspinous Ligament
NL	Nuchal Ligament
CL	Capsular Ligament
SPL	Spinous Process Length
FC	Facet
DC	Disc
EP	Endplate
A	Area
W	Width
H	Height
D	Depth
l	left or lower (depending on the context)
r	right
u	upper

Author details

Sabine Bauer[1*†] and Ivanna Kramer[1,2†]

1 Institute for Medical Engineering and Information Processing, University of Koblenz-Landau, Koblenz, Germany

2 Institute for Computer Visualistics, University of Koblenz-Landau, Koblenz, Germany

*Address all correspondence to: bauer@uni-koblenz.de

† These authors contributed equally.

References

[1] Bauer S. Basics of Multibody Systems: Presented by Practical Simulation Examples of Spine Models. In: Lopez-Ruiz R editor. Numerical Simulation. InTech; 2016. pp. 29-49. DOI: 10.5772/62864

[2] Galbusera F, Schmidt H, Neidlinger-Wilke C, Wilke HJ: The effect of degenerative morphological changes of the intervertebral disc on the lumbar spine biomechanics: a poroelastic finite element investigation. Comput Methods Biomech Biomed Engin. 2011;14(8):729-39. DOI: 10.1080/10255842.2010.493522

[3] Ryang Y, Pape H, Meyer B: Degenerative Lumbale Instabilität - Definition, klinische und radiologische Zeichen, Management. Die Wirbelsäule. 2017;01(02):101-116. isbn:2509-8241

[4] Bashkuev M, Reitmaier S, Schmidt H: Effect of disc degeneration on the mechanical behavior of the human lumbar spine: a probabilistic finite element study.The Spine Journal. 2018; 18(10):1910-1920. https://doi.org/10.1016/j.spinee.2018.05.046, isbn:1529-9430

[5] Mas Y, Gracia L, Ibarz E, Gabarre S, Pena D, Herrera A: Finite element simulation and clinical follow-up of lumbar spine biomechanics with dynamic fixations. PloS one. 2017;12 (11):1-19. https://doi.org/10.1371/journal.pone.0188328

[6] Schmidt H, Heuer F, Wilke H-J: Which axial and bending stiffnesses of posterior implants are required to design a flexible lumbar stabilization system?. Journal of Biomechanics. 2009; 42(1):48-54.isbn:0021-9290

[7] Wilke H-J, Heuer F, Schmidt H: Design optimization of a new posterior dynamic stabilization system. Journal of Biomechanics. 2008;41. DOI 10.1016/S0021- 9290(08)70312-9

[8] Goel VK, Grauer JN, Patel TCh, Biyani A, Sairyo K, Vishnubhotla S, Matyas A, Cowgill I, Shaw M, Long R, Dick D, Panjabi MM, Serhan H: Effects of charit´e artificial disc on the implanted and adjacent spinal segments mechanics using a hybrid testing protocol. Spine (Phila Pa 1976). 2005;30 (24):2755-64. DOI: 10.1097/01.brs.0000195897.17277.67.

[9] Dreischarf M, Zander T, ShiraziAdl A, Puttlitz C.M, Adam C.J, Chen C.S, Goel V.K, Kiapour A, Kim, Y.H, Labus K.M, Little J.P, Park W.M, Wang Y.H, Wilke H.J, Rohlmann A, Schmidt H: Comparison of eight published static finite element models of the intact lumbar spine: predictive power of models improves when combined together.Journal of biomechanics. 2014; 47:1757-1766

[10] Zander T, Dreischarf M, Timm AK, Baumann WW, Schmidt H: Impact of material and morphological parameters on the mechanical response of the lumbar spine - A finite element sensitivity study. J Biomech. 2017;53: 185-190. DOI: 10.1016/j.jbiomech.2016.12.014

[11] Panjabi MM, Duranceau J, Goel V, Oxland T, Takata K: Cervical human vertebrae. Quantitative three-dimensional anatomy of the middle and lower regions. Spine. 1991;16.8:861-9. DOI: 10.1097/00007632-199108000-00001

[12] Anderst WJ, Thorhauer ED, Lee JY, Donaldson WF, Kang JD: Cervical spine bone mineral density as a function of vertebral level and anatomic location. Spine J. 2011;11(7):659-67. DOI: 10.1016/j.spinee.2011.05.007

[13] White AA, Panjabi MM: Clinical biomechanics of the spine. 2nd ed. Philadelphia: Lippincott; 1990;752 p.

[14] van Lopik DW, Acar M. Development of a multi-body computational model of human head and neck. Proceedings of the Institution of Mechanical Engineers, Part K: Journal of Multi-body Dynamics. 2007;221(2): 175-197. DOI: 10.1243/14644193JMBD84

[15] Panjabi MM, Crisco JJ, Vasavada A, Oda T, Cholewicki J, Nibu K, Shin E: Mechanical properties of the human cervical spine as shown by three-dimensional load-displacement curves. Spine (Phila Pa 1976). 2001;26(24): 2692-700. DOI: 10.1097/00007632-200112150-00012

[16] Yoganandan N, Knowles SA, Maiman DJ, Pintar FA: Anatomic study of the morphology of human cervical facet joint. Spine (Phila Pa 1976). 2003; 28(20):2317-23. DOI: 10.1097/01.BRS.0000085356.89103.A5

[17] Panjabi MM, Oxland T, Takata K, Goel V, Duranceau J, Krag M: Articular facets of the human spine. Quantitative three-dimensional anatomy. Spine (Phila Pa 1976). 1993;18(10):1298-310. DOI: 10.1097/00007632-199308000-00009

[18] Yang KH, King AI: Mechanism of facet load transmission as a hypothesis for low-back pain. Spine (Phila Pa 1976). 1984;9(6):557-65. DOI: 10.1097/00007632-198409000-00005

[19] Jager, de MKJ: Mathematical head-neck models for acceleration impacts. Eindhoven: Eindhoven University of Technology; 1996. 143 p. https://doi.org/10.6100/IR460661

[20] Yoganandan N, Kumaresan S, Pintar FA: Biomechanics of the cervical spine Part 2. Cervical spine soft tissue responses and biomechanical modeling. Clin Biomech (Bristol, Avon). 2001;16 (1):1-27. DOI: 10.1016/s0268-0033(00) 00074-7

[21] Kadri, PA, Al-Mefty O: Anatomy of the nuchal ligament and its surgical applications. 2007;61:301–304. DOI: 10.1227/01.neu.0000303985.65117.ea

[22] Mattucci SF, Moulton JA, Chandrashekar N, Cronin DS: Strain rate dependent properties of younger human cervical spine ligaments. J Mech Behav Biomed Mater. 2012;10:216-26. DOI: 10.1016/j.jmbbm.2012.02.004

[23] Hueston S, Makola M, Mabe I, Goswami T: Cervical Spine Anthropometric and Finite Element Biomechanical Analysis. 2012. IntechOpen. https://openresearchlibrary.org, doi:10.5772/35524.

[24] Kumaresan S, Yoganandan N, Pintar FA, Maiman DJ, Goel VK: Contribution of disc degeneration to osteophyte formation in the cervical spine: a biomechanical investigation. J Orthop Res. 2001 Sep;19(5):977-84. DOI: 10.1016/S0736-0266(01)00010-9

[25] Tan SH, Teo EC, Chua HC: Quantitative three-dimensional anatomy of cervical, thoracic and lumbar vertebrae of Chinese Singaporeans. Eur Spine J. 2004;13(2): 137-46. DOI: 10.1007/s00586-003-0586-z

[26] Pooni J, Hukins D, Harris P, Hilton R, Davies K: Comparison of the structure of human intervertebral discs in the cervical, thoracic and lumbar regions of the spine. Surgical and Radiologic Anatomy. 2006;8:175-182. DOI: 10.5772/35524

[27] Skrzypiec DM, Pollintine P, Przybyla A, Dolan P, Adams MA: The internal mechanical properties of cervical intervertebral discs as revealed by stress profilometry. Eur Spine J. 2007 Oct;16(10):1701-9. DOI: 10.1007/s00586-007-0458-z

[28] Sellers WI, Crompton RH: Using sensitivity analysis to validate the

predictions of a biomechanical model of bite forces. Ann Anat. 2004;186(1):89-95. DOI: 10.1016/S0940-9602(04)80132-8.

[29] Wexler P, Anderson B, Gad S, Hakkinen B, Kamrin M, De Peyster A, Locey B, Pope C, Mehendale H, Shugart L: Encyclopedia of toxicology. 3rd ed. Waltham; Academic Press; 2005. 236-237p. DOI: 10.1177/1091581815586498

[30] Smith E, Szidarovszky F, Karnavas W, Bahill A. Sensitivity analysis, a powerful system validation technique. The Open Cybernetics Systemics Journal. 2008; 2(1). DOI: 10.2174/1874110X00802010039

[31] Taylor M: What is sensitivity analysis. Consortium YHE. University of York. 2009: 1-8.

[32] Qian G, Mahdi A: Sensitivity analysis methods in the biomedical sciences. Mathematical biosciences. 2020;323:108306. DOI: 10.1016/j.mbs.2020.108306

[33] Lehman S, Lawrence S: Three algorithms for interpreting models consisting of ordinary differential equations: sensitivity coefficients, sensitivity functions, global optimization. Mathematical Biosciences. 1982;62.1:107-122. DOI: 10.1016/0025-5564(82)90064-5

[34] Ghosh P. The biology of the intervertebral disc. 1st ed. Boca Raton, FL: CRC press; 1988. DOI: 10.1007/978-3-7091-1535-0

Chapter 6

A Numerical Simulator Based on Finite Element Method for Diffusion-Advection-Reaction Equation in High Contrast Domains

Hani Akbari

Abstract

Implementation of finite element method (FEM) needs special cares, particularly for essential boundary conditions that have an important effect on symmetry and number of unknowns in the linear systems. Moreover, avoiding numerical integration and using (off-line) calculated element integrals decrease the computational cost significantly. In this chapter we briefly present theoretical topics of FEM. Instead we focus on what is important (and how) to carefully implement FEM for equations that can be the core of a numerical simulator for a diffusion–advection-reaction problem. We consider general 2D and 3D domains, high contrast and heterogeneous diffusion coefficients and generalize the method to nonlinear parabolic equations. Although we use Matlab codes to simplify the explanation of the proposed method, we have implemented it in C++ to reveal the efficiency and examples are presented to admit it.

Keywords: finite element method, diffusion–advection-reaction equation, boundary condition, implementation

1. Introduction

In a domain Ω in $\mathbb{R}^n$, $n = 1, 2, 3$, consider a partial differential equation of the form of

$$\nabla \cdot (-\kappa \nabla u) + \gamma \cdot \nabla u + \mu u = f \quad \text{in } \Omega, \tag{1}$$

where the unknown function u, reaction coefficient μ and the given function f are scalar functions ($\Omega \to \mathbb{R}$). Diffusion coefficient κ and the (convergence free) advection coefficient γ are defined in Ω and give (normally piecewise constant) values in $\mathbb{R}^{n \times n}$ and $\mathbb{R}^n$, respectively. Eq. (1) is called a diffusion–advection-reaction equation and has immense applications in science and engineering. Transport of heat, momentum and energy, solid mechanics, CO_2 sequestration, computational fluid dynamics and biophysics are a few number of research fields that as a part of the solution strategy need to solve (1) numerically.

IntechOpen

After several steps, numerical solution of Eq. (1) is obtained by solving a linear system

$$\mathcal{A}x = b \tag{2}$$

where $\mathcal{A}$ and b are called stiffness matrix and right hand side vector, respectively. In practice each term of diffusion, advection or reaction is accumulated separately in the stiffness matrix.

Several challenges make finding the numerical solution of (1) difficult. 3 of the most important are 1) complex geometry of Ω, 2) heterogeneity and high contrast in coefficient κ that produce very ill-conditioned linear systems and 3) large scale domains. A very common numerical method to solve (1) is FEM that uses a mesh representing the complex geometries accurately and overcomes the first issue. We can employ powerful linear solvers such as multigrid or multiscale methods to resolve the effect of heterogeneity and high contrast in κ in the linear system. Finally a careful implementation in parallel machines reduces computational time of large scale simulations considerably. However, lots of effort and research are still needed to propose a method obtaining a reliable solution in a reasonable time for real problems. Other issues such as uncertainties in the data or nonlinear coupled system of equations should be addressed, as well.

Considering a positive definite κ we explain (and implement) how to obtain a finite element solution of (1) through the following steps.

1. In Matlab we generate arbitrary 2D and simple 3D domains. The main domain is divided into a collection of subdomains, called elements where each element includes some nodes. Numbered nodes and elements generate a mesh. Actually by a mesh we mean 2 data structures, `Cells` (integer valued) that stores identifier or index of nodes in each element and `Nodes` (real valued) that stores coordinates of each node. Many software such as Matlab, GID and Gmesh generate reliable meshes. Corresponding to each element e, a positive definite matrix (a constant or a 2×2 matrix in 2D or a 3×3 matrix in 3D), named κ_e, will be set to form κ. In each element e, we can also set μ_e and γ_e to form μ and γ in Eq. (1), respectively. See Section 2 for details.

2. A brief introduction to FEM is presented that covers weak formulation (Section 3), shape functions and reference elements (Section 4).

3. Evaluating of element integrals and preparing table of calculated integrals are discussed in Section 4.1. Then with help of affine mappings (Section 4.2) we use a linear combination of element integrals to accumulate into the stiffness matrix which is explained in Section 5. Note that different types of elements (for example triangles and rectangles in 2D) might exist in the mesh and we can have an unstructured mesh, generally. Hence element integrals should be considered for any type of elements exist in the mesh.

4. Efficient implementation of boundary conditions, particularly Dirichlet boundary condition is presented in Section 5.2. We show that how a correct but careless implementation of Dirichlet boundary conditions decreases both accuracy and efficiency.

5. After assembly of the linear system (2), we use Matlab functions to solve the linear system and plot the result. Linear solvers are the core of a numerical simulator and their efficiency has a direct impact on the overall simulation. We refer to [1] for further discussions and references.

6. Generalization to more complex equations such as Eq. (25) now becomes straightforward and solving strategy is given in Section 6.

We finish this chapter by introducing the necessary topics to have an independent and efficient simulator in Section 7. Most of the topics presented here are fully discussed in [1–3] which are of great value for further reading.

2. Mesh generation

Implementation of FEM is began with mesh generation which is dividing the main domain into several subdomains or elements. Each element is finite (finite

```
%% Part 1
model = createpde(1);
model.Geometry = multicylinder(2,[1 3 2],'ZOffset',[0 1 4]);
mesh = generateMesh(model,'Hmax',0.5,'GeometricOrder','linear');

figure
subplot 131
pdegplot(model,'FaceLabels','on','CellLabels','on','FaceAlpha',0.25);
subplot 132, pdeplot3D(model);

Nodes = mesh.Nodes;
Cells = mesh.Elements;

Nn = size(Nodes,2); % number of nodes
Nc = size(Cells,2); % number of cells or N_e!
dim = 3;            % dimension

CC2 = findElements(mesh,'region','Cell',2)'; % cell ID of middle region
% for boundary condition:
NFT = findNodes(mesh,'region','Face',6)';    % node ID of top face
NFB = findNodes(mesh,'region','Face',1)';    % node ID of bottom face

%% Part 2
kappa = ones(dim, Nc);
kappa(:,CC2) = 2;

global D Drr Dss Dtt Drs Drt Dst;
load LinearTetraElementIntegral 'D' 'Drr' 'Dss' 'Dtt' 'Drs' 'Drt' 'Dst';

[IM, JM, FFvec, DDvec] = calcMatrices(Nodes, Cells, kappa); % Listing. 3

A = sparse(IM,JM,DDvec+0.1*FFvec,Nn,Nn);

DirV = [ones(length(NFB),1); 10*ones(length(NFT),1)];
DirI = [NFB; NFT];                 % Index of nodes with Dirichlet BC
ResI = setdiff((1:Nn)', DirI); % Index of nodes with no Dirichlet BC

b = - A(:,DirI)*DirV;
A(DirI,:) = [];
A(:,DirI) = [];
b(DirI) = [];

sol = A\b;
u = zeros(Nn,1);
u(ResI) = sol;
u(DirI) = DirV;

subplot 133
pdeplot3D(model,'ColorMapData',u);
```

Figure 1.
*Creating a geometry and generating its mesh shown in **Figure 2**.*

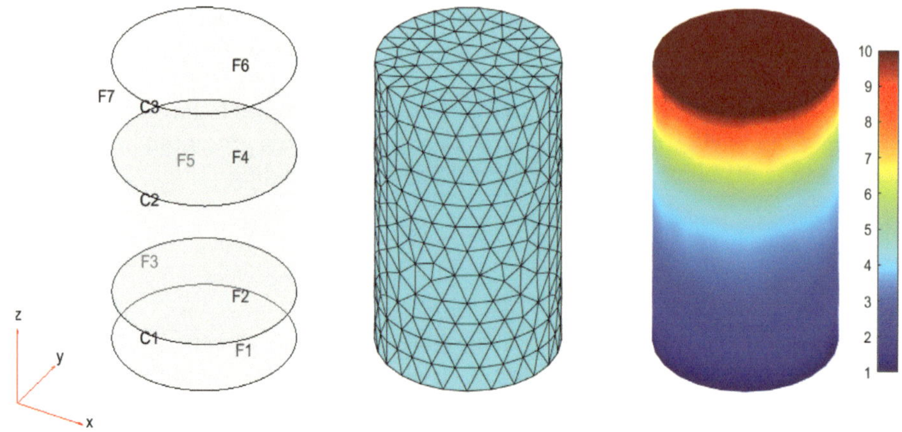

Figure 2.
A 3D domain consists of 3 stacked cylinders (left) and the generated mesh (middle) and the solution of Eq. (21) in right.

elements) and has a regular shape such as triangle or cuboid. Each element has a specific type, such as linear or bilinear, which is defined by shape functions and explained in Section 4. We prefer to use Lagrange triangles and parallelograms in 2D and tetrahedron and cuboid in 3D space as elements. The reason of such selection is to avoid numerical integration to evaluate definite integrals (of Lagrange elements) arise in FEM.

The main output of the mesh generation is two sets, nodes and elements that form the mesh. Nodes are not necessarily the vertices of the elements, for example in linear Crouzeix-Raviart element nodes are placed at edge midpoints or in quadratic Lagrange (triangle) elements midpoints and vertices together form the nodes.

In presence of complex geometries mesh generation can be a challenge for software mesh generators. For not very complicated geometry, Matlab, with a good quality, can generate triangular and tetrahedron mesh for 2D and 3D domains, respectively. We give an example, as shown in Part 1 of **Figure 1**.

First with `createpde (1)` we create a raw model considering one partial differential equation. Then we import or create geometry for this model. For our purposes 3 stacked cylinders with radius 2 and heights 1, 3 and 2 would be fine as shown in **Figure 2**. `pdegplot` shows how Matlab has specified regions (cylinders) and faces by numbering them. Then `generateMesh` generates a mesh with linear (or quadratic by default) tetrahedron elements and can be viewed by `pdeplot3D`. Smaller value for `Hmax` gives a finer mesh. Setting N_n and N_e for number of nodes and elements, respectively, `Nodes` that stores 3 Cartesian coordinates of each node is a matrix of size $3 \times N_n$. Since all elements are linear tetrahedron, `Cells` that stores index of nodes of each element, would be a $4 \times N_e$ matrix. For example the first element of the mesh is formed by 4 nodes, stored in `Cells(:,1)` and `Nodes(:,Cells(:,1))` returns 3D coordinates of that 4 nodes. So we can simply traverse over nodes and elements by their index. Moreover, we can extract element indices of a region with `findElements` or node indices of a region or a face by `findNodes`, where are necessary to set boundary conditions. For example we find node indices of bottom and top faces of the domain in Lines 21–22, since we will set boundary conditions on them.

3. Weak form

In a bounded domain Ω with Lipschitz boundary Γ, Green's formula says for two regular functions u and v we have

$$\int_\Omega \nabla \cdot (-\kappa \nabla u) v = \int_\Omega \kappa \nabla u \cdot \nabla v - \oint_\Gamma \nu \cdot \kappa \nabla u v \tag{3}$$

where ν is the (outward) normal vector. Regularity means that u and v have piecewise continuous (at least) first order partial derivatives to make the above integrals meaningful. So multiplying both side of (1) by v and applying Green's formula we obtain

$$\int_\Omega \kappa \nabla u \cdot \nabla v - \oint_\Gamma \nu \cdot \kappa \nabla u v + \int_\Omega \gamma \cdot \nabla u v + \int_\Omega \mu u v = \int_\Omega f v. \tag{4}$$

Approximation of functions in FEM is done by means of basis functions. Assume $\{\varphi_j\}_j, j = 1, \dots, N_n$ is a set of basis functions, introduced in Section 4, that we can write

$$u = \sum_{j=1}^{N_n} u_j \varphi_j, \tag{5}$$

and our goal is to find unknown coefficients $\{u_j\}_j$. Substituting (5) in (4) and setting $v = \varphi_i, i = 1, \cdots, N_n,$ which is called standard Galerkin method we obtain

$$\sum_{j=1}^{N_n} u_j \int_\Omega \kappa \nabla \varphi_j \cdot \nabla \varphi_i + \sum_{j=1}^{N_n} u_j \int_\Omega \gamma \cdot \nabla \varphi_j \varphi_i + \sum_{j=1}^{N_n} u_j \int_\Omega \mu \varphi_j \varphi_i + \oint_\Gamma \nu \cdot (-\kappa \nabla u) \varphi_i = \int_\Omega f \varphi_i \tag{6}$$

Eq. (6) consists of evaluation of (from left to right) diffusion, advection and reaction terms, boundary integral and right hand side that we calculate them in elements and then sum them up $\left(\int_\Omega = \sum_e \int_e\right)$ to assemble the linear system (2). u or its gradient in boundary integral are known, hence we do not write its expansion form.

4. Shape functions

In practice definite integrals of basis functions and their derivatives in (6) are evaluated in an element, hence restriction of basis functions over elements, called shape functions, contribute in computations. With help of affine mapping we map a typical element into a fixed element, called reference element and evaluate integrals in it. So only shape functions in reference element determine values of integrals in (6).

We use Lagrange elements which means corresponding to each node of an element a shape function is defined such that it is a polynomial in the element, takes value 1 at that specific node and 0 at other nodes of the element. Therefore a basis function in (5) becomes a continues function with value 1 at a specific node and 0 at neighboring nodes with trivial (zero) extension in the rest of the domain, as plotted in **Figure 3**. Note that the derivatives of Lagrange basis functions are not continues on the boundary of the elements.

Some examples of reference elements and their shape functions are presented in **Figure 4**. We see that in linear triangle element where nodes are the vertices of the triangle, the first shape function $S_1(r,s) = 1 - r - s$ corresponding to the first node $n_1 = (0,0)$ has value 1 at n_1 and 0 at $n_2 = (1,0)$ and $n_3 = (0,1)$. It is called a linear element since shape functions are combination of first order polynomials $\{1, r, s\}$.

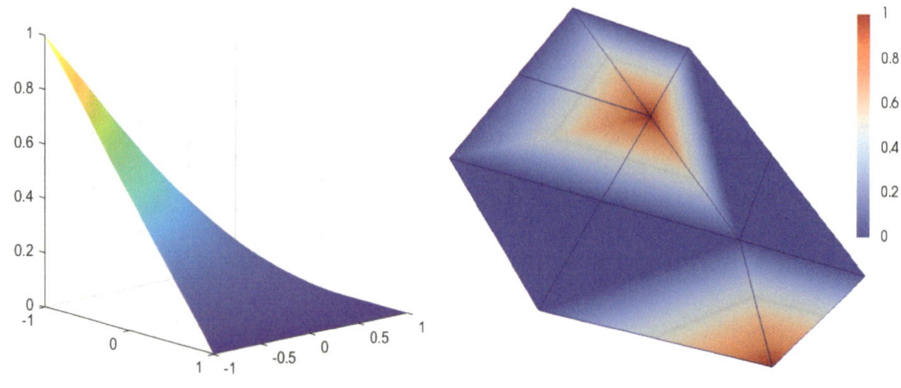

Figure 3.
Left) a shape function of reference bilinear quadrilateral. Right) 2 basis functions where their restrictions in a triangle is linear and in a quadrilateral is bilinear.

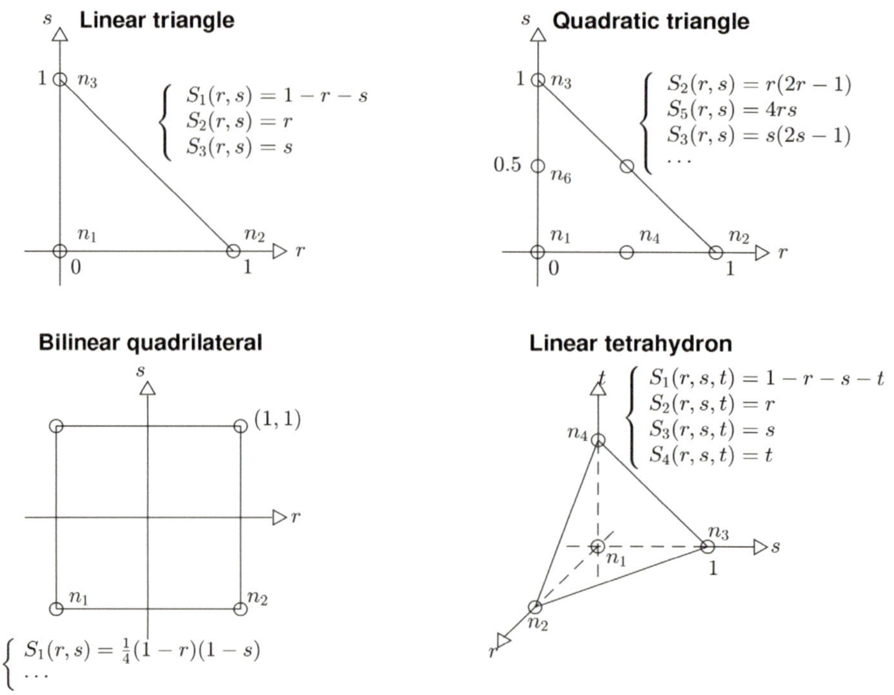

Figure 4.
Some of reference Lagrange elements and their shape functions.

Starting from n_1 nodes are numbered counterclockwise and we consider it for all Lagrange elements.

For Lagrange elements it is easy to find shape functions. For example to find S_1 for bilinear quadrilateral element where nodes are vertices of the square, it suffices to consider the line passing n_2 and n_3 which is $1 - r$ multiplied by the line passing n_3 and n_4 which is $1 - s$. So S_1 has to be of the form $\alpha(1 - r)(1 - s)$ which gives 0 for all nodes except $n_1 = (-1, -1)$. Substituting n_1 in equation of S_1, we can find α such that it gives 1 for n_1. This element is called bilinear since its shape functions are linear combination of $\{1, r, s, rs\}$.

For quadratic triangle element where vertices and midpoints form nodes, for shape function S_1 corresponding to n_1, we need the multiplication of two lines passing n_2 and n_6 which is $0.5 - r - s$ and n_3, n_4 and n_5 which is $1 - r - s$. So S_1 has to be of the form $\alpha(0.5 - r - s)(1 - r - s)$. Substituting n_1 in equation of S_1, we can find α such that it gives 1 for n_1. This element is called quadratic since its shape functions are linear combination of $\{1, r, s, rs, r^2, s^2\}$. Note that we first numbered nodes at vertices and then at edge midpoints.

Exercise 1 Find shape functions of linear tetrahedron as presented in **Figure 4**.

Exercise 2 Quadratic tetrahedron element has nodes at midpoint of edges of linear tetrahedron element, so it has 6 other nodes (numbered 5 to 10) in addition to vertices (numbered 1 to 4 similar to linear tetrahedron). Find the 10 shape functions. For example for node at origin we have $S_1 = 2(1 - r - s - t)(0.5 - r - s - t)$ or for node at $(0, 0, 1)$ we have $S_4 = 2t(t - 0.5)$.

4.1 Element integrals

Definite integrals of shape functions and their derivatives in reference elements, called element integrals, play an important role in assembly of stiffness matrix. Denoting the reference element by $\hat{e}$ which has $N_{\hat{e}}$ nodes and partial derivatives of shape functions by $\partial_r S$ which means $\frac{\partial S}{\partial r}$ we define the matrix D_{rr} (for diffusion term) such that its ijth entry is

$$D_{rr}^{ij} = \int_{\hat{e}} \partial_r S_i \ \partial_r S_j, \quad i,j = 1, \cdots, N_{\hat{e}}. \tag{7}$$

Similarly we define D_{rs} and D_{ss}. In 3D space we also have to define D_{rt}, D_{st} and D_{tt}. In **Figure 5** we show how to evaluate D_{rs} for linear tetrahedron element. For example D_{rs} and D_{st} for linear tetrahedron are

$$D_{rs} = \frac{1}{6}\begin{pmatrix} 1 & 0 & -1 & 0 \\ -1 & 0 & 1 & 0 \\ 0 & 0 & 0 & 0 \\ 0 & 0 & 0 & 0 \end{pmatrix}, \quad D_{st} = \frac{1}{6}\begin{pmatrix} 1 & 0 & 0 & -1 \\ 0 & 0 & 0 & 0 \\ -1 & 0 & 0 & 1 \\ 0 & 0 & 0 & 0 \end{pmatrix}.$$

Exercise 3 For quadratic tetrahedron element find D_{rs} as

$$D_{rs} = \frac{1}{30}\begin{pmatrix} 3 & 0 & 1 & 0 & -1 & -4 & -1 & 1 & 0 & 1 \\ 1 & 0 & -1 & 0 & -3 & 0 & 1 & 3 & 0 & -1 \\ 0 & 0 & 0 & 0 & 0 & 0 & 0 & 0 & 0 & 0 \\ 0 & 0 & 0 & 0 & 0 & 0 & 0 & 0 & 0 & 0 \\ -4 & 0 & 0 & 0 & 4 & 4 & 0 & -4 & 0 & 0 \\ -1 & 0 & -3 & 0 & 4 & 4 & 4 & -4 & 0 & -4 \\ -1 & 0 & 1 & 0 & 4 & 0 & 8 & -4 & 0 & -8 \\ 1 & 0 & 3 & 0 & -4 & -4 & -4 & 4 & 0 & 4 \\ 1 & 0 & -1 & 0 & -4 & 0 & -8 & 4 & 0 & 8 \\ 0 & 0 & 0 & 0 & 0 & 0 & 0 & 0 & 0 & 0 \end{pmatrix}$$

Other element integrals such as D_r (for advection term) and D (for reaction or right hand side terms) can be defined where their ijth entries are

$$D_r^{ij} = \int_{\hat{e}} \partial_r S_i \ S_j, \quad D^{ij} = \int_{\hat{e}} S_i \ S_j, \quad i,j = 1, \cdots, N_{\hat{e}}. \tag{8}$$

```
syms r s t;
Nn = 4;                % number of nodes of reference element
S1 = 1-r-s; S2 = r;  S3 = s;  S4 = t; % shape functions
S = [S1 S2 S3 S4];
Dxy = zeros(Nn,Nn);
for i=1:Nn
   for j=1:Nn
      Drs(i,j) = int(int(int(diff(S(i),r)*diff(S(j),s),t,...
                          [0 1-r-s]), s, [0 1-r]), r, [0 1]);
   end
end
```

Figure 5.
Element integrals D_{rs} *for linear tetrahedron.*

Note that D_{rr} and D are symmetric matrices while D_r is not. Moreover, since $D_{rs} = D_{sr}$ we calculate only one of them. All element integrals are evaluated once and saved for future use. In **Figure 1** we load element integrals of linear tetrahedron that we have calculated and saved in `LinearTetraElementIntegral.mat`.

4.2 Affine mapping on reference elements

So far we introduced shape functions and calculated element integrals, all in reference element. Now we explain how to map an arbitrary element in physical space into the reference element and evaluate integrals in (6) only in terms of element integrals and without any numerical integration. Clearly change of variable is necessary to evaluate integrals when we use a mapping which is explained in next subsection.

Consider an arbitrary triangle, say e, in 2D space with vertices $v_i = (x_i, y_i)$, $i = 1, 2, 3,$ and the reference linear triangle, $\hat{e}$. Set

$$T = \begin{pmatrix} x_2 - x_1 & x_3 - x_1 \\ y_2 - y_1 & y_3 - y_1 \end{pmatrix}, \quad \eta = \begin{pmatrix} x \\ y \end{pmatrix}, \quad \xi = \begin{pmatrix} r \\ s \end{pmatrix}, \quad \eta_o = \begin{pmatrix} x_1 \\ y_1 \end{pmatrix} \tag{9}$$

and define the affine (composition of a linear mapping and a translation) mapping $\hat{e} \to e$ with the rule

$$\xi \to \eta = T\xi + \eta_o. \tag{10}$$

We see that the affine mapping (10) maps the nodes of reference linear triangle, $\{n_i\}_{i=1,2,3}$, to the vertices of the triangle,

$$v_i = Tn_i + v_1, \quad i = 1, 2, 3. \tag{11}$$

Exercise 4 Show that the affine mapping in (9) and (10) also maps the reference quadratic triangle to an arbitrary triangle with 6 nodes (3 at vertices and 3 at edges midpoint).

Exercise 5 Show that the affine mapping,

$$\xi = \begin{pmatrix} r \\ s \end{pmatrix} \to \eta = \begin{pmatrix} x \\ y \end{pmatrix} = \frac{1}{2}\begin{pmatrix} x_2 - x_1 & x_4 - x_1 \\ y_2 - y_1 & y_4 - y_1 \end{pmatrix}\begin{pmatrix} r \\ s \end{pmatrix} + \frac{1}{2}\begin{pmatrix} x_1 + x_3 \\ y_1 + y_3 \end{pmatrix} \tag{12}$$

maps the reference bilinear quadrilateral to an arbitrary parallelogram. Remember in a parallelogram where v_1 and v_3 are opposite vertices, we have $x_1 + x_3 = x_2 + x_4$ and $y_1 + y_3 = y_2 + y_4$.

Exercise 6 Show that the affine mapping,

$$\xi = \begin{pmatrix} r \\ s \\ t \end{pmatrix} \rightarrow \eta = \begin{pmatrix} x \\ y \\ z \end{pmatrix} = \begin{pmatrix} x_2 - x_1 & x_3 - x_1 & x_4 - x_1 \\ y_2 - y_1 & y_3 - y_1 & y_4 - y_1 \\ z_2 - z_1 & z_3 - z_1 & z_4 - z_1 \end{pmatrix} \begin{pmatrix} r \\ s \\ t \end{pmatrix} + \begin{pmatrix} x_1 \\ y_1 \\ z_1 \end{pmatrix} \quad (13)$$

maps the reference linear and quadratic tetrahedron to an arbitrary tetrahedron.

4.3 Change of variables in integrals

Consider a scalar function (a typical basis function) $\varphi = \varphi(\eta)$, where $\eta = (x, y)$ in 2D space. Gradient of φ is $\nabla_\eta \varphi = \left(\partial_x \varphi, \partial_y \varphi\right)^t$, where t denotes the transpose of the vector. The chain rule and applying the affine mapping (10) on φ give

$$\nabla_\eta \varphi = T^{-t} \ \nabla_\xi \hat{\varphi} \quad (14)$$

where $\hat{\varphi}(\xi) = \varphi(\eta)$ and T^{-t} is the transpose of the inverse of T. Note that if φ is the restriction of a basis function in element e, then $\hat{\varphi}$ would be the corresponding shape function in $\hat{e}$. Recalling $d\eta = |T| d\xi$, we can write

$$\int_e \kappa_e \nabla_\eta \varphi_i \cdot \nabla_\eta \varphi_j d\eta = \int_e \left(\kappa_e \nabla_\eta \varphi_i\right)^t \nabla_\eta \varphi_j d\eta = \int_{\hat{e}} (\nabla_\xi S_i)^t T^{-1} \kappa_e^t T^{-t} \nabla_\xi S_j \ |T| d\xi. \quad (15)$$

Setting matrix $M_e = |T| T^{-1} \kappa_e^t T^{-t}$ and dropping subscript e as well as η and ξ from ∇, then we have

$$\begin{aligned} \int_e \kappa \nabla \varphi_i \cdot \nabla \varphi_j d\eta &= M_{11} \int_{\hat{e}} \partial_r S_i \ \partial_r S_j \ d\xi \\ &+ (M_{12} + M_{21}) \int_{\hat{e}} \partial_r S_i \ \partial_s S_j \ d\xi + M_{22} \int_{\hat{e}} \partial_s S_i \ \partial_s S_j \ d\xi \\ &= M_{11} D_{rr} + (M_{12} + M_{21}) D_{rs} + M_{22} D_{ss} \end{aligned} \quad (16)$$

Exercise 7 Show that in 3D space,

$$\begin{aligned} \int_e \kappa \nabla \varphi_i \cdot \nabla \varphi_j d\eta &= M_{11} D_{rr} + (M_{12} + M_{21}) D_{rs} + (M_{13} + M_{31}) D_{rt} \\ &+ M_{22} D_{ss} + (M_{23} + M_{32}) D_{st} \\ &+ M_{33} D_{tt}, \end{aligned} \quad (17)$$

where $M = |T| T^{-1} \kappa^t T^{-t}$.
Exercise 8 Referring to notation (8) show that

1. $$\int_e \varphi_i \ \varphi_j d\eta = |T| D \quad (18)$$

2. for a vector γ, defined for each element e in 3D space

$$\int_e \gamma \cdot \nabla \varphi_i \ \varphi_j d\eta = M_1 D_r + M_2 D_s + M_3 D_t \quad (19)$$

where vector $M = |T| T^{-1} \gamma$.

5. Assembly of the linear system

With help of element integrals, affine mapping and Eqs. (17)–(19) we can accumulate each term of (6) into the stiffness matrix of the linear system (2).

5.1 Diffusion term

Equation

$$\int_{\Omega} \kappa \nabla \varphi_j \cdot \nabla \varphi_i = \sum_e \int_e \kappa_e \nabla \varphi_j \cdot \nabla \varphi_i \tag{20}$$

suggests how to accumulate $\int_{\Omega} \kappa \nabla u \cdot \nabla v$ into the stiffness matrix:

1. traverse elements and in each element map $\int_e \kappa_e \nabla \varphi_i \cdot \nabla \varphi_j$ into the reference element

2. with help of element integrals and Eq. (17) evaluate the desired term

3. map the local matrix into the stiffness matrix.

Exercise 9 Follow the above steps to accumulate advection and reaction terms into the stiffness matrix.

In **Figure 6** we implemented accumulation of diffusion and reaction terms for linear or quadratic tetrahedron. Note that other elements only have a different affine mapping T and dimension, if we had elements in 2D space. We assume that κ is a diagonal matrix for each element, so a matrix storage of size $3 \times N_c$ can store κ of all elements. We also assume that reaction coefficient μ is constant, otherwise a vector of size N_c can store μ for all elements and should be an input argument. The evaluated integrals are saved such that the indices of row and column and value of the integral are set in `IM`, `JM` and `DDvec` (for diffusion term), respectively. So Line 33 of **Figure 1** is to assembly the stiffness matrix of the equation

$$\int_{\Omega} \kappa \nabla u \cdot \nabla v + 0.1 \int_{\Omega} uv = 0 \tag{21}$$

where κ is set in Lines 25–26. We also set boundary conditions at bottom and top faces of the boundary (with values 1 and 10, respectively) and explain how to implement it in next subsection to obtain the solution of the problem, plotted in **Figure 2**.

5.2 Boundary integral

Probably correct implementation of boundary integrals is the most important part of the FEM, since boundary conditions mainly determine the situation of physical problem. Neumann, Robin and Dirichlet are 3 types of boundary conditions that might be considered on different parts of the boundary. Although boundary conditions are set on the boundary, they only affect on boundary nodes.

Neumann boundary condition. $\nu \cdot (-\kappa \nabla u) = g$ is a Neumann boundary condition, so boundary integral in (6) becomes $g \oint \varphi_i$. If g is zero, then nothing has to be done for the boundary integrals. In fact incorporating only diffusion–advection-

```
function [IM, JM, FFvec, DDvec] = calcMatrices(Nodes, Cells, kappa)

% Nn = number of nodes ( Nn = size(Nodes,2)), Ne = number of elements.
% Then A1 = sparse(IM,JM,DDvec,Nn,Nn) and A2 = sparse(IM,JM,FFvec,Nn,Nn)
% assembly the diffusion and reaction terms, respectively.
% kappa is considered a diagonal matrix for each element (3*Nc)

  global D Drr Drs Drt Dss Dst Dtt; % saved element integrals

  % ln is the number of nodes of each cell (4 for linear tetrahedron)
  ln = size(Cells,1);
  ln2 = ln*ln;
  ne = size(Cells,2); % number of elements
  NN = ln2*ne;

  IM = zeros(NN,1);
  JM = zeros(NN,1);
  FFvec = zeros(NN,1);
  DDvec = zeros(NN,1);

  lk = 0;
  for K=1:ne     % traverse elements

      enodes = Cells(:,K); % (global) index of the nodes of element K
      x = Nodes(1,enodes); % x-coordinate of the nodes of element K
      y = Nodes(2,enodes); % y-coordinate
      z = Nodes(3,enodes); % z-coordinate

      T = [[x(2)-x(1) x(3)-x(1) x(4)-x(1)];
           [y(2)-y(1) y(3)-y(1) y(4)-y(1)];
           [z(2)-z(1) z(3)-z(1) z(4)-z(1)]];

      detT = det(T);
      TTinv = inv(T)';
      M = detT*(TTinv')*diag(kappa(:,K))*TTinv;

      locDD = M(1,1)*Drr + (M(1,2) + M(2,1))*Drs + (M(1,3)+M(3,1))*Drt + ...
              M(2,2)*Dss + (M(2,3) + M(3,2))*Dst + ...
              M(3,3)*Dtt;

      locFF = detT*D;

      % use indices of nodes to map local matrices to global index
      locI = repmat(enodes,1,4);
      IM(lk+1:lk+ln2) = reshape(locI',[],1);
      JM(lk+1:lk+ln2) = locI(:);
      DDvec(lk+1:lk+ln2) = locDD(:);
      FFvec(lk+1:lk+ln2) = locFF(:);

      lk = lk + ln2;
  end % end for
end % end function
```

Figure 6.
Assembly of diffusion–reaction terms for (linear or quadratic) tetrahedron.

reaction terms and right hand side function means that we have considered a pure Neumann problem and this is exactly what we do in practice. After that we add requested boundary conditions to the linear system of (2). For example if g is non-zero on part of the boundary, say Γ_N, then $-g\oint \varphi_i$ is known for indices that lie on Γ_N and should be added to the ith entry of b in (2).

Dirichlet boundary condition. $u = g$ is a Dirichlet boundary condition, which means value of u at some nodes is known. So if we decompose node indices into 2 sets, say U and K, then we can write

$$u = \sum_{j \in U} u_j \varphi_j + \sum_{j \in K} u_j \varphi_j, \tag{22}$$

where U and K include indices of unknown and known values of u, respectively. If with permutation vector $[U, K]$ we reorder rows and columns of the linear system in (2), then we can write

$$\mathcal{A}u = \begin{pmatrix} \mathcal{A}_{11} & \mathcal{A}_{12} \\ \mathcal{A}_{21} & \mathcal{A}_{22} \end{pmatrix} \begin{pmatrix} u_U \\ u_K \end{pmatrix} = \begin{pmatrix} b_U \\ b_K \end{pmatrix} \tag{23}$$

The first line gives a smaller linear system

$$\mathcal{A}_{11} u_U = b_U - \mathcal{A}_{12} u_K \tag{24}$$

which preserves the symmetry or positive definiteness of the original problem. In Lines 35–42 of **Figure 1** we showed how to implement it in Matlab. Setting values 1 and 10 for bottom and top faces, we expect a smooth solution starting from 1 at the bottom and reaching to 10 at the top as shown in **Figure 2**.

The second approach to set Dirichlet condition that does not use permutation of the linear system is setting $\mathcal{A}_{21}$ and $\mathcal{A}_{22}$ in (23) to zero and identity matrices, receptively and solving the modified linear system. Although this approach gives the correct solution theoretically, it is very error-prone numerically as shown in a test case in **Figure 7** and explained as follows.

A domain with cuboid elements is refined in the middle and along the x-axis to simulate a fracture in the domain. We solve $\nabla \cdot (-\kappa \nabla u) = 0$ where κ in fracture is 100 times larger than rest of the domain. Setting 10^6 and 5×10^6 as Dirichlet boundary condition on left and right faces (along x-axis) of the domain, respectively, a correct solution should be started from 10^6 and monotonically reached to 5×10^6. The linear system is symmetric positive definite and preconditioned conjugate gradient method is employed to solve it. The correct solution shown in left image of **Figure 7** is obtained by (24). However, the second approach gives the nonphysical (wrong) solution along the fracture as shown in right. Moreover, the correct solution is obtained 3 times faster than the wrong one, mainly because the second approach violates the symmetry (but preserves the positive definiteness) of the linear system, due to setting $A_{21} = 0$.

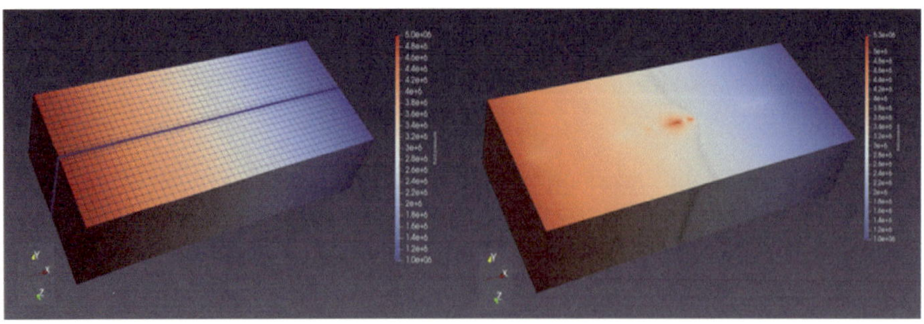

Figure 7.
Left) Result of correct implementation of Dirchlet BC by using Eq. (24). Right) nonphysical (wrong) result due to using the modified linear system (the second approach to implement Dirchlet BC).

On the other hand reduction in size of the linear system by using (24), has great advantage in applications that large number of nodes have Dirichlet (more precisely essential) boundary condition. For example, in pore scale modeling and to approximate the permeability of a sandstone model, we had to set Dirichlet boundary condition for 75% of the nodes. Since the model included almost 10 million elements, solving a linear system with 2.5 million unknowns gave a significant speed up in computational time, in addition to significant improvement in accuracy.

Exercise 10 Robin boundary condition is a combination of Neumann and Dirichlet conditions which states $u - \beta\nu \cdot \kappa\nabla u = g$ on Γ_R. β is a non-zero function (normally a constant number) on Γ_R. Considering $-\nu \cdot \kappa\nabla u = (g - u)/\beta$, explain how to implement it.

5.3 Right hand side term

Function f in (1) is the source or sink term and hence normally is defined on a very small part of the domain. It can be defined as a constant number over an element or can be approximated by its nodal values in the form $f = \sum f_j\varphi_j$ and therefore right hand side term becomes $\sum_{j=1}^{N_n} f_j \int \varphi_j\varphi_i$ with possibly too many $f_j = 0$. We prefer nodal value approximation of f, since in applications that need to solve parabolic equation, f has a value in all elements; see $q^{[k]}$ in Eqs. (27) and (28).

6. Generalization to parabolic equation

In applications such as simulation of compressible flows we need to solve the time dependent equation

$$\frac{\partial\rho(u)}{\partial t} + \nabla \cdot (-\rho(u)\kappa\nabla u) = f(u), \quad \text{in } \Omega \tag{25}$$

where ρ and f are nonlinear functions of u and boundary condition and initial value are provided. A fully implicit time discretization of (25) gives

$$\frac{\rho^{n+1} - \rho^n}{\Delta t} + \nabla \cdot \left(-\rho^{n+1}\kappa\nabla u^{n+1}\right) = f^{n+1}$$

which simplifies to

$$G(u) = \frac{1}{\Delta t}\rho + \nabla \cdot (-\rho\kappa\nabla u) - f(u) - \frac{1}{\Delta t}\rho^n = 0. \tag{26}$$

In Eq. (26), the unknown u-dependent variables ρ and f should be evaluated at the current time t^{n+1}, while ρ^n is known from the previous time step.

Newton–Raphson linearization is the most common method to solve the nonlinear Eq. (26) which generates a sequence $\{u^{[k]}\}$, $k = 0, 1, 2, \ldots$ by solving the following equation

$$\nabla \cdot \left(-\rho^{[k]}\kappa\nabla u\right) + \mu^{[k]}u = q^{[k]}, \tag{27}$$

where

$$\mu^{[k]} = \frac{1}{\Delta t}\frac{d\rho}{du}\bigg|_{u^{[k]}} - \frac{df}{du}\bigg|_{u^{[k]}}, \quad q^{[k]} = \mu^{[k]}u^{[k]} + f^{[k]} - \frac{1}{\Delta t}\rho^n. \tag{28}$$

Having bounded derivatives and a good starting point $u^{[0]}$ (which is normally chosen u^n), the resulted sequence $u^{[k]}$ in Eq. (27) converges quadratically to u^{n+1}, the solution of Eq. (26). Note that (26) is a diffusion–reaction equation which is completely explained how to solve and (core) codes were provided, as well.

7. Conclusion and future works

In this introductory chapter we presented the framework of FEM in brief (but effective) and implemented a numerical solver for diffusion–advection-reaction equation. Accumulation of different terms and setting boundary conditions correctly as well as evaluating of definite integrals without numerical integration were explained and their codes were also given. Finally we showed that nonlinear parabolic equations can be solved by combining of Newton method and diffusion–reaction equation where the latter was explained in this chapter.

However, to have a reliable and efficient simulator several topics and challenges should be addressed. Assuming correct mathematical model, mesh generation and assigning (measured or suggested) values to the coefficients of the problem are very important to make close the numerical model and its solution to the real problem [4, 5]. In a real problem, 2D and 3D elements appear together. For example 2D elements are employed to model faults in a 3D reservoir. Moreover, the generated mesh is normally unstructured and different types of elements are included in the mesh. Hence to assembly the linear system, traversing elements and nodes should have been implemented very efficiently.

Solving linear systems is normally the most time consuming part of the simulation and efficient implementation of linear solvers particularly in parallel machines, are of great interest [6]. Multigrid and multiscale methods, particularly their algebraic form, have attracted interest to solve large scale linear systems since they have shown good scalability in addition to resolving low frequency parts of the error in solving linear systems [7, 8]. Standard Galerkin method that we used in this chapter is not a conservative scheme hence modifications or other methods such as discontinues Galerkin method would be necessary to solve problems in computational fluid dynamics [9, 10]. Proposing and implementing of advanced numerical algorithms to linearize and solve a system of nonlinear coupled initial-boundary value problem in a large scale domain become necessary. At last comparison with real data (if available) and quantifying uncertainties might force us to revisit all steps of the simulation again.

Acknowledgements

Hani Akbari would like to thank Prof. Lars K. Nielsen, scientific director at Novo Nordisk Foundation Center for Bio-Sustainability. Hani Akbari is grateful to DTU-Biosustain since this work was completed when he was a postdoc at DTU-Biosustain.

Author details

Hani Akbari
Shamim Institute of HPC, Scientific Visualization and Machine Learning, Mashhad, Iran

*Address all correspondence to: hani.akbari@shamimhpc.ir

References

[1] Trangenstein J. Numerical Solution of Elliptic and Parabolic Partial Differential Equations. Cambridge University Press; 2013

[2] M. Larson, F. Bengzon, The Finite Element Method: Theory, Implementation, and Applications, Springer, 2013

[3] Ern A, Guermond J. Theory and Practice of Finite Elements. Springer; 2004

[4] Frey P, George PL. Mesh Generation: Application to Finite Elements. second ed. John Wiley & Sons; 2013

[5] Edelsbrunner H. Geometry and Topology for Mesh Generation. Cambridge University Press; 2001

[6] Saad Y. Iterative methods for sparse linear systems. Second ed. SIAM; 2000

[7] Notay Y. Aggregation-based algebraic multigrid for convection-diffusion equations. SIAM J. Sci. Comput. 2012;**34**:A2288-A2316

[8] Akbari H, Pereira F. An algebraic multiscale solver with local Robin boundary value problems for flows in high-contrast media. Journal of Engineering Mathematics. 2020;**123**: 109-128

[9] Zienkiewicz OC, Taylor RL, Nithiarasu P. The Finite Element Method for Fluid Dynamics. 7th ed. Butterworth-Heinemann; 2013

[10] Hesthaven J, Warburton T. Nodal Discontinuous Galerkin Methods. Analysis, and Applications, Springer: Algorithms; 2008

Section 3

Diffusion Phenomena Numerical Analysis

Chapter 7

A Modified Spectral Relaxation Method for Some Emden-Fowler Equations

Gerald Tendayi Marewo

Abstract

In this chapter, we present a modified version of the spectral relaxation method for solving singular initial value problems for some Emden-Fowler equations. This study was motivated by the several applications that these equations have in Science. The first step of the method of solution makes use of linearisation to solve the model problem on a small subinterval of the problem domain. This subinterval contains a singularity at the initial instant. The first step is combined with using the spectral relaxation method to recursively solve the model problem on the rest of the problem domain. We make use of examples to demonstrate that the method is reliable, accurate and computationally efficient. The numerical solutions that are obtained in this chapter are in good agreement with other solutions in the literature.

Keywords: Emden-Fowler equations, Lane-Emden equations, singular initial value problem, spectral relaxation method, numerical method

1. Introduction

The singular initial value problem

$$\begin{aligned} &\frac{d^2y}{dx^2}+\frac{\gamma}{x}\frac{dy}{dx}+r(x,y)=s(x), 0<x,\gamma>0 \\ &y(0)=\alpha, \frac{dy}{dx}(0)=0, \alpha\in\mathbb{R} \end{aligned} \tag{1}$$

for the Lane-Emden Eq. (1) models several phenomena such as the thermal behaviour of a spherical cloud of gas acting under the mutual attraction of its molecules [1], the temperature variation of a self gravitating star, the kinetics of combustion [2], thermal explosion in a rectangular slab [3] and the density distribution in isothermal gas spheres [4]. Moreover, Eq. (1) has been used many a time as a benchmark for new methods.

A particular case of Eq. (1) is the Emden-Fowler equation of the first kind:

$$\frac{d^2y}{dx^2}+\frac{2}{x}\frac{dy}{dx}+y^m=0, y(0)=1, \frac{dy}{dx}(0)=0, m\in\mathbb{N} \tag{2}$$

As mentioned in [5], Eq. (2) represents the dimensionless form of the governing equation for the gravitational potential of a Newtonian self-gravitating, spherically symmetric, polytropic fluid. The equation provides a useful approximation for stars.

A more general form for Eq. (2) is the Emden-Fowler equation

$$\frac{d^2y}{dx^2} + \frac{\gamma}{x}\frac{dy}{dx} + f(x)g(y) = 0 \tag{3}$$

which can be written as

$$(py')' + qy = h(x, y, y') \text{ where}()' = \frac{d}{dx}(), \tag{4}$$

an equation which was discussed in [6]. An existence result for the solution is given therein under certain conditions on $p(x), q(x)$ and $h(x, y, y')$.

Exact solutions are available for particular cases of Eq. (3) [7], but not for the general case according to the best of our knowledge. This is motivation enough for seeking approximate solutions. To this end several approximate analytical methods were used by other researchers to solve Eq. (3). Van Gorder [8] made use of the Homotopy analysis method (HAM) and its variant, the Optimal homotopy analysis method, to solve a boundary value problem for the Lane-Emden equation of the second kind. The two respective analytical solutions that they obtained were in strong agreement. The Homotopy perturbation method (HPM) is another variant of the HAM that was used by Chowdhury and Hashim [9] to solve an initial value problem for Eq. (3). Their analytical solutions were the same as those that were obtained by Wazwaz [10] using the Adomian decomposition method (ADM). Chowdhury and Hashim observed that the HPM was less computationally expensive than the ADM. Wazwaz [11] made use of the variational iteration method (VIM) to solve both initial value problems and boundary value problems for Eq. (3) and for some inhomogeneous Emden-Fowler equations. The results that they obtained demonstrated the reliability and effectiveness of the VIM.

Some numerical methods have been used by other researchers to approximate solutions to Eq. (3). Many of these numerical methods fall in the class of *collocation methods*. Examples of these collocation methods include but are not limited to the Chebyshev wavelet finite difference method (CWFDM) [12], the Haar wavelet collocation method (HWCM) [13], the Taylor wavelet method (TWM) [14] and the Radial basis function - differential quadrature method (RDF-DQM) [15]. One distinct feature of these collocation methods is the choice of collocation points for discretizing the problem domain. Another distinct feature is the choice of basis functions that are used either for constructing numerical solutions or for numerical differentiation. The CWFDM makes use of Chebyshev-Gauss-Lobatto collocation points and Chebyshev wavelet finite difference basis functions. The HWCM makes use of the collocation points

$$x_j = (j - 0.5)/2M, j = 1, \cdots, 2M$$

on the problem domain $[0, L]$, and the method uses integrals of Haar wavelets as basis functions. The TWM uses roots of shifted Legendre polynomials as collocation points, and as basis functions the method uses Taylor wavelets which are special functions that defined in terms of Taylor polynomials. Convergence results for the Taylor wavelet solution were presented in [14]. The RDF-DQM uses collocation points

$$x_j = \frac{2}{L}\left(1 - \cos\left(\frac{j-1}{N-1}\right)\right), j = 1, \cdots, N$$

on the problem domain $[0, L]$ and the method makes use of Radial basis functions. Unlike making use of collocation methods for solving Eq. (3) Van Gorder and Vajravelu [1] used the Runge–Kutta-Felhberg 4-5 (RKF45) method to validate the analytical solutions that they obtained from using the HAM and from using the traditional power series method. The RKF45 method is an embedded Runge–Kutta-pair which makes use of an adaptive stepsize to control the method and to ensure stability properties such as A-stability. See [16] for more details on the RKF45 method.

In this chapter we make use of a modified version of the spectral relation method (SRM) to solve an initial value problem for Eq. (3). We denote our method by MSRM. The SRM was successfully used to solve fluid flow problems by for example Motsa [17], Motsa *et al.* [18], Shateyi *et al.* [19] and Gangadhar *et al.* [20] to mention a few. In [17–20] the SRM was shown to be accurate, computationally efficient and easy to implement. Moreover, the SRM was applied only after transforming the governing partial differential differential equations to ordinary differential equations. However, not so long ago the SRM was modified in such a way that it was directly applicable to partial differential equations. See for example [21]. The SRM was used to solve other types of problems such as hyperchaotic systems [22]. It is to the best of our knowledge that the MSRM has not been used in existing literature. We chose the MSRM because it is not computationally intensive and it is easy to implement.

In Section 2 we describe the MSRM for the model problem. In Section 3 we make use of examples to demonstrate the accuracy and computational efficiency of the MSRM. Section 4 concludes this chapter.

2. The MSRM for the model problem

We seek an approximate solution to

$$y'' + \frac{\gamma}{x} y' + f(x)g(y) = 0, 0 < x \leq L, y(0) = \alpha, y'(0) = 0 \tag{5}$$

where $\gamma > 0, L, \alpha$ are given constants and f and g are given functions.

We follow the idea behind the solution method by Ramos [23], where the singularity at $x = 0$ is isolated in a sufficiently small subinterval $I_\varepsilon = [0, \varepsilon]$ of $I = [0, L]$ where $\varepsilon > 0$. The point $x = \varepsilon$ splits interval I into two subintervals: I_ε and $I - I_\varepsilon = [\varepsilon, L]$. A linearisation method is used to solve Eq. (5) restricted to I_ε, i.e., *near the singularity* at $x = 0$. In order to improve the accuracy of the method on the subinterval $I - I_\varepsilon$ we form a partition

$$I - I_\varepsilon = \cup_{m=1}^{M} I_m \text{ where } I_m = [x_{m-1}, x_m], x_0 = \varepsilon \text{ and } x_M = L \tag{6}$$

The method by Ramos proceeds by using the same linearisation method on the subintervals I_m, $(m = 1, \cdots, M)$ of $I - I_\varepsilon$, i.e. *away from the singularity*. To this end, at the interface x_{m-1} of I_{m-1} and I_m we make use the solution of Eq. (3) restricted to I_{m-1} to generate the initial conditions for Eq. (3) restricted to I_m. This ensures continuity of the solution. In this chapter we avoid linearisation on $I - I_\varepsilon$ by making use of the SRM on the subintervals of $I - I_\varepsilon$. However, as was done in [23] we make use of linearisation on I_ε. This approach results in the MSRM. A detailed description of the MSRM is given in Sections 2.1 and 2.2.

2.1 Near the singularity

Let $\varepsilon \in (0, L)$ be a sufficiently small number. Restrict problem (5) to $[0, \varepsilon]$ and re-arrange to get

$$y'' = \underbrace{-\left[\frac{\gamma}{x}y' + f(x)g(y)\right]}_{u(x,y,y')}, 0 < x \le \varepsilon, y(0) = \alpha, y'(0) = 0 \tag{7}$$

If we Taylor expand u about $\boldsymbol{\varepsilon}_0 = \left(\varepsilon, \underbrace{y(0)}_{y_0}, \underbrace{y'(0)}_{y'_0}\right)$ and neglect higher order terms we get

$$u(x,y,y') = \underbrace{u(\boldsymbol{\varepsilon}_0)}_{u_0} + \underbrace{u_y(\boldsymbol{\varepsilon}_0)}_{H_0}(y - y_0) + \underbrace{u_{y'}(\boldsymbol{\varepsilon}_0)}_{G_0}(y' - y'_0) + \underbrace{u_x(\boldsymbol{\varepsilon}_0)}_{L_0}(x - \varepsilon)$$

where u_s denotes $\frac{\partial u}{\partial s}$. Consequently, Eq. (7) can be replaced by

$$y'' = u_0 + H_0(y - y_0) + G_0(y' - y'_0) + L_0(x - \varepsilon), 0 \le x \le \varepsilon, y(0) = \alpha, y'(0) = 0, \tag{8}$$

where the differential equation now holds at $x = 0$ because the singularity there is no longer present. If $H_0 \neq 0$ and $R_0 := (G_0/2)^2 + H_0 > 0$, then problem (8) has analytical solution [23]

$$y(x) = A_0 \exp\left(\lambda_0^+(x - \varepsilon)\right) + B_0 \exp\left(\lambda_0^-(x - \varepsilon)\right) + C_0(x - \varepsilon) + D_0 \tag{9}$$

for $0 \le x \le \varepsilon$, where $\lambda_0^\pm = G_0/2 \pm \sqrt{R_0}, C_0 = -L_0/H_0$,

$$D_0 = -(G_0C_0 + P_0)/H_0, P_0 = u_0 - H_0y_0 - G_0y'_0,$$

$$A_0 = \frac{\exp\left(\lambda_0^+\varepsilon\right)}{\lambda_0^+ - \lambda_0^-}\left(y'_0 - C_0 - \lambda_0^-\left(y_0 + C_0\varepsilon - D_0\right)\right),$$

$$B_0 = \frac{\exp\left(\lambda_0^-\varepsilon\right)}{\lambda_0^+ - \lambda_0^-}\left(\lambda_0^+\left(y_0 + C_0\varepsilon - D_0\right) - \left(y'_0 - C_0\right)\right)$$

Thus

$$y(\varepsilon) = \underbrace{A_0 + B_0 + D_0}_{\alpha_0} \text{ and } y'(\varepsilon) = \underbrace{\lambda_0^+A_0 + \lambda_0^-B_0 + C_0}_{\alpha'_0} \tag{10}$$

We take α_0 and α'_0 as initial values for problem (7) restricted to $[\varepsilon, L]$ in the next section.

2.2 Away from the singularity

We seek $y(x)$ satisfying

$$y'' + \frac{\gamma}{x}y' + f(x)g(y) = 0, \varepsilon \le x \le L, y(\varepsilon) = \alpha_0, y'(\varepsilon) = \alpha'_0 \tag{11}$$

In this section we begin by describing the SRM for problem (11). In practical applications it is usually important to obtain a solution to (11) which possesses a prescibed degree of accuracy. To this end we make use of the SRM on a partition of the problem domain $[\varepsilon, L]$. This is our last task in this section.

The first step of the SRM for (11) is to let $v = y$ and $w = v'$ so that we obtain the equivalent problem

$$v' = w, v(\varepsilon) = \alpha_0, \tag{12}$$

$$w' + \frac{\gamma}{x} w + f(x)g(v) = 0, w(\varepsilon) = \alpha_0' \tag{13}$$

for $\varepsilon \leq x \leq L$ which upon making use of the change of variable

$$x(\eta) = \frac{L+\varepsilon}{2} + \frac{L-\varepsilon}{2}\eta$$

becomes

$$\beta \frac{dv}{d\eta} = w, v(-1) = \alpha_0 \tag{14}$$

$$\beta \frac{dw}{d\eta} + \frac{\gamma}{x(\eta)} w + f(x(\eta))g(v) = 0, w(-1) = \alpha_0' \tag{15}$$

for $-1 \leq \eta \leq 1$ where $\beta = 2/(\mathrm{L} - \varepsilon)$. As described in [19] the next step of the SRM mimicks the Gauss–Seidel method for linear systems and it yields the iteration

$$\beta \frac{dv_{r+1}}{d\eta} = \underbrace{w_r}_{R^{(1)}(\eta)}, v_{r+1}(\eta_N) = \alpha_0 \tag{16}$$

$$\beta \frac{dw_{r+1}}{d\eta} + \frac{\gamma}{x} w_{r+1} = \underbrace{-f(x(\eta))g(v_{r+1})}_{R^{(2)}(\eta)}, w_{r+1}(\eta_N) = \alpha_0' \tag{17}$$

for $-1 = \eta_{\mathrm{N}} \leq \eta \leq \eta_0 = 1$ where $r = 0, 1, \cdots$ and on $[-1, 1]$ we formed a grid consisting of the Chebyshev-Gauss-Lobatto collocation points

$$\eta_i = \cos\left(\frac{\pi i}{N}\right), \qquad i = 0, 1, \cdots, N. \tag{18}$$

If the initial approximations v_0 and w_0 to v and w, respectively, are prescribed then Eqs. (16) and (17) generate sequences $\{v_r\}_{r=1}^{\infty}$ and $\{w_r\}_{r=1}^{\infty}$ of consecutive approximations. To this end we assume that v_r and w_r are known at the end of the rth iteration. Once Eq. (16) is solved for v_{r+1} the right hand side $R^{(2)}$ of Eq. (17) becomes known and we solve this equation for w_{r+1}. Since v_{r+1} and w_{r+1} are now known, we proceed in a similar manner to compute v_{r+2} and w_{r+2}, and so on. As done in [19] we solve Eqs. (16) and (17) by making use of Chebyshev differentiation [24] to obtain

$$\underbrace{\hat{D}}_{A_1} \mathbf{v}_{r+1} = \mathbf{R}^{(1)}, v_{r+1}(\eta_N) = \alpha_0, \tag{19}$$

$$\underbrace{(\hat{D} + \gamma \mathrm{diag}[1/x(\eta_0), \cdots, 1/x(\eta_N)])}_{A_2} \mathbf{w}_{r+1} = \mathbf{R}^{(2)}, w_{r+1}(\eta_N) = \alpha_0' \tag{20}$$

where $\hat{D} = \beta D, D$ is the $N \times N$ Chebyshev differentiation matrix,

$$\mathrm{diag}[b_0, \cdots, b_N] = \begin{pmatrix} b_0 & & \\ & \ddots & \\ & & b_N \end{pmatrix} \tag{21}$$

is a diagonal matrix, and $\mathbf{R}^{(i)} = \left[R^{(i)}(\eta_0), \cdots R^{(i)}(\eta_N)\right]^T$ with $i = 1, 2$. Moreover,

$$\mathbf{v}_r = [v_r(\eta_0), \cdots, v_r(\eta_N)]^T, \mathbf{w}_r = [w_r(\eta_0), \cdots, w_r(\eta_N)]^T, \quad r = 0, 1, \cdots \tag{22}$$

and $[\,]^T$ denotes matrix transpose.

We prescribe v_0 and w_0 by requiring that

$$v_0(\varepsilon) \equiv y_0(\varepsilon) = \alpha_0 \text{ and } w_0(\varepsilon) \equiv y_0'(\varepsilon) = \alpha_0'$$

Moreover, we assume that

$$\lim_{r \to \infty} v_r = v \text{ and } \lim_{r \to \infty} w_r = w$$

The initial conditions

$$v_{r+1}(\eta_N) = \alpha_0 \text{ and } w_{r+1}(\eta_N) = \alpha_0'$$

are included in the iterative scheme consisting of Eqs. (19) and (20) as shown below.

$$\left(\frac{A_1}{0 \;\cdots\; 0 \;1}\right)\begin{pmatrix} v_{r+1}(\eta_0) \\ \vdots \\ v_{r+1}(\eta_{N-1}) \\ \hline v_{r+1}(\eta_N) \end{pmatrix} = \begin{pmatrix} R^{(1)}(\eta_0) \\ \vdots \\ R^{(1)}(\eta_{N-1}) \\ \hline \alpha_0 \end{pmatrix} \tag{23}$$

$$\left(\frac{A_2}{0 \cdots 0 \;1}\right)\begin{pmatrix} w_{r+1}(\eta_0) \\ \vdots \\ w_{r+1}(\eta_{N-1}) \\ \hline w_{r+1}(\eta_N) \end{pmatrix} = \begin{pmatrix} R^{(2)}(\eta_0) \\ \vdots \\ R^{(2)}(\eta_{N-1}) \\ \hline \alpha_0' \end{pmatrix} \tag{24}$$

The larger L is the less reliable is the SRM. As a workaround to this problem is we subdivide interval $[\varepsilon, L]$ into a disjoint union of non-overlapping subintervals as detailed in Eq. (6). Given the the model problem on I_1, we use the SRM to compute estimate $\tilde{y}$ to y on I_1. We make use of $\tilde{y}$ to compute initial values for the problem on I_2. We repeat this procedure for the problem on I_2 and continue in a similar manner until we exhaust $[\varepsilon, L]$. Shown in Algorithm 2.2 is an outline of the MSRM for problem (5).

Algorithm 1: Putting the MSRM together.

---

1. Let $[0, L] = I_\varepsilon \cup [\varepsilon, L]$ where $I_\varepsilon := [0, \varepsilon]$ for some sufficiently small number $\varepsilon > 0$.

2. Replace nonlinear Eq. (5) with linear Eq. (8) on I_ε and compute solution $y_\varepsilon(x)$ to Eq. (7) on I_ε.

3. Set $\alpha_0 = y_\varepsilon(\varepsilon)$, set $\alpha_0' = y_\varepsilon'(\varepsilon)$ and let $\varepsilon = x_0 < x_1 < \cdots < x_M = L$.

4. **for** $m = 1, 2, \cdots, M$ **do**

5. Define $I_m := [x_{m-1}, x_m]$ and make use of the SRM to solve

$$y'' = u(x, y, y'), x \in I_m, y(x_{m-1}) = \alpha_0, y'(x_{m-1}) = \alpha'_0$$

for the estimate $\tilde{y}$ to y on I_m.

6. Set $\alpha_0 = \tilde{y}(x_m)$ and set $\alpha'_0 = \tilde{y}'(x_m)$.

7. **end**

2.3 Examples

In this section we make use of some examples to investigate the accuracy and computational effeciency of the MSRM.

Example 1 We look for a numerical solution to the problem

$$y'' + \frac{1}{x}y' + 3y^5 - y^3 = 0, 0 < x \leq 10 y(0) = 1, y'(0) = 0 \tag{25}$$

which Wazwaz [11] solved using the VIM and obtained the approximate. analytical solution

$$y(x) = \frac{1}{\sqrt{1 + x^2}} \tag{26}$$

When we apply the MSRM on (25) we get the following components for constructing the numerical solution.

1. Coefficients

$$u_0 = -2, L_0 = 0, H_0 = -12, G_0 = -10^4 \tag{27}$$

for problem (8).

2. Initial values

$$y(\varepsilon) := \alpha_0 = 9.999999926 \times 10^{-1} \text{ and } y'(\varepsilon) := \alpha'_0 = -1.264241093 \times 10^{-4} \tag{28}$$

for problem (11).

3. Entries

$$R_i^{(1)} = w(\eta_i) \text{ and } R_i^{(2)} = -\left(3v_{r+1}^5(\eta_i) - v_{r+1}^3(\eta_i)\right), i = 0, \cdots, N$$

in the vectors on the right hand sides of Eqs. (19) and (20).

The MSRM generates a numerical solution to problem (25) which is plotted together with the analytical solution (26) in **Figure 1(a)**. **Figure 1(a)** shows a good agreement between the numerical and analytical solutions. For a more detailed comparion of the two solutions see **Table 1**. **Table 1** and **Figure 1(b)** show that the absolute error of the numerical solution is no more than 0.5×10^{-6}. Thus the numerical solution agrees with the analytical solution in the first 6 decimal places.

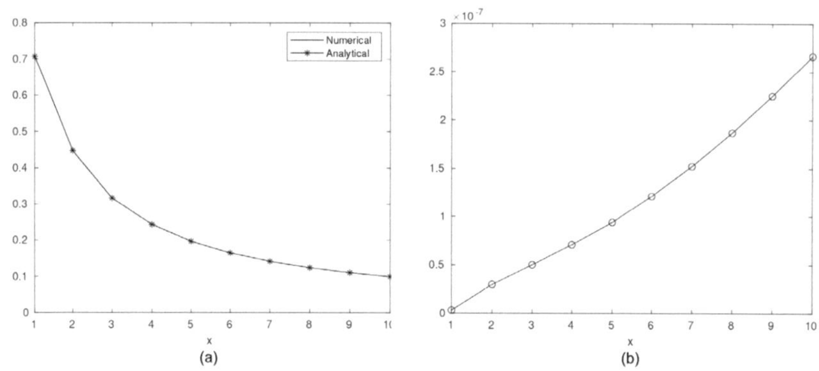

Figure 1.
(a) Solutions to problem (25). (b) Absolute error of the MSRM solution to (25).

x	MSRM	Solution (26)	Absolute error
1	0.70710678	0.70710678	3.1×10^{-9}
2	0.44721363	0.44721360	3×10^{-8}
3	0.31622782	0.31622777	5×10^{-8}
4	0.24253570	0.24253563	7.1×10^{-8}
5	0.19611623	0.19611614	9.4×10^{-8}
6	0.16439911	0.16439899	1.2×10^{-7}
7	0.14142151	0.14142136	1.5×10^{-7}
8	0.12403492	0.12403473	1.9×10^{-7}
9	0.11043175	0.11043153	2.3×10^{-7}
10	0.09950399	0.09950372	2.7×10^{-7}

Table 1.
Comparison of numerical values of y with solution (26).

This degree of accuracy was achieved by the MSRM upon choosing $N = 60, M = 10$, setting the maximum number of iterations as $r_{max} = 20$ and imposing the stopping criterion

$$\max\left\{\|\mathbf{v}_{r+1} - \mathbf{v}_r\|_\infty, \|\mathbf{w}_{r+1} - \mathbf{w}_r\|_\infty\right\} \leq \varepsilon, r = 0, 1, \cdots$$

for the iterative method, where $\varepsilon = 10^{-6}$ is the tolerance and $\|\mathbf{w}\|_\infty = \max_{0 \leq i \leq N}|w_i|$ is the infinity norm. It took only $r = 5$ iterations for the MSRM to achive the given degree of accuracy.

Example 2 We consider the initial value problem

$$y'' + \frac{5}{x}y' + 8\left(e^y + 2e^{y/2}\right) = 0, 0 < x \leq 10, y(0) = y'(0) = 0 \qquad (29)$$

that was solved by Wazwaz [10] using the ADM to obtain the approximate analytical solution

$$y(x) = -2\ln\left(1 + x^2\right) \qquad (30)$$

Applying the MSRM on (29) produces the following building blocks for constructing the numerical solution.

1. Coefficients

$$u_0 = -24, L_0 = 0, H_0 = -16, G_0 = -5 \times 10^4 \tag{31}$$

for problem (8).

2. Initial values

$$y(\varepsilon) := \alpha_0 = -3.846468396 \times 10^{-8} \text{ and } y'(\varepsilon) := \alpha_0' = -4.767657761 \times 10^{-4} \tag{32}$$

for problem (11).

3. Elements

$$R_i^{(1)} = w(\eta_i) \text{ and } R_i^{(2)} = -8\left(e^{v_{r+1}(\eta_i)} + 2e^{v_{r+1}(\eta_i)/2}\right), i = 0, \cdots, N$$

of the vectors on the right hand sides of Eqs. (19) and (20).

A comparison of the MSRM solution to problem (29) with the analytical solution (30) is shown in **Figure 2(a)**. The graph suggests that the two solutions are exactly the same. However, a closer look at the two solutions is provided in **Table 2** and we observe that the analytical and numerical solutions are slightly different. A plot of the absolute error of the MSRM solution over a grid on the problem domain $[0, 10]$ is shown in **Figure 2(b)**. We observe that the absolute error does not exceed 0.5×10^{-7}. Hence the MSRM solution and the analytical solution agree to within 7 decimal places. For the MSRM to achieve this degree of accuracy we chose $N = 60, M = 10, r_{max} = 20$ and $\varepsilon = 10^{-6}$. We observed that the MSRM stopped at iteration $r = 5$ for problem (29).

Example 3 We seek a numerical solution to

$$y'' + \frac{2}{x}y' - 6y - 4y\ln y = , 0 < x \leq 1, y(0) = 1, y'(0) = 0, \tag{33}$$

where

$$y(x) - e^{x^2} \tag{34}$$

is the exact solution [25]. We restrict the problem to a relatively small interval $[0, 1]$ because the solution (34) grows rapidly.

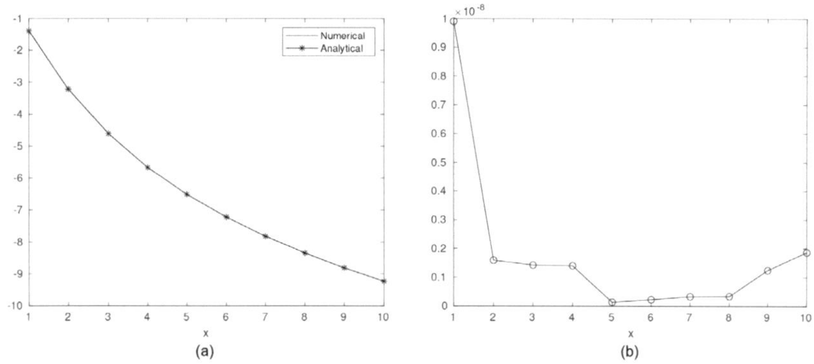

Figure 2.
(a) Solutions to problem (29). (b) Absolute error of the MSRM solution to (29).

x	MSRM	Solution (34)	Absolute error
1	−1.38629437	−1.38629436	9.9×10^{-9}
2	−3.21887583	−3.21887582	1.6×10^{-9}
3	−4.60517018	−4.60517019	1.4×10^{-9}
4	−5.66642669	−5.66642669	1.4×10^{-9}
5	−6.51619308	−6.51619308	1.4×10^{-10}
6	−7.22183583	−7.22183583	2.3×10^{-10}
7	−7.82404601	−7.82404601	3.4×10^{-10}
8	−8.34877454	−8.34877454	3.4×10^{-10}
9	−8.81343849	−8.81343849	1.2×10^{-9}
10	−9.23024103	−9.23024103	1.9×10^{-9}

Table 2.
Comparison of numerical values of y with solution (30).

The MSRM for (34) gives the following components for constructing the numerical solution.

1. Coefficients

$$u_0 = 6, L_0 = 0, H_0 = 10 \text{ and } G_0 = -2 \times 10^4 \tag{35}$$

for problem (8).

2. Initial values

$$\alpha_0 = 1.000000017 \text{ and } \alpha_0' = 2.593994191 \times 10^{-04} \tag{36}$$

for problem (11).

3. Right hand sides

$$R_i^{(1)} = w_r(\eta_i) \text{ and } R_i^{(2)} = 6v_{r+1}(\eta_i) + 4v_{r+1}(\eta_i)\ln(v_{r+1}(\eta_i)), i = 0, \cdots, N \tag{37}$$

for linear systems (19) and (20).

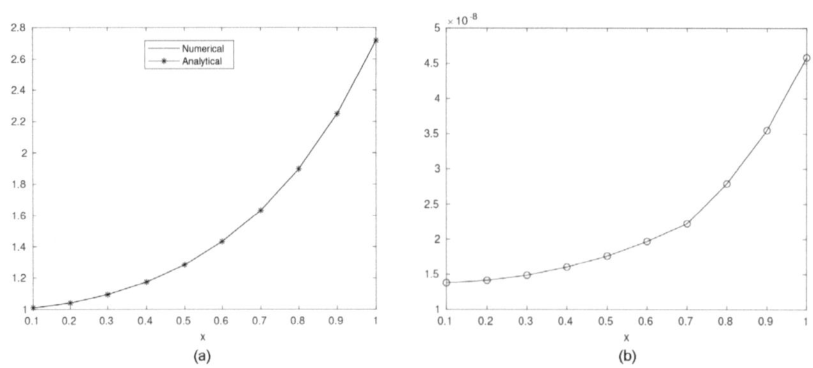

Figure 3.
(a) Solutions to problem (33). (b) Absolute error of MSRM solution to (33).

x	MSRM	Solution (34)	Absolute error
0.1	1.01005018	1.01005017	1.4×10^{-8}
0.2	1.04081079	1.04081077	1.4×10^{-8}
0.3	1.09417430	1.09417428	1.5×10^{-8}
0.4	1.17351089	1.17351087	1.6×10^{-8}
0.5	1.28402543	1.28402542	1.8×10^{-8}
0.6	1.43332943	1.43332941	2×10^{-8}
0.7	1.63231624	1.63231622	2.2×10^{-8}
0.8	1.89648091	1.89648088	2.8×10^{-8}
0.9	2.24790802	2.24790799	3.6×10^{-8}
1	2.71828187	2.71828183	4.6×10^{-8}

Table 3.
Comparison of numerical values of y with solution (34).

Figure 3(a) shows that the numerical solution agrees well with the analytical solution (34). For a closer look at how the numerical and analytical solutions compare see **Table 3**. We observe that the MSRM solution agrees with the analytical solution (34) on the problem domain $[0, 1]$ to within at least 7 decimal places. A plot of the absolute error at these grid points on $[0, 1]$ is shown in **Figure 3(b)**. We observe that the absolute error in the MSRM increases as me move away from the singular point $x = 0$, but the absolute error never exceeds 0.5×10^{-7}. This degree of accuracy is achieved with $N = 60, M = 10, r_{max} = 20$ and $\varepsilon = 10^{-6}$. Moreover, only 6 iterations were required to achieve this degree of accuracy.

3. Conclusions

In this chapter we presented a modified spectral relaxation method (denoted by MSRM) for solving singular initial value problems for some Emden-Fowler equations. We made use of some examples of the model problem to demonstrate that the MSRM is reliable, accurate and computationally efficient. The method provided a reliable treatment of the singular point. The MSRM solutions were compared with analytical solutions that were obtained using other methods, i.e., the Variational iteration method and the Adomian decomposition method. There was agreement between the solutions that were compared in the first 6 decimal places. A possible way of increasing the degree of accuracy of the MSRM would be to increase the tolerance for the method. This and other ways for optimizing the method could consistute future work. In all the examples that were considered, it took at most 6 iterations for the MSRM to converge. Hence the method exhibited rapid convergence.

Author details

Gerald Tendayi Marewo
North-West University, Potchefstroom, South Africa

*Address all correspondence to: gerald.marewo@nwu.ac.za

References

[1] Van Gorder R.A., Vajravelu K., Analytic and numerical solutions to the Lane–Emden equation, Physics Letters A, 372 (2008) 6060-6065.

[2] Šmarda Z., Khan Y., An efficient computational approach to solving singular initial value problems for Lane-Emden type equations, Journal of Computational and Applied Mathematics 290 (2015) 65-73.

[3] Van Gorder R.A., Analytic and numerical solutions to the Lane-Emden equation, New Astronomy 16 (2011) 492-497.

[4] Momoniat E., Harley C., Approximate implicit solution of a Lane-Emden equation, New Astronomy 11 (2006) 520-526.

[5] Căruntu B., Bota C., Approximate polynomial solutions of the nonlinear Lane-Emden type equations arising in astrophysics using the squared remainder minimization method, Computer Physics Communications 184 (2013) 1643-1648.

[6] Răb M., Bounds for solutions of the equation $(pu')' + qu = h(x, u, u')$, Arch. Math.2, Scripta Fac.Sci.Nat.UJEP Brunessis, X1, 1975, 79-84.

[7] Torres-Córdoba R., Martínez-García E.A., Exact analytic solution of an unsolvable class of first Lane-Emden equation for polytropic gas sphere, New Astronomy 82 (2021) 101458.

[8] Van Gorder R.A., Relation between Lane-Emden solutions and radial solutions to the elliptic Heavenly equation on a disk, New Astronomy 37 (2015) 42-47.

[9] Chowdhury M.S.H., Hashim I., Solutions of Emden-Fowler equations by homotopy-perturbation method, Nonlinear analysis: real world applications 10 (2009) 104-115.

[10] Wazwaz A.M., Adomian decomposition method for a reliable treatment of the Emden-Fowler equation, Applied Mathematics and Computation 161 (2005) 543-560.

[11] Wazwaz A.M., A reliable treatment of singular Emden-Fowler initial value problems and boundary value problems, Applied Mathematics and Computation 217 (2011) 10387-10395.

[12] Kazemi Nasab A., Kılıçman A., Pashazadeh Atabakan Z., Leong W.J., A numerical approach for solving singular nonlinear Lane-Emden type equations arising in astrophysics, New Astronomy 34 (2015) 178-186.

[13] Singh R., Garg H., Guleria V., Haar wavelet collocation method for Lane-Emden equations with Dirichlet, Neumann and Neumann-Robin boundary conditions, Journal of Computational and Applied Mathematics 346 (2019) 150-161.

[14] Gümgüm S., Taylor wavelet solution of linear and nonlinear Lane-Emden equations, Applied Numerical Mathematics 158 (2020) 44-53.

[15] Parand K., Hashemi S., RBF-DQ method for solving non-linear differential equations of Lane-Emden type, Ain Shams Engineering Journal (2018) 9, 615-629.

[16] Iserles A., A first course in the numerical analysis of differential euqations, Cambridge, 2009.

[17] Motsa S.S., A new spectral relaxation method for similarity variable nonlinear boundary layer flow systems, Chemical Engineering Communications Volume 201, 2014 - Issue 2.

[18] Motsa S.S., Dlamini P.G., Khumalo M., Spectral relaxation method and spectral quasilinearization

method for solving unsteady boundary layer flow problems, Advances in Mathematical Physics, vol. 2014, Article ID 341964, 12 pages, 2014.

[19] Shateyi S., Marewo G.T., Numerical analysis of unsteady MHD flow near a stagnation point of a two-dimensional porous body with heat and mass transfer, thermal radiation, and chemical reaction, Boundary Value Problems 2014, 2014:218.

[20] Gangadhar K., Kannan T., Sakthivel G., DasaradhaRamaiah K., Unsteady free convective boundary layer flow of a nanofluid past a stretching surface using a spectral relaxation method, International journal of ambient energy 2020, vol. 41, no. 6, 609-616. https://doi.org/10.1080/01430750.2018.1472648

[21] Motsa S.S., Animasaun I.L., Unsteady Boundary Layer Flow over a Vertical Surface due to Impulsive and Buoyancy in the Presence of Thermal-Diffusion and Diffusion-Thermo using Bivariate Spectral Relaxation Method, Journal of Applied Fluid Mechanics, Vol. 9, No. 5, pp. 2605-2619, 2016.

[22] Motsa S.S., Dlamini P.G., Khumalo M., Solving hyperchaotic systems using the spectral relaxation method, Abstract and Applied Analysis Volume 2012, Article ID 203461, 18 pages. https://doi.org/10.1155/2012/203461.

[23] Ramos J.I., Linearization techniques for singular initial-value problems of ordinary differential equations, Applied Mathematics and Computation 161 (2005) 525-542.

[24] Trefethen L.N., Spectral Methods in MATLAB, SIAM, 2000.

[25] Karimi Dizicheh A., Salahshour S., Ahmadian A., Balaeanu D., A novel algorithm based on the Legendre wavelets spectral technique for solving Lane-Emden equations, Applied Numerical Mathematics 153 (2020) 443-456. https://doi.org/10.1155/2014/341964.

Section 4

Numerical Forecasting Solutions

Chapter 8

Numerical Modeling of Soil Water Flow and Nitrogen Dynamics in a Tomato Field Irrigated with Municipal Wastewater

Ali Erfani Agah

Abstract

Because of water scarcity, reduction of annual rainfall and the use of wastewater in agriculture, there is a need for research to evaluate the potential impacts of using such sources on hydraulic soil properties and groundwater quality. Nitrate loss from the area under cultivation and regular use of fertilizer and wastewater is a major reason for non-point source contamination on agricultural lands. Numerical model, Hydrus-1D used to simulate soil nitrate in soil cultivated with tomato-crop during the growing period, in North-East Iran. A randomized completely blocked design with five irrigation treatments with different sources of nitrogen was applied. Comparison between simulated and measured soil moisture content shows that the model can follow the temporal variation of soil water content. However, some over estimation of the measured data was observed during the simulation period. To evaluate the Hydrus model performance with respect to nitrogen transport and transformations, the simulated nitrogen concentrations (NH4-N and NO3-N) are compared for different treatments at different depths of soil profile, (7.5, 22.5, 37.5, 52.5 and 120 cm from soil surface). It takes about 4 days to convert 90% of urea into ammonium and it takes about 70 days to convert 90% of ammonium into nitrate. However, urea concentrations decreased with time between irrigations as a result of hydrolysis. As expected, at 3.73 days, the urea was concentrated near the surface, immediately after fertigation. Ammonium remained concentrated in the immediate in the top soil at all times for all treatments. There was only slight movement, because of soil adsorption and subsequent fast nitrification and/or root uptake. In contrast to ammonium, nitrate moved continuously downwards during the 28-day simulation period, as nitrate is not adsorbed, whereas denitrification was assumed negligible. Leaching percentages were smaller for nitrate wastewater compared to nitrate- fertilizer, and manure. Base on simulation results treated municipal wastewater by an aerated lagoon can be used as a valuable source of irrigation without causing contamination of groundwater.

Keywords: Irrigation, Wastewater, Tomato, Nitrate leaching, Hydrus-1D

1. Introduction

Irrigation with wastewater is one of the best options to reduce the stress on limited fresh water available today and to meet the nutrient requirement of crops.

There is potential for these nutrients present in recycled water to be used as a fertilizer source when the water is recycled as an irrigation source for agriculture [1]. Nitrogen is a valuable nutrient contained in wastewater [2]. Various studies confirm that municipal wastewater can be useful as an additional water resource for irrigation [3–7]. Some researchers have shown that the best way to use wastewater after treatment is in agriculture [8].

Understanding the behavior of nitrogen in the soil system helps to maximize crop production while reducing the impacts of N fertilization on the environment. Nitrogen applied as fertilizer or wastewater may be: utilized and stored in the plant; stored as organic nitrogen in the soil; volatilized as ammonia, nitrogen gas or nitrous oxide; lost in runoff; or leached to the groundwater as nitrate [9, 10]. The main processes response for nitrogen transport and transformations in the soil are mass transport of inorganic nitrogen forms, commonly described by the general convection–dispersion equations and both chemical and biological reactions [11].

Nitrate is one of the nitrogen compounds most susceptible to leaching. Three kinds of soil transformation of the N contained in wastewater are important. The first of theses in mineralization:

$$\text{Organic} - N \rightarrow NH_4^+ \tag{1}$$

Mineralization occurs in soil as microorganisms, both aerobic and anaerobic convert organic nitrogen to inorganic forms. After wastewater application to soil, organic N quickly converts to ammonium nitrogen and then to nitrate nitrogen [12]. The sequel to mineralization is nitrification:

$$NH_4^+ + 3/2O_2 \rightarrow NO_2^- + H_2O + 2H^+ \tag{2}$$

$$No_2^- + 1/2O_2 \rightarrow NO_3^- \tag{3}$$

Microbial activity is also responsible for the two steps of nitrification. Nitrosomonas convert ammonium to nitrite. The second step of nitrification occurs through Nitrobacter species, which convert nitrite to nitrate. This step rapidly follows ammonium conversion to nitrite, and consequently, nitrite concentrations are normally low in soils.

Another important nitrogen transformation in soils is denitrification.

Nitrate, which is the end product of the nitrification process in aerobic soils, it can undergo reduction to NO_2 and finally to N_2 when the soil oxygen content is low and decomposable organic materials are present to furnish energy for the process. The sequence of products is:

$$NO_3^- \text{——>} NO_2 \text{——>} NO \text{——>} N_2O \text{——>} N_2 \tag{4}$$

This process, which is done by a group of bacteria, is called desalination or denitrification. Denitrification occurs under oxygen-limiting conditions when anaerobic bacteria use nitrate in respiration in the presence of carbon sources such as organic matter. NO_2 and N_2 are both gaseous and emitted from the soil. Factors influencing denitrification control include oxygen restriction and the presence of organic matter. In the case of wastewater irrigation, only wastewater with a high BOD can be a source of organic matter for denitrification [12].

The amount of soil nitrogen losses through denitrification depends on the type of soil and irrigation management applied in the field and may vary between zero and 90% of applied nitrogen [12].

Mineralization and nitrification processes convert the organic N and NH_4^+ into NH_4^+ and NO_3^{2-} respectively, which are absorbed and utilized by crops and termed as available nitrogen [13–15]. Nitrate is highly mobile and leachable. It has been established that excessive application of nitrogen leads to nitrate pollution of groundwater and surface water [16, 17]. Leaching of NO_3^{2-} below the root zone can be affected by a range of factors, including fertilizer application rates and the timing of applications [18].

Computer models are tools used in science to approximate natural phenomena. Therefore, models that predict flow and transport processes in soils are increasingly being applied to address practical problems. The use of simulation models allows extrapolation, in time and space, of data from leaching experiments and monitoring studies. More recently, computer simulation tools have been applied to predict the fate and transport of contaminants for risk evaluation [19].

In this study we present results of field experimentation and numerical simulations on a loamy soil cultivated with tomato plants, which were used to evaluate the performance of the different component of water and nitrogen dynamic in the soil. Model parameters were either solely derived from laboratory measurements or optimized by the inverse simulation method. The objective of this study was to determine the difference in concentration of nitrate in soil water below the root zone (about 1.5(m) for plots treated with (1) municipal wastewater (2) manure and (3) commercial chemical fertilizers, using HYDRUS-1D model, [31] at the research station of Mashhad in north-east of Iran. Field data, collected on a loamy soil cultivated with tomato plants, were used to evaluate the performance of the different component of water and nitrogen dynamic in the soil.

2. Material and methods

2.1 Site description and measurements

The weather data (daily maximum and minimum temperature, wind speed, humidity, sunshine hour and rainfall data) was collected from metrological station installed 2006 and 2007 at the Mashhad research station site, (36° 13′ latitude, 59°38′ longitude) in Northern east Iran. A soil profile pit was excavated to 120 cm depth and soil samples at different soil texture layers were sampled on 20 March 2009 before tomato sowing and basic properties, including soil water retention and saturated hydraulic conductivity were measured. The soil consists of heterogeneous layers with a deep groundwater ground water table (far below 80 m) and is characterized as sandy loam top soil (0–40 cm) over sandy clay loam (40–65) over sandy clay (65-120 cm). From the rooting depth (120 cm) of tomato crop 100 soil samples were collected and analyzed for various physical and chemical parameters before starting the experiment (**Table 1**).

TDR probes and ceramic cup tensiometers were installed at 0–20, 20–40, 40–60 and 60–100 cm soil layers in the investigated area. Water content measurements were taken daily starting in January 2009 and concluded in October 2010. TDR data will be used to assess estimate of shallow soil water content at soil profile. The irrigation scheduling was based on the soil moisture deficit in the root zone at each irrigation event (difference between root zone soil water at field capacity and at irrigation time) with intervals of 10 days. The characteristics of water and wastewater are summarized in **Table 2**. Total irrigation depth during this period was 23.9 cm.

Physical properties of soil			
Parameters	**Soil layers, cm**		
	0–40	**40–65**	**65–120**
Clay %	11.1	19.4	34.9
Silt %	39.4	35.9	23.8
Sand %	50.5	44.7	42.3
Textural class	Sandy loam	sandy clay loam	sandy clay
Bulk density, g/cm^3	1.55	1.43	1.35
FC, (vol. %)	21.34	27.61	28.22
PWP, (vol. %)	7.18	9.87	11.18
Chemical properties of soil			
pH	7.54	7.54	7.54
EC (ds/m)	1.53	1.53	1.50
Organic carbon (%)	1.82	1.79	1.71
Organic matter, g/100 g soil	1.19	0.67	0.63
Total nitrogen, g/100 g soil	0.059	0.060	0.054
Available P (mg/kg)	54.2	54.2	48

Table 1.
Physical and chemical properties of soil at initial condition.

Parameter	Unit	Well Water	Wastewater	Standard value
PH	—	7.9	7.9	6–8.5
EC	dS/m	0.58	1.1	—
Na^+	mg/l	63.7	85.8	—
K^+	mg/l	2.3	24.6	—
Ca^{2+}	mg/l	24	36.5	—
Mg^{2+}	mg/l	15	18.2	100
NH_3-N	mg/l	—	32	
NO_3-N	mg/l	—	0	
Org – N	mg/l	—	14	
P	mg/l	0.18	2.8	—
Cl	mg/l	12.4	118	600
SAR	mg/l	2.45	2.87	—
B	mg/l	0.86	0.9	1
Alkanity	—	—	670	
Hardness	—	—	189	
Total N of Coliform	—	—	1000	—
BOD_5	mg/l		115	100
COD	mg/l		145	200

Table 2.
Physico-chemical characteristics of water and treated wastewater.

The rainfall at the same period was 11.48 cm and reached 42.84 cm for the whole simulation period of one year.

2.2 Experimental design

Plots were irrigated with either well water or wastewater in a random complete block design (RCBD), with four replications according to the following treatments:

T1 - Irrigation by treated wastewater during all growing season, (%100 wastewater).

T2 - Alternate irrigation by treated wastewater and well water.

1. Alternate irrigation of tomato with wastewater and well water during the growing season, (%75 wastewater +%25 well water).

2. Alternate irrigation of tomato with wastewater and well water during the growing season, (%50 wastewater +%50 well water).

3. Alternate irrigation of tomato with wastewater and well water during the growing season, (%25 wastewater +%75 well water).

T3 -Irrigation with well water plus application animal manure.

T4 - Irrigation with well water plus application of fertilizer.

T5 - Irrigation with well water only.

To obtain these ratios were used in the operation of the irrigation turn. (**Table 3**).

The experiments were carried out on 20 plots, and each experiment included five irrigated furrows 4 m in width and 4.2 m in length (along the crop rows). Each plot consisted of 5 crop rows with a plant row spacing of 75 cm. The plots (T3) were grazing prior sowing with 3000 kg ha^{-1} or (3 kg m^{-2}) animal manure. The chemical analysis of animal manure has been showed in **Table 4**.

The plots (T4) were fertilized based on soil sample tests with 300 kg ha^{-1} or (30 gr m^{-2})of triple super phosphate, broadcast at seedbed preparation, and 110 kg ha^{-1} of net nitrogen or (200 kg ha^{-1}) of urea at tillage time (at two equal section) and 6 weeks after panting. In this study, the total manure was applied prior sowing and

Irrigation turns	**Mixing ratios of wastewater**				
	1%	**2%**	**3%**	**4%**	**5%**
	100s	**75**	**50**	**25**	**0**
First	Wastewater	Well water	Well water	Well water	Well water
Second	Wastewater	Wastewater	Wastewater	Well water	Well water
Third	Wastewater	Wastewater	Well water	Well water	Well water
Fourth	Wastewater	Wastewater	Wastewater	Wastewater	Well water
Fifth	Wastewater	Well water	Well water	Well water	Well water
Sixth	Wastewater	Wastewater	Wastewater	Well water	Well water
Seventh	Wastewater	Wastewater	Well water	Well water	Well water
Eighth	Wastewater	Wastewater	Wastewater	Wastewater	Well water
Ninth	Wastewater	Wastewater	Wastewater	Well water	Well water
Tenth	Wastewater	Well water	Well water	Wastewater	Well water

Table 3.
The proportions of water and wastewater.

pH	EC	N	P	K	OC	C:N	Fe	Mn	Cu	Zn
—	ds/m^{-1}		%			—		Mgkg^{-1}		
7.73	13.6	2.4	1.02	0.81	61	25.4	1611	72	4	54

Table 4.
Chemical analyses of animal manure.

for chemical fertilizer, 50% N and total P fertilizers were applied to the sowing seeds. Tomato was seeded on first week of May of each year in the plots at a plant spacing of 75 cm; Weed, diseases and insect control were uniformly managed during the growing season. After planting, irrigation was applied as required with well water until green stage and then treatments and irrigation applied as required during the growing season.

2.3 Model selection: HYDRUS 1D

The HYDRUS-1D software package uses numerical methods to solve the Richards' equation for saturated–unsaturated water flow and the convection–dispersion equation for solute transport [20]. In this study, we used HYDRUS-1D to analyze water flow and nitrogen transport through tomato field irrigated with wastewater and soil surface management strategies. The measured data used are taken from completed research projects in field study. The data measurements were realized by [3, 21, 22] and were combined with additional measurements. Before simulation, the model was calibrated with field data.

2.4 Boundary conditions

As the all the plots were at field capacity during the transplantation, therefore, the initial condition for volumetric soil water content was between 0.2–0.3 $cm^3\ cm^{-3}$ for all simulations. The upper boundary soil condition was the atmospheric boundary with a surface layer at which rainfall and evaporation occurred. The upper and lower soil boundary conditions (BC) for solute transport were considered as flux BC and zero concentration gradient.

2.5 Soil hydraulic properties

In this model, soil hydraulic properties, concerning soil moisture retention characteristics, $\theta(h)$, and saturated hydraulic conductivity, $K_{sat,}$ were measured in the field. The parameters of the [23] model were evaluated by fitting on $\theta(h)$ data using the Curve RETC code. The average values of Van Genuchten parameters for study at different soil depths are given in **Table 5**.

2.6 Parameter values

Initial soil water contents for tomato in different soil depths were 0.20–0.30 $cm^3\ cm^{-3}$ (giving a mean value of 0.27 $cm^3\ cm^{-3}$). Transport parameters were the model inputs. They were modified to calibrate the model. The modified longitudinal dispersivity and molecular diffusion coefficients of NO_3–N in free water (Do) were used as/set at 1.0 cm and 1.65 $cm^2\ d^{-1}$), respectively. Urea and $NO3^-$ were assumed to be present only in the dissolved phase (i.e., Kd = 0 $cm^3\ g^{-1}$) soil. The first-order decay coefficient, μ, for urea, representing hydrolysis, was set at 0.38 day^{-1}. Again, similar values were used in the literature, for example by [24] and

Depth	θ_r	θ_s	a	n	m	K_{sat}
(cm)	(cm^3 cm^{-3})	(cm^3 cm^{-3})	cm^{-1}	—	—	cm $hr.^{-1}$
25	0.067	0.412	0.0073	1.86	0.414	3.896
60	0.095	0.375	0.0075	1.317	0.5310	2.160
85	0.185	0.421	0.0068	1.657	0.3916	0.131
110	0.048	0.473	0.0298	1.751	0.4259	18.922

Note: Van Genuchten model: $\theta(h) = \theta_r + (\theta_s - \theta_r)/\left[1+(a|h|)^n\right]^m$ ***, m = 1–1/n.***

$K(h) = K_{sat}Se^{1/2}[1-(1-Se^{1/m})^m]^2$ with $Se = (\theta - \theta_r)/(\theta_s - \theta_r)$, θ_s and θ_r are saturated and residual water content, respectively, K_{sat} is saturated hydraulic conductivity, a and n empirical parameters.

Table 5.
Parameters of Van Genuchten equation for the soil moisture retention characteristics and the hydraulic conductivity function.

by [25] who all considered hydrolysis to be in the range of 0.36 or 0.38 to 0.56 day^{-1}. Nitrification from NO_4^+ to NO_3^- was modeled using the rate coefficient of 0.2 day^{-1}, which represents the center of the range of values reported in the literature, e.g., 0.2 day^{-1} ([24, 26], 0.02–0.5 day^{-1} [27], 0.226–0.432 day^{-1} [28], 0.15–0.25 day^{-1} [25], and 0.24–0.72 day^{-1} [29]. It is further assumed that the maximum rooting depth increases logistically with time (increases from 2 cm at germination at 60 cm at harvest), and that there is an exponential root distribution with depth. Uptake of nitrate and ammonium is by passive uptake only. That means that, e.g. NO_3-N uptake at a given time and depth is equal to the water uptake multiply nitrate concentration in neglected. But assume that the maximum allowed concentration for solute is (50 ppm N 550 mgN/L). Assume the soil profile is initially solute free.

2.7 Model testing

The model was evaluated by comparing measured and simulated values over time and depth using both qualitative and quantitative procedures. The qualitative procedures consisted of visually comparison between measured and simulated values over time and depth. For quantitative procedures, statistical analysis were used to calculate the average error (AE), the root mean square error (RMSE), the coefficient of residual mass (CRM), and the modeling efficiency (EF) between the measured and simulated values of water content in the soil during the study period [30–32].

The (AE) is the average difference between the simulated and the measured values. The AE with a positive or negative sign indicates whether the model tends to overestimate or underestimate the measured values. The RMSE statistical index shows the mean difference between simulated and observational data. The RMSE coefficient is equal to the variance of the remaining error and the lower the value, the higher the accuracy of the model. In the Nash-Sutcliffe criterion (EF), the numerical value of one indicates the complete conformity of the simulated and observational data. The CRM also shows the difference between experimental and estimated values. Positive CRM values indicate that the proposed model estimates the values less than its actual value, and vice versa. In the most optimal case, the RMSE and CRM values are equal to zero, in which case the proposed model estimates the values with the highest possible accuracy. Wilmott agreement statistical index (d) with a value it is between zero and one that the value of one indicates the best fit. The value of Willmott's index (*d*) reflects the degree of agreement, and *d* = 1 indicates perfect agreement between the measured and simulated values.

The closer the calculated values are to zero, the better the approximation of the simulated data to the field data [33]. The optimum values of AE, RMSE, EF and

CRM criteria are 0, 0, 1, and 0, respectively. Positive values of CRM indicate that the model underestimates the measurements and negative values for CRM indicate a tendency to overestimate them. If EF is less than zero, the models' predicted values are worse than simply using the observed mean. The average error and root mean square error are calculated as outlined in [33]:

The average error is defined as:

$$\textit{Average error}\,(AE) = \frac{\left(\sum_{i=1}^{n} S_i - Q_i\right)}{n} \tag{5}$$

The Root Mean Square Error is defined as:

$$\textit{Root mean square error}\,(RMSE) = \sqrt{\frac{\left(\sum_{i=1}^{n} S_i - Q_i\right)^2}{n}} \tag{6}$$

The Nash-Sutcliffe Efficiency is defined as:

$$EF = \left[1 - \frac{\sum_{i=1}^{n} (S_i - O_i)^2}{\sum_{i=1}^{n} \left(O_i - \overline{O_i}\right)^2}\right] \tag{7}$$

The coefficient of residual mass (CRM) is defined as:

$$CRM = \left[\frac{\sum_{i=1}^{n} (Q_i - S_i)}{\sum_{i=1}^{n} (O_i)}\right] \tag{8}$$

Where n is the number of observations, $\overline{O}_i$ is the average of the observed values and S_i and O_i are the simulated and measured values, respectively [34].

3. Results and discussion

3.1 Water flow simulation

Simulated soil water content in the soil profile are shown in **Figure 1**. However, the results in this section are presented to have an idea about the water regime in the soil profile with respect to the day of planting and harvesting the tomato. There were several rainfall events during the simulation period; however, more rainfall events were registered in the first part of the simulation period. As a result, the soil water content showed at all depths fluctuated more frequently in the first part of the simulation compared to the second and third part.

HYDRUS 1-D was also compared to the water content from field data collected from each treatment (by TDR) and simulated data for the soil profile over the growing season. The following observations are based on visual assessment of model fit compared with observed values of moisture contents of soil. The simulated and measured water contents at 20, 40, 60 and 100 cm are shown in **Figure 2A** and **B**, **Figure 3C** and **Figure 4D**, respectively. The predicted water

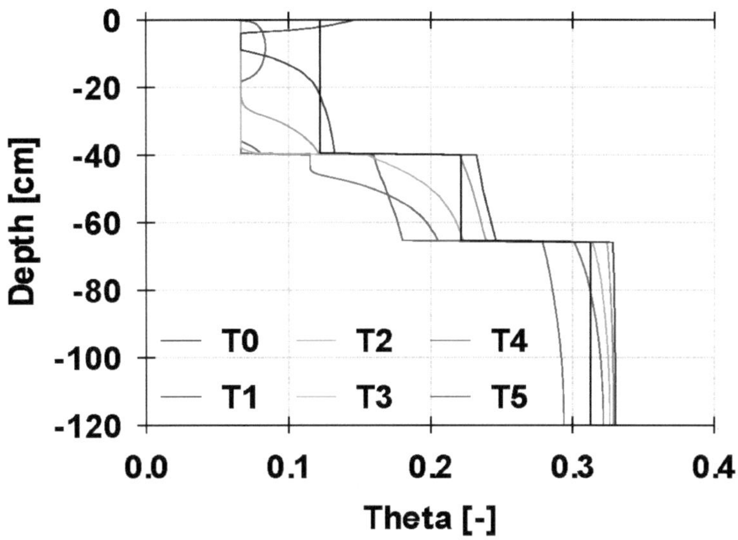

Figure 1.
Simulated soil water content at different times of the experiment in the soil profile. With (T0=50 days, T1=150 days, T2=200 days, T3= 240 days, T4=300 days and T5=339 days)..

contents at 20 cm depth agree well with the measured values during growing season. The simulation closely match the measured moisture dynamics, except in the (wet) spring and winter of 2010 when the model at times underestimates the soil water content in the top soil. The simulated water contents did not agree well with the measured data at 40, 60 and 100 cm depths, the response of the model was lower than measured, especially deeper in the soil profile. At all depths, a close agreement between the measured and simulated data was registered during (wet) winter period. The difference between simulated and measured water contents varied with depth from −0.045 to 0.152 $cm^3\ cm^{-3}$. For the deeper positions (40, 60 and 100 cm), the model systematically underestimates the measured water content by 0.04 to 0.110 cm^3/cm^3 over the entire growing season at deeper depths, potentially due to under-estimation in the amount of free drainage and an over-estimation of the soil porosity, although the dynamics (water depletion in summer, replenishment in winter) is well simulated. Given that the underestimation is not just limited to the growing season, but is also evident in winter periods when there is little evapotranspiration and the entire soil profile is draining suggests that the problem is not with the crop parameters or evapotranspiration, but rather with the soil hydraulic properties of the deeper soil horizons: the parameters of the van Genuchten-Mualem K-h-θ relationship control the equilibrium water contents in winter ('field capacity').

The statistical criteria of quantitative model evaluation between simulated and measured soil water content are summarized in **Table 6**. Overall, the values calculated demonstrate a good correlation of the model to field data. The results of the simulations may be affected by the value of the saturated hydraulic conductivity (Ks). Therefore, optimizing this parameter for all the three layers using inverse modeling of the Hydrus-1D, would slightly improve the simulation results. So the predicted water contents at −40, −60 and -100 cm are indeed much closer to the measured value, and this parameter change does not affect the (good) match observed for -20 cm. For further improvement, other hydraulic parameters (e.g. θs and α) also should be optimized. In addition, changing the matric pressure head may lead to good results.

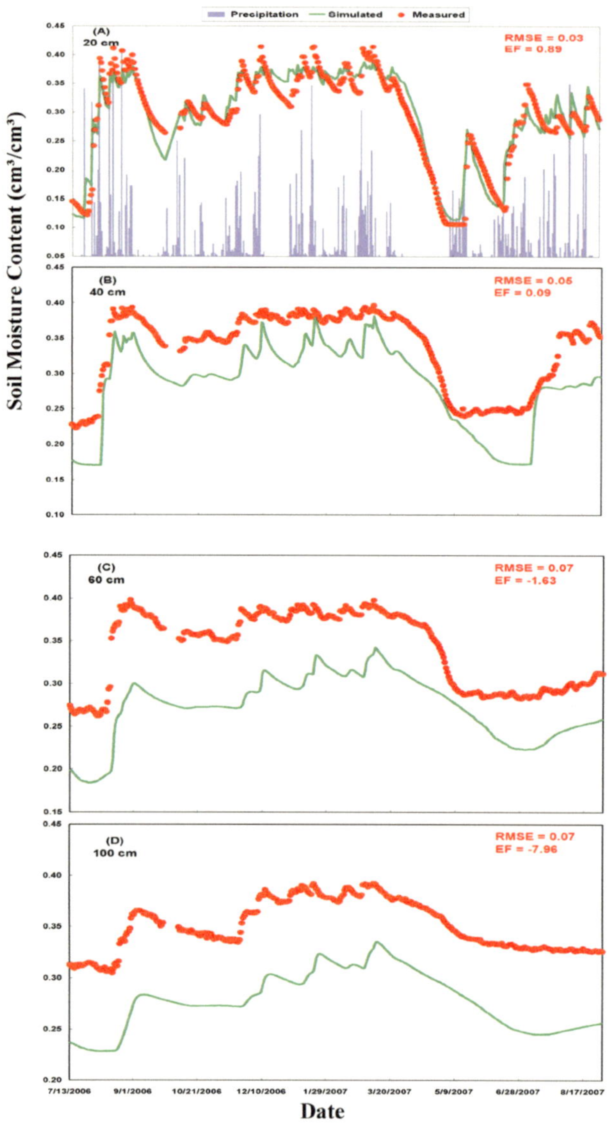

Figure 2.
Daily rainfall (solid bars, A) [cm], and measured (circles) and simulated (lines) soil water contents at 20 cm (A), 40 cm (B), 60 cm (C) and 100 cm (D) depths.

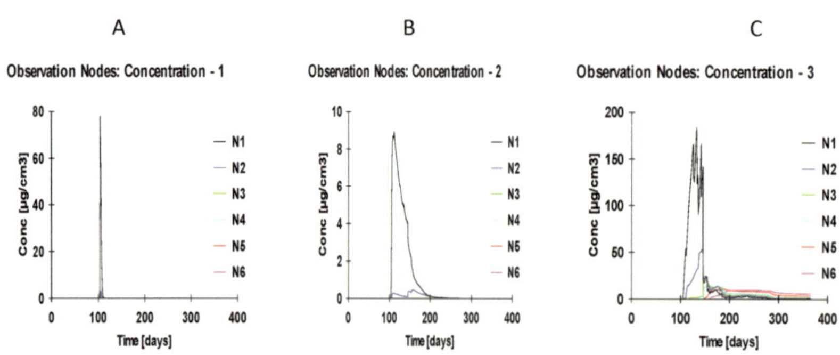

Figure 3.
Variation of concentration Urea, Ammonium and Nitate in different observational nodes of experiment in the soil profile. With (N1=20cm , N2=40cm, N3=60cm, N4=100cm and N5=120cm).

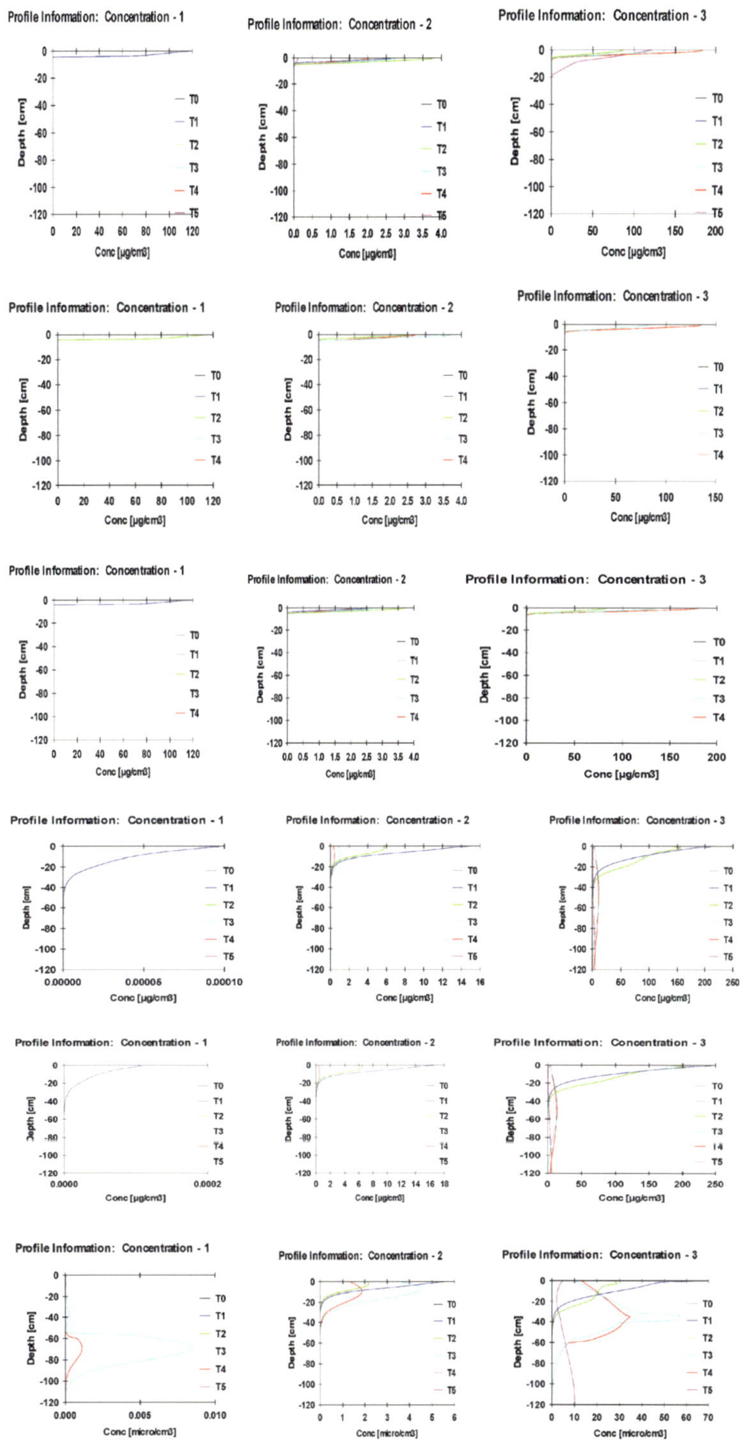

Figure 4.
Variation of concentration urea, ammonium and Nitate with depths for different times and depths of experiment for different treatment. With (T0=50 days, T1=150 days, T2=200 days, T3= 240 days , T4=300 days and T5=339 days).

3.2 Fate of nitrogen sources

We determined the effects of different sources of nitrogen on the soil distribution of urea, ammonium, and nitrate during of growing season tomato field.

Depth	Statistical criteria				
(cm)	n	AE ($cm^3 cm^{-3}$)	RMSE (%)	EF	CRM
20	365	−0.078	0.115	0.953	0.277
40	365	−0.094	0.115	0.958	0.303
60	365	−0.085	0.106	0.964	0.274
100	365	−0.032	0.058	0.989	0.098

Note: n = number of measurements, AE = average error, RMSE = relative root mean square error, EF = modeling efficient and CRM= coefficient of residual mass.

Table 6.
Statistical criteria for the simulated and measured soil water content.

N balance components Applied – Inflow kg N ha^{-1}	Losses – Outflow kg N ha^{-1}			Corn grain yield
	Crop uptake		Leaching	kg ha^{-1}
N applied by wastewater	Ammonium-N	Nitrate-N	Nitrate-N	
T1 = 100	5.3	47	-	82510
T2 = 75	5.2	47		81830
T2 = 50	4.1	39		7812
T2 = 25	1.7	23		63253
T3 = N-manure =100	4.2	43	21	78956
T4 = N-fertilizer =90	5.8	44	25	77582
T4 = split application	4.9	43	23	78962
T5 = well water = 0	—	—	—	51254

Table 7.
Components of nitrogen balance at the end of simulation period in kg N ha^{-1} for a soil depth of 150 cm.

To evaluate the Hydrus model performance with respect to nitrogen transport and transformations, the simulated nitrogen concentrations (NH4-N and NO3-N) are compared for different treatments at different depths of soil profile, (7.5, 22.5, 37.5, 52.5 and 120 cm from soil surface). **Figure 3 (A–C)** gives the daily variations in the simulated Urea-N, NH4-N and NO3-N concentrations respectively. It takes about 4 days to convert 90% of urea into ammonium and it takes about 70 days to convert 90% of ammonium into nitrate. Urea fertilizer is easily dissolved in water and transferred to the soil. After fertilization, urea is hydrolysed in the soil a urea concentration decreased over time between irrigations and ammonium is formed and then, during the nitration process by bacteria in the soil, convert ammonium to nitrite and then to nitrate. Immediately after fertigation, at 3.73 days, the urea was concentrated near the soil surface.

For all treatments ammonium accumulated in the topsoil immediately (**Figure 4**). Because of soil adsorption and subsequent fast nitrification and/or root uptake, there was only a slight movement of ammonium in the soil profile. The results obtained in this study indicated that nitrate moved continuously downwards during the 28-day of growing season simulation. Also, nitrate is easily exposed to leaching due to its high mobility and is not adsorbed to the soil, therefore denitrification was assumed negligible.

Nitrogen is applied to the soil solution by fertilizer application, treated wastewater irrigation and animal manure.

Ammonium is usually ionically exchanged and stabilized in the surface of clay minerals. It is found in small amounts in soil solution and can be retained by the negative charges of clay mineral particles and organic particles. Therefore, the mobility of ammonium ions is lower than that of nitrate ions. The NH4-N then transformed to nitrate by the nitrification process, which is the most soluble form of nitrogen in the soil for uptake by crops.

Table 7 shows the different components of nitrogen balance during the simulation period. Slightly smaller leaching percentages were computed for the urea–ammonium–nitrate wastewater compared to the nitrate- fertilizer and manure. Fertilizer use efficiency ranged from 54% (treatmentT4) to 84.9% (treatment T1). The results reported from nitrogen balance components show that nitrate leaching losses (0%, 23% and 25%) in treatments T1, T3, T4 respectively, and mainly occurred during the winter period. The reduced level of leaching is explained by low amount of drainage water, low nitrogen concentration of irrigation wastewater and excessive nitrogen uptake by the crop. Since the nitrate transport through the soil profile and out into field drains or deep groundwater, is usually controlled by water movement. The effect of irrigating different on the grain yield of tomato was also significant ($P < 5\%$) (**Table 7**). The results showed an increase in the mean of fresh and dry forage yield (8.25% fresh forage and 23.14% of dry forage) (**Table 7**). Because treated wastewater is an important source of plant nutrients and can be reused for irrigation to increase forage crop production.

4. Conclusions

The HYDRUS-1D software was performed to simulated water flow and nitrogen transport in tomato crop soil for wastewater irrigation and fertilization. Based on the study carried out in the field, the ability of the model to predict the moisture in the soil at various depths is accurate. This can be due to an acceptable method in the simulation model.

The results reported from nitrogen balance components show that nitrate leaching losses (0%, 23% and 25%) in treatments T1, T3, T4 respectively and mainly occurred during the winter period. The reduced level of leaching is explained by low amount of drainage water, and excessive nitrogen uptake by the crop. Since the nitrate transport through the soil profile and out into field drains or deep groundwater, is usually controlled by water movement. It was fund that the slightly smaller leaching percentages for the urea–ammonium–nitrate wastewater compared to the nitrate- fertilizer and manure. Fertilizer use efficiency ranged from 54% (treatment T4) to 84.9% (treatment T1). Based on these results we conclude that nitrogen from wastewater has smaller nitrate leaching compared to nitrogen from animal manure and commonly fertilizer. Nevertheless, our simulation results provide guidance on the appropriate fertigation strategy for use of waste water in irrigation.

Author details

Ali Erfani Agah
Department of Water Engineering, Faculty of Agriculture, Shahrood University of Technology, Shahrood, Iran

*Address all correspondence to: ali.erfani68@gmail.com

References

[1] Erfani Agah A Environmental aspects use of treated municipal wastewater in irrigation. Published and presented in Iranian National Committee on Irrigation and Drainage (IRNCID) In, 1996.

[2] Erfani Agah A, Haghnia G, Alizade A. Effect of irrigation by treated wastewater on the yield and quality of tomato. Journal of Agricultural Sciences and Technology. 2001;**15**:65-70

[3] Erfani Agah A, Haghnia G, Alizade A. Effect of irrigation with wastewater on yield and lettuce qualitative and soil properties Esfahan Agric Tec And Sci Natural Resourses J. 2002;**5**:72-77

[4] Kaboosi K. The assessment of treated wastewater quality and the effects of mid-term irrigation on soil physical and chemical properties (case study: Bandargaz-treated wastewater) applied water science 7:2385-2396. 2017. DOI: 10.1007/s13201-016-0420-5

[5] Kurian M, Reddy VR, Dietz T, Brdjanovic D. Wastewater re-use for peri-urban agriculture: A viable option for adaptive water management? Sustainability Science. 2013;**8**:47-59. DOI: 10.1007/s11625-012-0178-0

[6] Lyu SD, Chen WP, Zhang WL, Fan YP, Jiao WT. Wastewater reclamation and reuse in China: Opportunities and challenges Journal of Environmental Sciences. 2016;**39**:86-96. DOI: 10.1016/j.jes.2015.11.012

[7] Lavrnic S, Zapater-Pereyra M, Mancini ML. Water Scarcity and Wastewater Reuse Standards in Southern Europe: Focus on Agriculture Water Air and Soil Pollution. 2017:228. DOI: 10.1007/s11270-017-3425-2

[8] Pescod MB (1992) Wastewater treatment and use in agriculture.. In: Drainage FIa (ed). Rome, p 47

[9] Mtshali JS, Tiruneh AT, Fadiran AO. Characterization of sewage sludge generated from wastewater treatment plants in Swaziland in relation to agricultural uses. Res Environ. 2014;**4**: 190-199

[10] Kominko H, Gorazda K, Wzorek Z. The possibility of Organo-mineral fertilizer production from sewage sludge. Waste and Biomass Valorization. 2017;**8**:1781-1791

[11] Hazen and Sawyer (2010) Literature review of nitrogen fate and transport modeling. Florida Department of Health .Division of Environmental Health. Bureau of Onsite Sewage Programs 4042 Bald Cypress Way Bin #A-08. Tallahassee, FL 32399-1713.

[12] Pettygrove, Asano (1990) California State Water Resources Control Board., "Irrigation with reclaimed municipal wastewater – guidance manual". Edited by: G. Stuart Pettygrove and Takashi Asano. Prepared by: Department of Land, Air and Water Resources. University of California. Pub: Lewis Publishers.

[13] Robertson GP, Groffman PM. Nitrogen transformations. In: Paul EA, editor. Soil Microbiology, Ecology and Biochemistry. Fourth ed. Burlington, Massachusetts, USA: Academic Press; 2015. pp. 421-446

[14] Kim DY, Burger JA. Nitrogen transformations and soil processes in a wastewater-irrigated, mature Appalachian hardwood forest Forest ecology and management 90:1-11. 1997. DOI: https://doi.org/10.1016/S0378-1127(96)03889-3

[15] Wlodarczyk T, Kotowska U. Nitrogen Transformations and Redox Potential Changes in Irrigated Organic Soil. Lublin: Institute of Agrophysics PAS; 2005

[16] Almasri MN. Kaluarachchi JJ. Modeling nitrate contamination of

groundwater in agricultural watersheds Journal of Hydrology. 2007;**343**:211-229. DOI: 10.1016/j.jhydrol.2007.06.016

[17] Waskom R. Best management practices for irrigation management. Colorado State Univ. Coop. Ext. Bul. 1994;**XCM-173**

[18] Mekala C, Nambi IM. Experimental and simulation studies on nitrogen dynamics in unsaturated and saturated soil using HYDRUS-2D Procedia technology 25:122-129. 2016. DOI: https://doi.org/10.1016/j.protcy.2016.08.089

[19] Šimůnek J, van Genuchten MT, Šejna M. Recent Developments and Applications of the HYDRUS Computer Software Packages Vadose Zone Journal. 2016:15. DOI: 10.2136/vzj2016.04.0033

[20] Šimůnek J, Šejna M, Saito H, M VG (2013) The Hydrus-1D Software Package for Simulating the Movement of Water, Heat, and Multiple Solutes in Variably Saturated Media, Version 4.17, HYDRUS Software Series 3,

[21] Erfani Agah A, Wyseure G. Experimental investigation of water flow and solute transport in unsaturated columns International Journal of AgriScience. 2013;**3**:228-239

[22] Erfani Agah A, Meire P, Deckere Ed (2016) Simulation of Phosphorus Transport in Soil Under Municipal Wastewater Application Using Hydrus-1D. In: Larramendy ML, Soloneski S (eds) Soil Contamination - Current Consequences and Further Solutions. InTech, Rijeka, p Ch. 09. doi:10.5772/66214

[23] Van Genuchten MT (1980) A closed-form equation for predicting the hydraulic conductivity of unsaturated soils soil science Society of America Journal 44:892-898

[24] Hanson BR, Simunek J, Hopmans JW. Evaluation of urea-ammonium-nitrate fertigation with drip irrigation using numerical modeling Agricultural Water Management. 2006;**86**:102-113. DOI: 10.1016/j.agwat.2006.06.013

[25] Ling G, El-Kadi A. A lumped parameter model for nitrogen transformation in the unsaturated zone. Water Resources Research. 1998;**34**:203-212. DOI: 10.1029/97WR02683

[26] Johnsson H, Bergstrom L, Jansson P-E, Paustian K. Simulated nitrogen dynamics and losses in a layered agricultural soil agriculture. Ecosystems & Environment. 1987;**18**:333-356. DOI: https://doi.org/10.1016/0167-8809(87)90099-5

[27] Lotse EG, Jabro JD, Simmons KE, Baker DE. Simulation of nitrogen dynamics and leaching from arable soils Journal of Contaminant Hydrology. 1992;**10**:183-196. DOI: https://doi.org/10.1016/0169-7722(92)90060-R

[28] Selim HM, Iskandar IK. Modeling nitrogen transport and transformations in soils: 1. Theoretical considerations. Soil Science. 1981;**131**:233-241

[29] Misra C, Nielsen DR, Biggar JW. Nitrogen transformations in soil during leaching; I. theoretical Considerations1 soil science Society of America Journal 38:289-293. 1974. DOI: 10.2136/sssaj1974.03615995003800020024x

[30] Vanclooster M, Ducheyne S, Dust M, Vereecken H. Evaluation of pesticide dynamics of the WAVE model. Agric Water Manage. 2000;**44**:371-388

[31] Loague K, Green RE. Statistical and graphical methods for evaluating solute transport models: Overview and application. Validation of flow and transport models for the unsaturated zone. J Contam Hydrol. 1991;7:51-73

[32] Antonopoulos V. Simulation of soil moisture dynamics on irrigated cotton in semi-arid climates. Agric Water Manage. 1997;**34**:233-246

[33] Olinski AJ. Verification of Drainmod ver. 5.1 for estimating water balance and nitrogen transport through soils in southern Ontario. M.Sc. Thesis. University of Guelph, Guelph. Ontario. 2005;**3**:24-27

[34] Moriasi DN, Arnold JG, Van Liew MW, Bingner RL, Harmel RD. Veith TL. Model evaluation guidelines for systematic quantification of accuracy in watershed simulations Transactions of the Asabe. 2007;**50**:885-900

Chapter 9

Determination of Values Range of Physical Quantities and Existence Parameters of Normal Spherical Detonation by the Method of Numerical Simulation

Myron Polatayko

Abstract

Using elements of the theory of classical detonation and previously obtained relations for spherical waves, the author tried to establish the range of admissible values of temperature, Mach numbers, and specific hydrogen content in the gas mixture of the possible existence of normal spherical detonation. The work took into account the critical values of the parameters associated with the kinetics of the chemical reaction at the front of the blast wave and the parameters that determine the intensity of the shock transition (minimum and maximum Mach number) for a given reacting medium. Using the example of the interaction of hydrogen and oxygen in a hydrogen-oxygen mixture, it was possible to graphically determine the range of values of the main physical quantities and parameters - the critical temperature, the detonation temperature of the quiescent medium, and the specific hydrogen content in the mixture required for spherical detonation. Mathematical modeling of the process was carried out at a fixed value of the pressure of the gaseous medium.

Keywords: detonation wave, critical temperature, range of permissible values, process modeling

1. Introduction

Explosions are widely used in many areas of science and engineering, and their models are applied to elucidate various physical phenomena. Moreover, the unexpected explosions in industry and everyday life often result in catastrophes with numerous human losses, which invokes the intensive study of a supersonic burning nowadays. Those researches are carried out using both analytical methods [1] and numerical simulations [2, 3]. This work aims at studying the range of parameters needed for a normal spherical detonation in a gas mixture to take place. It is the kind of detonation that precedes the plane (classical) detonation, but emerges at lower shock wave velocities [4]. The spherical wave produced by a strong point explosion corresponds to the initial stage of the whole detonation process and transforms gradually into the classical variant. In gaseous explosive mixtures, the detonation regime of the explosive transformation is possible only at certain concentrations of the combustible gas, depending on the chemical composition of the mixture, pressure and temperature. A decrease in pressure leads to the appearance of a pulsating detonation front, and subsequently to the formation of the so-called spin detonation, in which the three-shock wave configurations arising at the detonation wave front rotate along a helical line. With a further decrease in pressure, the supersonic combustion process dies out. At present, the reasons for the onset and existence of pulsating detonation [5] have not been fully investigated. It is hoped that in the near future this issue will be resolved after a detailed study of spherical detonation waves [6] and volumetric detonation.

In the earlier work [7], the model for the transition of an explosion spherical wave to the Chapman–Jouguet regime was proposed. In the other work [8], the concept of the critical temperature at the wave front was introduced as a basic criterion for the transformation of a shock wave to the detonation one. In this work, using the example of a gaseous hydrogen-oxygen mixture, an attempt is made to graphically determine the ranges of physical parameters and quantities at which spherical detonation is probable.

2. Critical values of parameters related to the chemical reaction kinetics

The classical theory considers detonation waves with sharp front edge. In its framework, the chemical transformations are assumed to begin right after a jump-like increase of the pressure. Actually, the process develops somewhat differently [9]. The temperature and pressure profiles behind the shock front of a detonation wave are schematically shown in **Figure 1**. After the shock transition (1–2), the vibrational and rotational degrees of freedom of gas molecules become excited (2–3), which is accompanied by a temperature reduction.

Then the induction period (3–4) takes place, the duration of which can be equal to more than 90% of the whole chemical reaction time (3–5), if the activation energy of the process is sufficiently high (E = 20 ÷ 40 kcal/mol). In the stationary detonation regime (the Chapman–Jouguet regime), profile 1–5 does not change in time. The reaction zone adjoins the region of non-stationary flow, rarefaction wave (5–6), the profile of which can change.

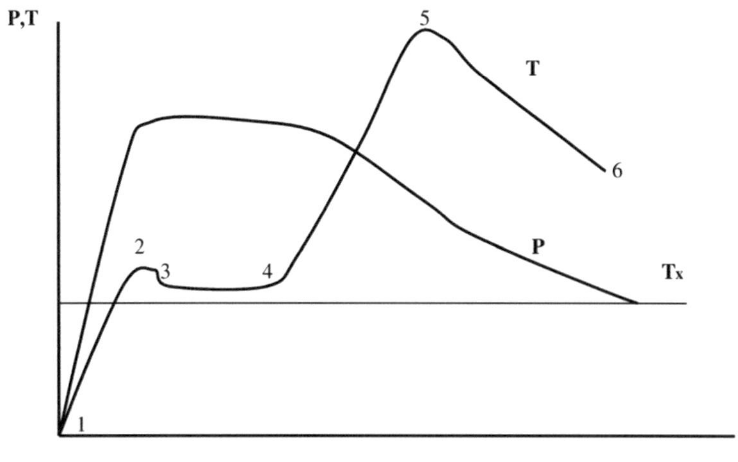

Figure 1.
Schematic diagrams of the pressure, P, and gas temperature, T, profiles behind the shock wave front [9] under the condition $T_2 \geq T_x$, where T_2 is the temperature at point 2.

It turns out that, in the case of a hydrogen-oxygen mixture compressed by a shock wave, a lot of free radicals emerge in section 2–4 [10, 11], with their concentration reaching $10^{12} \div 10^{15} cm^{-3}$. Rapid chain transformations start just from those initial centers [10] and run following the Lewis scheme. In this case, we have

$$OH + H_2 = H_2O + H, \tag{1}$$

$$H + O_2 = OH + O. \tag{2}$$

Note also that the temperature T_2, at which the branching probability δ equals unity,

$$\delta = 1, \tag{3}$$

is critical: the process becomes considerably accelerated, and the rapid chain reaction takes place. Under the indicated conditions, according to the results of work [4], the equality

$$T_2 = T_x, \tag{4}$$

plays the role of a criterion for qualitative variations in the kinetics of the interaction between hydrogen and oxygen. The Lewis scheme ignites the detonation mechanism, although the process itself can run in a certain different way, following a different scenario, in which the reaction rate is higher by an order of magnitude. From the chemical viewpoint, we have already stated the fact that, in order to obtain the supersonic burning at the shock wave front, it is necessary to reach the temperature T_x in the medium, at which the branching probability δ equals unity. How can T_x be determined? In work [4], the relation between the key parameters of a chemical reaction at the shock wave front, on the one hand, and the physical quantities that characterize the process of shock transition, on the other hand, was obtained,

$$T_x^2 = \frac{2.5 \times 10^5 Q T_0 (\gamma - 1)\left(2\gamma M^2 - \gamma + 1\right)\left(2 + (\gamma - 1)M^2\right)^2}{4\gamma^2(\gamma + 1)M^6 K^* P_0} \times \exp\left(-\frac{E_2}{K^* T_x}\right), \tag{5}$$

where M is the Mach number (it reflects the shock transition intensity); P_0 the initial, before the explosion ignition (at 293 K), pressure of the gas mixture reckoned in mm Hg units; E_2 the activation energy of the branching reaction (2); K^* the gas constant; Q the combustion energy per gas mole; and γ the adiabatic index for the given gas mixture. For the hydrogen-oxygen mixture, the corresponding numerical values are [12]: $\gamma = 1.4$, $Q = 286.5$ kJ/mol, $K^* = 8.31$ J/mol/K, $E_2 = 16 \times 10^3 \times 4.19$ J/mol, and $T_0 = 293$ K. Then Eq. (5) reads

$$T_x{}^2 = \frac{5.38 \times 10^{10}\left(2 + 0.4M^2\right)^2\left(2.8M^2 - 0.4\right)}{P_0 M^6} \times e^{-8067/T_x} \tag{6}$$

It is evident that the temperature T_x is different for different Mach numbers. Formula (6) describes the functional dependence of the critical temperature T_x on the Mach number M for the given initial pressure P_0 and allows one to compare its value with the real temperature

$$T_2 = \frac{\left(2\gamma M^2 - \gamma + 1\right)\left(2 + (\gamma - 1)M^2\right)}{(\gamma + 1)^2 M^2} \times T_1, \tag{7}$$

where T_1 is the temperature of the medium in front of the wave front. Hence, in our case, the important criterion,

$$T_2 \geq T_x, \tag{8}$$

has to be satisfied for the detonation to take place as a real process.

3. Elements of the hydrodynamic theory of detonation: limiting parameters dependent on the Mach number minimum and maximum

The detonation process of explosive materials is considered as a cumulative action of the shock wave and the chemical reaction, when the shock compression initiates the reaction, and the reaction energy maintains the detonation process afterward. The hydrodynamic theory [13] enables one to evaluate the size of a chemical reaction zone and the values of medium parameters in the chemical reaction zone (at the interface with the detonation products). The classical theory considers a plane detonation front,

$$d = \Delta t\left(D - v_g\right), \tag{9}$$

where d is the chemical reaction zone width, Δt the reaction duration, D the shock wave velocity, and v_g the gas velocity behind the reaction front (the Jouguet point). To be exact, in a real situation (**Figure 2**), there exists some shock transition interval (1–2) before the temperature T_2 is achieved, which is not taken into account in this case. One can see in **Figure 2** that front (3–3) separates the chemical reaction zone from detonation products. This means that the substance being suddenly compressed by the shock wave burns out completely within the time inter Δt.

The theory is based on two important postulates: (1) the whole substance compressed by the shock wave burns out, and (2) the combustion energy is enough to maintain the shock wave velocity to be constant ($D = const$). According to the theory, the pressure P_3 and the density ρ_3 in the chemical reaction zone at the interface with detonation products (the Jouguet point), are connected with each other by the following relations [13]:

$$P_3 = \frac{P_1 + \rho_1 D^2}{1+\gamma}, \tag{10}$$

$$\frac{\rho_3}{\rho_1} = \frac{D^2(\gamma+1)}{b_1{}^2 + \gamma D^2}, \tag{11}$$

Where P_3 is the pressure at front (3–3) separating the reaction zone from the reaction products, P_1 the pressure in front of the shock wave front, ρ_1 the gas density in front of the wave front, D the wave velocity, γ the adiabatic index, ρ_3 the medium density at the wave front (3–3), and b_1 the sound velocity in the motionless medium in front of the front. From the Mendeleev–Clapeyron equation

$$PV = \frac{m}{\mu} K^* T \Rightarrow T = \frac{P\mu}{\rho K^*}, \tag{12}$$

substituting the values of P_3 (10) and ρ_3 (11), we determine the temperature T_3 at the Jouguet point,

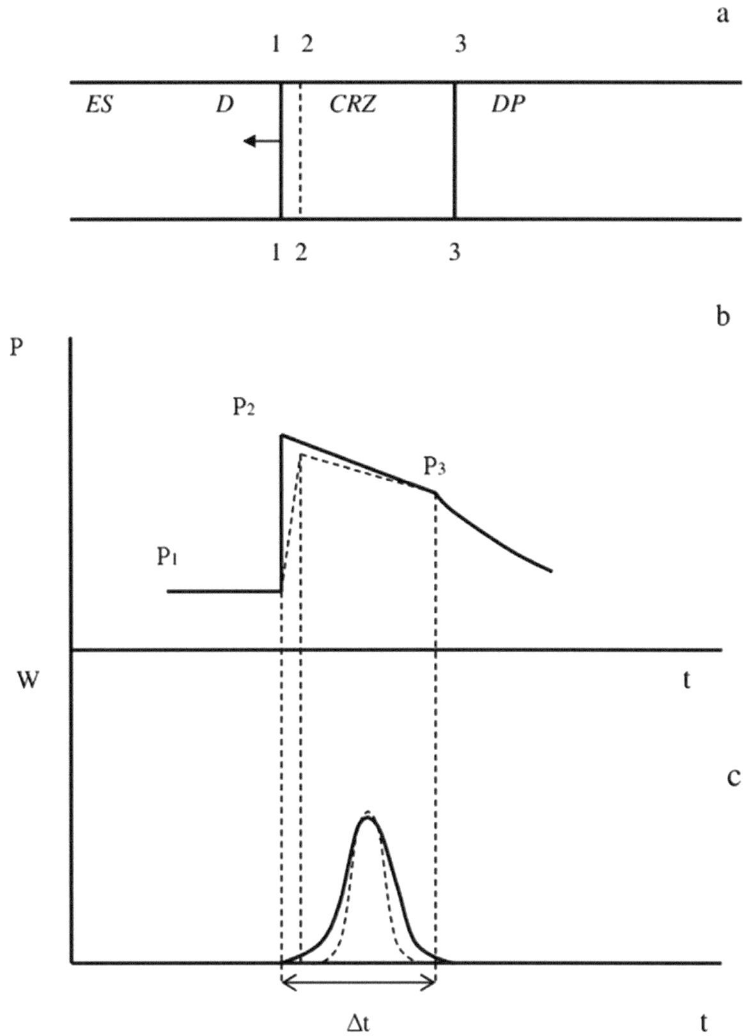

Figure 2.
Schematic structure of a plane detonation wave: explosive substance (ES), detonation products (DP), and chemical reaction zone (CRZ) (a). The pressure changing in time: in front of the wave front P_1, at the wave front P_2, and in the chemical reaction zone (the Jouguet point, P_3) (b). Reaction rates (c).

$$T_3 = \frac{\rho_1 D^2}{\gamma+1} \times \frac{\mu}{K^*} \frac{(\gamma D^2 + {b_1}^2)}{\rho_1 D^2 (\gamma+1)} = \frac{\mu(\gamma D^2 + {b_1}^2)}{K^*(\gamma+1)^2} = \frac{\mu {b_1}^2}{K^*} \frac{(\gamma M^2+1)}{(\gamma+1)^2} = T_1 \gamma \times \frac{(\gamma M^2+1)}{(\gamma+1)^2}, \tag{13}$$

where $D = b_1 M$. Here, we took into account that

$$P_3 \approx \frac{\rho_1 D^2}{\gamma+1}, \tag{14}$$

when $\frac{P_3}{P_1} \gg 1$, and that

$${b_1}^2 = \gamma \frac{K^* T_1}{\mu}, \tag{15}$$

where μ is the molar mass. It is evident that if we consider the detonation and the support of a chemical reaction by the shock wave, the following condition has to be satisfied:

$$T_3 > T_2; \tag{16}$$

or, in a wider sense (**Figure 3**),

$$T_3 > T_2 > T_x \tag{17}$$

Let us analyze inequality (16) in detail. From the theory of shock waves [13], it is known that the temperature at point 2 in **Figure 3** is determined by relation (7). Therefore, inequality (16) can be transformed as follows:

$$T_1\gamma\frac{\gamma M^2 + 1}{(\gamma+1)^2} > T_1\frac{(2\gamma M^2 - \gamma + 1)(2 + (\gamma-1)M^2)}{(\gamma+1)^2 M^2} \tag{18}$$

or

$$\gamma(\gamma M^2 + 1) - \frac{(2\gamma M^2 - \gamma + 1)(2 + (\gamma-1)M^2)}{M^2} > 0, \tag{19}$$

since $T_1 > 0$ and $\gamma > 0$. By solving the equation

$$\gamma(\gamma M^2 + 1) - \frac{(2\gamma M^2 - \gamma + 1)(2 + (\gamma-1)M^2)}{M^2} = 0 \tag{20}$$

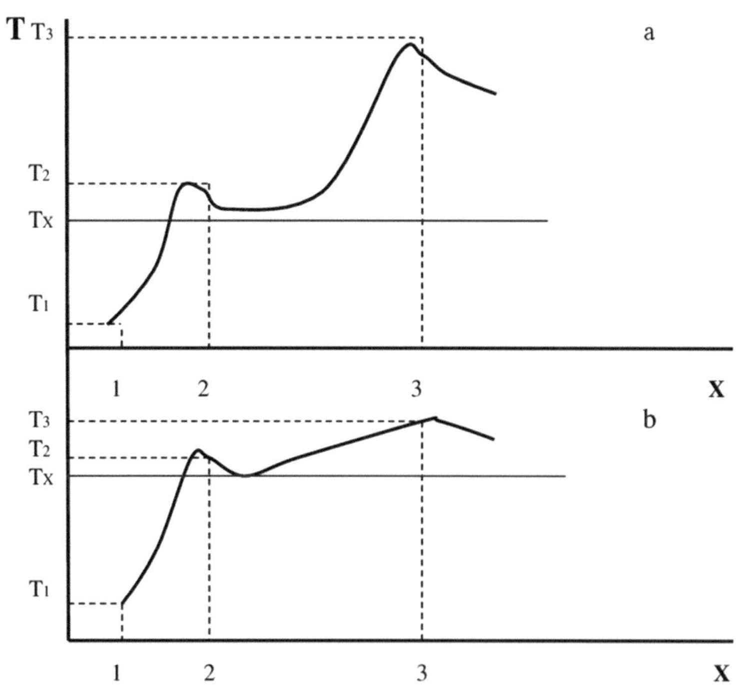

Figure 3.
Schematic diagrams for the temperature profiles behind the wave front: the general case (a) and the case where $T_3 \approx T_2 \approx T_x$ (b) corresponding to the limiting detonation process.

with respect to M (keeping in mind that $M > 0$), we obtain

$$M^4(2\gamma - \gamma^2) + M^2(\gamma^2 - 5\gamma + 1) + 2\gamma - 2 = 0 \tag{21}$$

or, substituting the corresponding γ -value,

$$0.84M^4 - 4.04M^2 + 0.8 = 0. \tag{22}$$

The positive roots of this equation are $M_1 = 2.145$ and $M_2 = 0.455$. For shock transitions, the most interesting is the first root, $M_1 = 2.145 \approx 2.2$. On the basis of inequality (18), we may assert that the detonation process is not possible for all shock waves, but only for those with the Mach number $M > 2.2$. Owing to hydrodynamic reasons, there is no detonation for waves with $M < 2.2$. While analyzing Eq. (20), it is expedient to admit that the temperature equality [14]

$$T_3 \approx T_2 \approx T_x \tag{23}$$

describes the lower temperature limit of the detonation (**Figure 3**). In so doing, we took into account that the rate of the chemical reaction decreases together with the temperature in the chemical reaction zone. At the same time, according to the hydrodynamic theory, the amount of the substance that was compressed by the shock wave and interacted under its action has to remain at the previous level. This circumstance inevitably results in the time growth for the active reaction phase, and, as a consequence of the process continuity, gives rise to a considerable reduction of the induction period (interval 3–4 in **Figure 1**). In this connection, there emerges a possibility for the detonation wave to create a gas layer with an approximately identical temperature, and the Mach number $M \approx 2.2$ should be considered as the lower detonation limit.

In order to determine the upper limit of the detonation wave emergence by initiating an explosion in reacting gas media, let us use the model describing the continuous transition of a spherical explosion wave into the Chapman–Jouguet regime [4]. For the normal spherical detonation, it can be determined from the formula

$$M = \left[\frac{(\gamma + 1)^2(\gamma - 1)Qc}{4\gamma^2 K^* T_1}\right]^{\frac{1}{2}}, \tag{24}$$

derived in work [4]. All quantities in this formula are known, except for the parameter c, the specific content of the burned out gas (hydrogen). The intensity of a detonation wave can be controlled by changing, mainly, two parameters: c (in the numerator) and T_1 (in the denominator). In our case, all hydrogen compressed by the shock wave burns out. The values of coefficient c are confined within the interval $0.66 \geq c > 0$. Let we have the stoichiometric mixture of hydrogen with oxygen ($c = 0.66 = max$), and the medium temperature $T_1 = -100^{\circ}C \approx 173K = min$. We suppose that a further temperature decrease will result in changes of the adiabatic index γ and the physical properties of the reacting mixture [15, 16], i.e. let formula (24) be valid for real gases at $T_1 \geq 173K$. In this case, we obtain a rough estimate for the Mach number maximum, $M_{max} = 6.2$. Note again that a strong explosion takes place in a cooled down medium. In this case, $M_{max} = 6.2$. Hence, we estimated the interval of possible Mach numbers for the normal spherical detonation of the hydrogen-oxygen gas mixture under real conditions:

$$6.2 \geq M \geq 2.2. \tag{25}$$

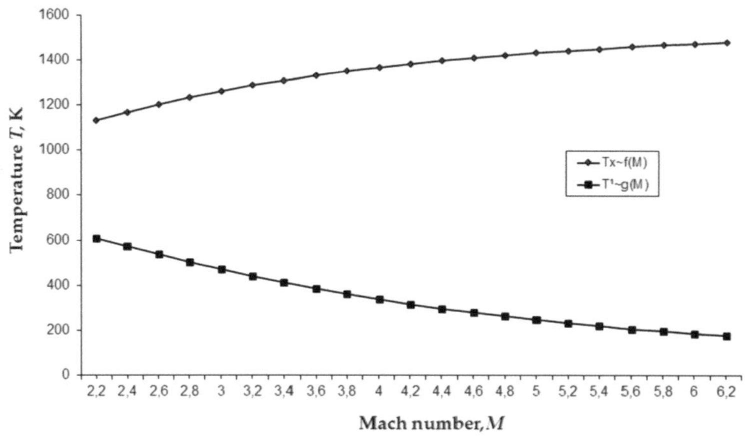

Figure 4.
Dependences of the critical temperature T_x and the detonation temperature in the motionless medium, T^1, on the Mach number M at the fixed pressure P = 60 mm Hg.

In view of relation (6), let us plot the dependence of the critical temperature on the Mach number, T_x (M) (**Figure 4**). Since the Mach number range was found, we will calculate the critical temperature T_x for every M from the indicated interval with an increment of 0.2 and the fixed initial pressure P_0 (see **Table 1**). Transcendental equations were solved using the "Consortium Scilab (Inria, Enpc)" software package with the "Scilab-4.1.2" code. When solving equations, only roots with real values that have physical meaning should be taken into account (the procedure was applied in [4]).

The larger the Mach number, the higher is the critical temperature. However, at $M \geq 5$, the critical temperature growth becomes a little slower. At the lower limit $M = 2.2$, $T_x = 1130\ K$, and, at the upper limit $M = 6.2$, $T_x = 1479\ K$. Hence, in a hydrogen-oxygen mixture, the critical temperature T_x for the allowable values of Mach number M accepts values from the following interval:

$$1479\ K \geq T_x \geq 1130\ K. \tag{26}$$

Figure 4 also exhibits the dependence of the detonation temperature T^1 on the Mach number M, which can easily be obtained [4] by substituting the critical temperature T_x into relation (7):

$$T^1 = \frac{(\gamma+1)^2 M^2 T_x}{(2\gamma M^2 - \gamma + 1)(2 + (\gamma - 1)M^2)}. \tag{27}$$

From Eqs. (25) and (26), it follows that the detonation temperature for a motionless medium falls within the interval

M	2.2	2.4	2.6	2.8	3	3.2	3.4	3.6	3.8	4
T_x, [K]	1130	1166	1201	1233	1260	1286	1309	1329	1349	1365
T^1, [K]	609	572	537	503	470	440	412	385	360	337
4.2	4.4	4.6	4.8	5	5.2	5.4	5.6	5.8	6	6.2
1381	1396	1408	1420	1431	1441	1450	1458	1466	1473	1479
316	297	279	262	247	233	219	207	196	186	176

Table 1.
Values of critical temperature T_x and detonation temperature T^1 depending on the Mach number M (P_o = 60 mm Hg).

$$609\,K \geq T^1 \geq 176\,K. \tag{28}$$

This is the minimum temperature in front of the shock wave that makes the detonation possible.

4. Results of calculations and their discussion

Using the intervals obtained above for some physical quantities, let us graphically determine the region of existence for the normal spherical detonation. The upper limit of the hydrogen content in the mixture is confined in our case by the value $c = 0.66$. Above this value, the chemical reactions resulting from the interaction between hydrogen and oxygen in the mixture, which were considered in work [4], become more complicated, and this circumstance may result in different values of critical temperature. Further researches of this issue are required. Below, the choice $T_1 = 800\,K$ for the upper limit of the medium temperature is explained in detail.

With regard for dependence (24) of the Mach number M on the temperature of a motionless medium T_1 and the hydrogen content c, let us plot the dependences $M(c)$, $T_x(c)$, and $T^1(c)$ at fixed T_1 and P_0. We proceed from the plots of the dependence $M(c)$ at $T_1 = const$ exhibited in **Figure 5** for T_1 -temperatures in the interval $800\,K \geq T_1 \geq 173\,K$ (see **Table 2**). The lower curve corresponds to the gas mixture temperature $T_1 = 800\,K$, and the upper one to $T_1 = 173\,K$. According to expression (24), this family of curves has a power dependence on the hydrogen content in the mixture, c, with a power exponent of 0.5. Let us fix the maximum content of burned out hydrogen, $c = 0.66$, which corresponds to the stoichiometric composition of hydrogen-oxygen mixture, and draw a vertical line. The Mach number corresponding to its intersection with the mentioned family of curves changes from $M = 2.8$ at point 4 to $M = 6.2$ at point 5. Another important detail should be emphasized. Four dashed lines are drawn in **Figure 5**. Two horizontal

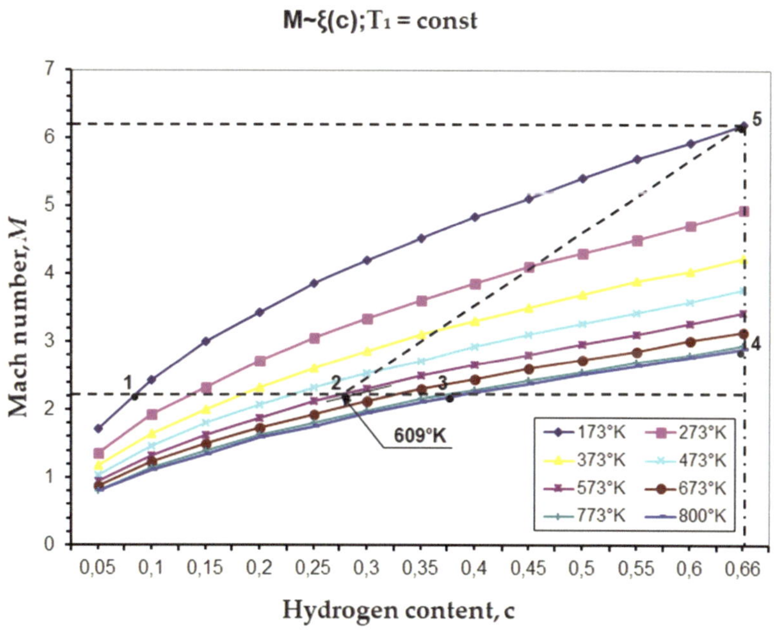

Figure 5.
Dependence of the Mach number M on the hydrogen content c in the gas mixture ($P_0 = 60$ mm Hg) for various temperatures in the motionless medium, T_1.

	c	0.05	0.1	0.15	0.2	0.25	0.3	0.35	0.4	0.45	0.5	0.55	0.6	0.66
173°K	M	1.71	2.42	3	3.42	3.85	4.19	4.51	4.84	5.1	5.41	5.7	5.93	6.2
273°K		1.36	1.93	2.32	2.72	3.05	3.34	3.6	3.85	4.1	4.31	4.5	4.72	4.95
373°K		1.17	1.65	2	2.33	2.6	2.85	3.1	3.3	3.5	3.69	3.9	4.04	4.23
473°K		1.03	1.46	1.8	2.07	2.32	2.53	2.72	2.93	3.1	3.27	3.42	3.59	3.76
573°K		0.94	1.33	1.62	1.88	2.12	2.3	2.5	2.66	2.81	2.97	3.1	3.26	3.42
673°K		0.87	1.23	1.5	1.74	1.93	2.13	2.3	2.45	2.6	2.74	2.86	3.01	3.15
773°K		0.81	1.14	1.39	1.62	1.8	1.98	2.16	2.29	2.43	2.56	2.7	2.81	2.94
800°K		0.8	1.11	1.34	1.59	1.75	1.95	2.11	2.25	2.38	2.52	2.64	2.76	2.89

Table 2.
Data of the Mach number M, depending on the specific content of hydrogen c at $T_1 = const$, $P_0 = 60\ mm\ Hg$.

ones confine the region of allowable Mach numbers corresponding to the normal spherical detonation. The first line corresponds to the minimum $M_{min} = 2.2$, and the second one to the maximum $M_{max} = 6.2$. The third dashed vertical line corresponds to the stoichiometric composition of the hydrogen-oxygen mixture and is the optimal variant for the detonation. The fourth line will be discussed below.

Let us consider points 1 to 5 in **Figure 5** separately. Segment 1–2 corresponds to the lower limit of the shock wave velocity $M_{min} = 2.2$, but the medium temperature for the segment points turns out lower than the detonation one (**Figure 4**). Therefore, the detonation is impossible in this case. The region of the probable detonation for this Mach number is restricted to segment 2–3, because the temperature of motionless medium reaches the detonation temperature values here. On the basis of **Figure 4**, it is also possible to draw a conclusion that, for the medium temperature $T_1 = 173\ K$, the detonation is possible if $M = M_{max} = 6.2$ (point 5 in **Figure 5**). In other words, for the chosen temperature, $800\ K \geq T_1 \geq 173\ K$, and hydrogen content, $0.66 \geq c \geq 0.075$, intervals, the region of the probable detonation is bounded by segments 2–3, 3–4, 4–5, and 5–2. Segment 5–2 is presented in **Figure 5** schematically by a straight line. In the general case, in view of the nonlinear dependence $M(c, T_1)$, this segment is curvilinear.

Let us derive the functional dependence $T_x(c)$ by substituting the $M(c)$ dependence (Eq. (24)) into Eq. (5). To make transformations simpler, let us rewrite Eq. (24) in a slightly different form,

$$M = [\eta c]^{\frac{1}{2}}, \tag{29}$$

where

$$\eta = \frac{(\gamma+1)^2(\gamma-1)Q}{4\gamma^2 K^* T_1}. \tag{30}$$

Then we obtain the following transcendental equation for the critical temperature T_x:

$$T_x^2 = \frac{2.5\times10^5 Q T_0(\gamma-1)(2\gamma\eta c-\gamma+1)(2+(\gamma-1)\eta c)^2}{4\gamma^2(\gamma+1)K^* P_0 \eta^3 c^3} \times \exp\left(-\frac{E_2}{K^* T_x}\right) \tag{31}$$

or, taking Eq. (30) into account,

$$T_x^2 = \frac{2.5\times10^5 T_0 T_1(2\gamma\eta c-\gamma+1)(2+(\gamma-1)\eta c)^2}{(\gamma+1)^3 P_0 \eta^2 c^3} \exp\left(-\frac{E_2}{K^* T_x}\right). \tag{32}$$

In **Figure 6**, using expression (32) and **Table 3**, we plotted the dependences $T_x(c)$ at $P_0 = const$ and $T_1 = const$. By the form, they are similar to the previous plots (**Figure 5**) and confirm the conclusions made for points 1 to 5.

More interesting is the dependence of the detonation temperature in the motionless medium on the hydrogen content, $T^1(c)$, at $P_0 = const$ and $T_1 = const$. It can be determined from relation (5) with regard for Eqs. (7) and (4):

$$(T^1)^2 = \frac{2.5 \times 10^5 Q T_0 (\gamma - 1)(\gamma + 1)^3}{4\gamma^2 K^* P_0 (2\gamma M^2 - \gamma + 1) M^2} \times \exp\left(- \frac{E_2 (\gamma + 1)^2 M^2}{K^* T^1 (2\gamma M^2 - \gamma + 1)(2 + (\gamma - 1) M^2)} \right). \tag{33}$$

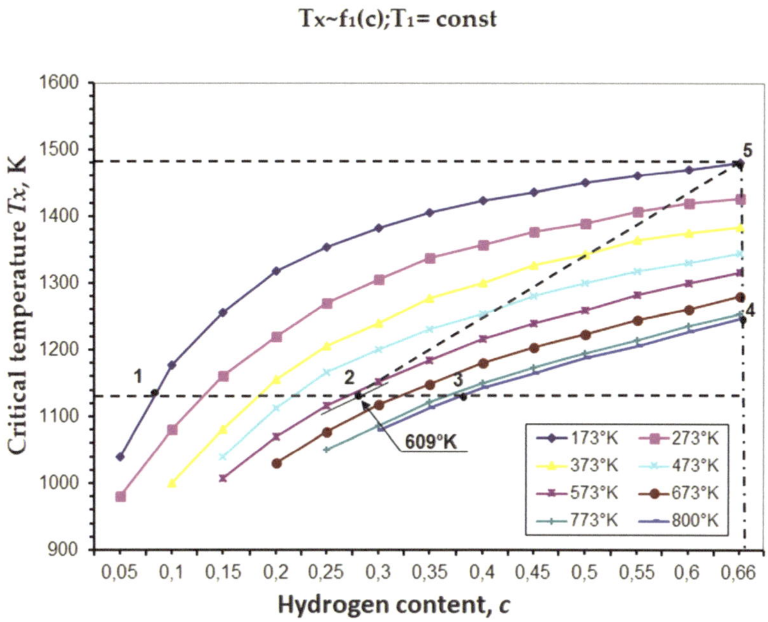

Figure 6.
Dependences of the critical temperature T_x on the hydrogen content c in the gas mixture (P = 60 mm Hg) for various temperatures in the motionless medium, T_1.

	c	0.05	0.1	0.15	0.2	0.25	0.3	0.35	0.4	0.45	0.5	0.55	0.6	0.66
173°K	T_x	1040	1176	1256	1317	1354	1382	1405	1423	1436	1450	1460	1470	1480
273°K		980	1080	1160	1220	1270	1305	1338	1358	1376	1390	1408	1420	1427
373°K			1000	1080	1155	1205	1240	1276	1300	1327	1343	1364	1375	1384
473°K				1040	1113	1166	1200	1230	1254	1280	1300	1318	1330	1344
573°K				1008	1070	1116	1152	1184	1216	1240	1259	1282	1300	1316
673°K					1030	1076	1118	1148	1180	1204	1223	1244	1260	1280
773°K						1050	1085	1122	1150	1174	1195	1215	1236	1253
800°K							1078	1112	1142	1164	1187	1205	1226	1246

Table 3.
Data showing the functional dependence of the critical temperature T_x on the specific content of hydrogen c at $T_1 = const$, $P_0 = 60$ mm Hg.

Making allowance for Eqs. (29) and (30), relation (33) can be simplified to the following form:

$$(T^1)^2 = \frac{2.5 \times 10^5 T_0 T_1 (\gamma + 1)}{c P_0 (2\gamma\eta c - \gamma + 1)} \times \exp\left(-\frac{E_2 (\gamma + 1)^2 \eta c}{K^* T^1 (2\gamma\eta c - \gamma + 1)(2 + (\gamma - 1)\eta c)}\right). \tag{34}$$

The corresponding family of curves is shown in **Figure 7**, according to **Table 4**.

While analyzing the plots, the attention should be drawn to the following facts.

(i) Every temperature T_1 of the gas mixture is associated with a specific dependence

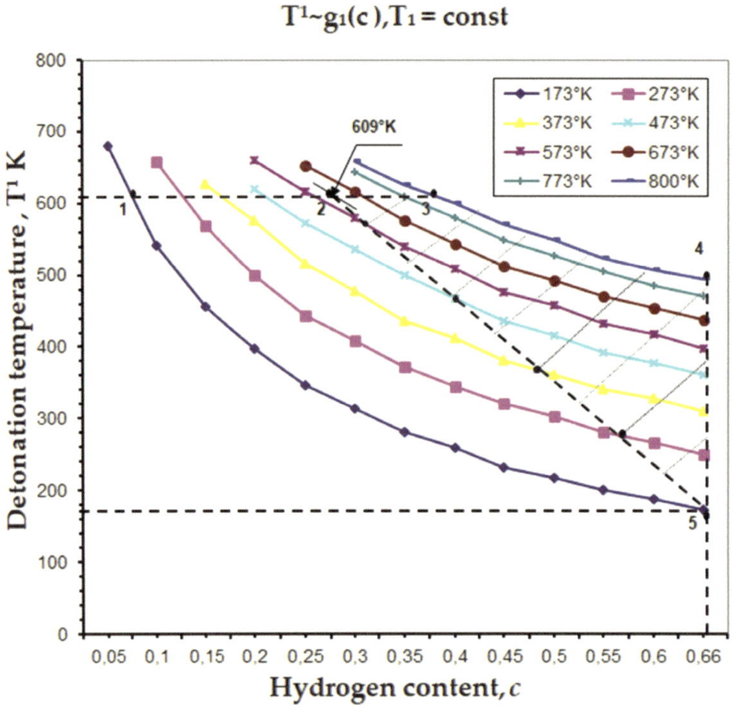

Figure 7.
Diagrams of the dependence of the detonation temperature T^1 on the hydrogen content c in an explosive gas mixture $H_2 + O_2$ (P = 60 mm Hg) for different temperatures in a stationary environment, T_1.

	c	0.05	0.1	0.15	0.2	0.25	0.3	0.35	0.4	0.45	0.5	0.55	0.6	0.66
173°K	T^1	680	542	456	398	346	313	280	258	232	217	200	188	173
273°K			658	568	500	442	409	372	345	320	302	280	266	250
373°K				626	576	516	477	436	412	381	360	340	328	310
473°K					620	572	535	500	467	436	416	392	378	360
573°K					660	616	580	540	508	476	458	432	418	398
673°K						652	616	576	543	512	492	470	454	438
773°K							644	609	580	548	526	504	484	470
800°K							658	625	600	570	548	523	507	493

Table 4.
Values of the detonation temperature T^1, depending on the specific content of hydrogen c at T_1 = const, P_o = 60 mm Hg.

$T^1(c)$. (ii) As the hydrogen content in the mixture grows, the temperature of the detonation T^1 in the motionless medium drastically decreases, which is especially appreciable at low temperatures. (iii) Let us draw a horizontal line that intersects the family of curves (for example, let it be the dashed line $T^1 = 173\ K$). At the point of its intersection with the curve corresponding to the same temperature of the motionless medium (in our case, this is $T_1 = 173\ K$), the detonation condition $T_2 = T_x$ (point 5) is satisfied. Detonation becomes probable, because the current temperature of motionless medium reaches the detonation temperature for this medium ($T_1 = T^1$). (iv) The intersection points of any horizontal line (see item iii) correspond to the critical hydrogen contents in the mixture, below which the detonation is impossible. The dashed line connecting points 2 and 5 in **Figure** 7 corresponds to the condition $T_2 \geq T_x$ for the whole family of curves.

Let the hydrogen content in the mixture change from 0.075 (point 1) to 0.66 (point 4). Then, on the basis of the plots shown in **Figure** 7, one may assert the following.

1. As was indicated above, a temperature lower than $T^1 \approx 173 \div 176\ K$ can give rise to a variation in the physical properties of the reacting mixture. Then the proposed formulas will produce erroneous results. The horizontal dashed line that passes through point 5 corresponds to this temperature, and point 5 testifies to the explosion with the maximum Mach number $M_{max} = 6.2$.

2. According to the plot of the functional dependence $T^1(c)$ at $T_1 = 273\ K$ (see Eq. (34) and **Figure** 7), the detonation is possible if the hydrogen content in the mixture is not lower than 0.57.

3. Physical restrictions imposed by the minimum Mach number $M_{min} = 2.2$ bring about the existence of the upper limit for the detonation temperature, $T^1 = 609\ K$. Points of both segments 1–2 and 2–3 correspond to the allowable values of Mach number. However, the detonation is possible only for the points on segment 2–3, because the main condition $T_2 \geq T_x$ is satisfied at $T_1 \geq 609\ K$. Whence it follows that $c = 0.27$ is the minimum hydrogen content in the mixture, below which the detonation is mpossible even at very high temperatures.

4. Experimental results testify that, if the temperature of a gas mixture is higher than $T_1 = 800\ K$, the spontaneous ignition takes place, which can transform into the detonation, if the hydrogen content in the mixture is not lower than 0.37. Therefore, this temperature is a kind of upper limit, to which the hydrogen-oxygen mixture can be heated.

From the reasons given above, it follows that the region of spherical supersonic burning is bounded by segments 2–3, 3–4, 4–5, and 5–2 in **Figure** 7. For the illustrative purpose, it was hatched.

5. Conclusions

The dependences between the temperature, Mach number, the hydrogen content in the hydrogen-oxygen mixture as the main parameters characterizing the process of transformation of a shock wave into a detonation one and affecting the chemical reactions between reacting components are studied. On the basis of relations obtained earlier [4], the conditions are found, under which the probability of a chain branching reaches unity ($\delta = 1$), and a fast chain reaction is started. The existence of the critical temperature T_x at the front of a shock wave, above which the detonation takes place, is substantiated, as well as the functional dependence (5) of the critical temperature on the Mach number. In author's opinion, the latter should be taken as a basis, while studying the processes of spherical detonation. Summarizing the results of work [4], the ondition $T_2 \geq T_x$ is found, which connects the kinetics of a chemical reaction with the detonation in a gas mixture. On the basis of the relations of the hydrodynamic theory of detonation, the region of possible values for the temperatures at the shock wave front, T_2, and in the chemical reaction zone, T_3, is determined. The equality $T_x \approx T_2 \approx T_3$ which couples them, corresponds to the lower limit, at which the detonation is possible. The minimum and maximum values of Mach number in reacting gas media are also determined, which enables the process of supersonic burning to be analyzed in more details and the region of physical parameters and quantities (the critical emperature, the temperature of detonation in the motionless medium, and the hydrogen content in the mixture), at which the spherical detonation is probable, to be indicated. The latter is llustrated, by using the hydrogen-oxygen mixture as an example.

To summarize, it should be noted that this paper is final in a cycle of works devoted to the study of the whole process of normal spherical detonation.

Acknowledgements

The author expresses his sincere gratitude to Yu.L. Birkovoi, the former head of the technological department of microelectronics of the known, in the past, production association "Rodon", as well as to the whole team of this department, for a long-term fruitful cooperation.

Nomenclature

Basic designations:

H_2	hydrogen molecule
O_2	oxygen molecule
H_2O	water molecule
O	oxygen atom
H	hydrogen atom
OH	compound of an oxygen atom with a hydrogen atom
T_x	critical temperature
T_2	temperature at the shock front
δ	branching probability
D	shock wave velocity, detonation velocity
M	Mach number
P, T, ρ	pressure, temperature, density of the medium
ES	explosive substance
DP	detonation products
CRZ	chemical reaction zone
γ	adiabatic index
Q	*combustion energy of one mole of combustible gas*
μ	*molar mass*
K^*	*universal gas constant*
E_2	the activation energy of the branching reaction

d	the chemical reaction zone width
Δt	the reaction duration
v_g	the gas velocity behind the reaction front
b_1	the sound velocity in the motionless medium in front of the front
m	gas mass
T_3	temperature at the Jouguet point
c	coefficient of flammable gas content in the mixture
W	chemical reaction rate
T^1	detonation temperature
M_{max}	maximum Mach number
M_{min}	minimum Mach number
$f(x)$	function of the variable x
$\exp(x)$	exponential function
Δx	increment of variable x
$\ll$	much less
$\gg$	much more
$\approx$	almost equal
$[a;b]$	line segment

Author details

Myron Polatayko
Kyiv National University, Kyiv, Ukraine

*Address all correspondence to: pmm.miron@gmail.com

References

[1] Orlenko L. P, editor. Physics of Explosion. 3rd ed. Moscow: Fizmatlit; 2004. V.1. 832 p.

[2] Mader, C. Numerical Simulation of Detonation. 1st ed. Berkeley: University of California; 1979. 485 p.

[3] Fedorov A. V, Tropin D. A, and Bedarev I. A. Mathematical modeling of the suppression of the detonation of a hydrogen-oxygen mixture by inert particles. Fiz. Goren. Vzryva. 2010; 46 (3):103–115.

[4] Myron Polatayko (November 5th 2018). Determination of the Velocity of the Detonation Wave and the Conditions for the Appearance of Spherical Detonation during the Interaction of Hydrogen with Oxygen [Online First], IntechOpen, DOI: 10.5772/intechopen.81792. Available from: https://www.intechopen.com/ online- first/determination-of-the-velocity-of-the-detonation-wave-and-the-conditions-for-the- appearance-of-spheri

[5] Kasimov A.R., Faria L.M., Rosales R. R. On a Theoretical Prediction of the Dynamics of Pulsating and Cellular Detonations in Gases. Combustion and Explosion. 2016; 9 (2): 42–50

[6] Myron Polatayko. Features of Spherical Detonation in Explosive Gas Environments, Public Science Framework, Physics Journal. 2019; 5(1): 1–7. Available from: http://www.aiscie nce.org/journal/allissues/pj

[7] Polatayko M. M. Determination of detonation wave velocity in an explosive gas mixture. Ukr. Fiz. Zh. 2012; 57(6): 606–611.

[8] Polatayko, M.M. Conditions for the Occurrence of Spherical Detonation in a Hydrogen- Oxygen Mixture. Ukr. Fiz. Zh.. 2013; 58 (10): 963–968.

[9] Soloukhin R. I. Shock Waves and Detonations in Gases. Baltimore: Mono Books; 1966. 176p.

[10] Semenov N. N. Chemical Kinetics and Chain Reactions. Oxford: Clarendon Press; 1935. 480 p.

[11] Semenov N. N. Some Problems of Chemical Kinetics and Reactivity. Princeton: Princeton Univ. Press; 1959. V.1. 254 p.

[12] Matveev A. N. Molecular Physics. Moscow: Mir Publishers; 1985. 450 p.

[13] Sysoev N. N, and Shugaev F. V. Shock Waves in Gases and Condensed Media. Moscow: Moscow State University; 1987. 133 p.

[14] Polatayko M. M. Possibility of Normal Spherical Detonation in a Hydrogen-Oxygen Gas Mixture: Allowable Temperature, Mach Number, and Hydrogen Content. Ukr. Fiz. Zh. 2014; 59(10):980–988.

[15] Ravdel A. A, Ponomareva A. M, editors. A Brief Handbook of Physico-Chemical Values. Eighth ed. Leningrad: Khimiya; 1983. 232 p.

[16] Morachevsky A. G., Sladkov I. B. Physico-Chemical Properties of Molecular Inorganic Compounds. Leningrad: Khimiya; 1987. 192p.

Chapter 10

On Statistical Assessments of Racial/Ethnic Inequalities in Cigarette Purchase Price among Daily Smokers in the United States: Non-Hispanic Whites Pay Least

Julia N. Soulakova and Trung Ha

Abstract

We discuss statistical methods suitable for comparing multiple populations versus one reference population and consider two common problems: (1) detecting all significant mean differences and (2) demonstrating that all mean differences are significant. Discussed methods include the Bonferroni approach (both problems), Min test (problem 2), and Strassburger-Bretz-Hochberg (SBH) confidence interval for estimating the smallest mean difference (problem 2). They illustrate the methods using the pooled 2010–2015 Tobacco Use Supplement to the Current Population Survey (TUS-CPS) data on the cigarette purchase price (per pack) reported by adult daily smokers (n = 34,728). The goal was to show that among seven considered racial/ethnic groups of daily smokers, non-Hispanic (NH) Whites paid least for cigarettes (on average). We used the design-based multiple linear regression to derive the estimates and raw p-values. The Min test supported the study goal. Likewise, SBH lower 95% confidence interval bound was $0.08, indicating that the other racial/ethnic groups of daily smokers paid at least eight cents more for a pack of cigarettes (on average) than did non-Hispanic Whites. However, Bonferroni method (that was originally proposed for problem 1) failed to support the study goal. The study highlights the importance of choosing the right statistical method for a given problem.

Keywords: balanced repeated replications, complex survey, multiple comparisons, statistical multiple-testing problems

1. Introduction

In this chapter, we discuss statistical methods for comparing multiple populations relative to one population (termed "reference"). These types of multiple comparisons commonly arise in behavioral science, for example, when multiple racial/ethnic groups are compared to non-Hispanic (NH) White smokers in terms of tobacco-use-related behaviors [1–4]. When the statistical parameter of interest is the mean difference, the most common study goal is one of the following two goals. **Goal 1** is to detect all significant mean differences among the considered populations (versus the reference population), that is, to draw an individual conclusion

IntechOpen

regarding significance of each mean difference. **Goal 2** is to demonstrate that all mean differences among the considered ones are significant. Note that if one assessed Goal 1 and concluded that each mean difference was significant then s/he has (indirectly) assessed Goal 2 as well. Other more intricate study goals, such as the ones arising in pharmaceutical statistics which involve a hierarchical structure among the primary and secondary end points, were addressed elsewhere and are outside of the scope of this chapter [5–11].

We discuss how Goals 1 and 2 can be assessed in a study of racial and ethnic disparities, where Hispanic (H) population and five non-Hispanic populations such as American Indian/Alaska Native (AIAN), Asian (ASIAN), Black/African American (BAA), Hawaiian/Pacific Islander (HPI), and Multiracial (MULT), are compared to non-Hispanic White (W) population in terms of the mean differences.

$$\mu_{AIAN} - \mu_W, \mu_{ASIAN} - \mu_W, \mu_{BAA} - \mu_W, \mu_H - \mu_W, \mu_{HPI} - \mu_W, \text{and } \mu_{MULT} - \mu_W, \quad (1)$$

where μ_{AIAN}, μ_{ASIAN}, μ_{BAA}, μ_H, μ_{HPI}, $\mu_{MULT,}$ and μ_W denote, respectively, the mean responses for AIAN, ASIAN, BAA, H, HPI, MULT, and W populations. Furthermore, suppose that each positive mean difference in Eq. (1) corresponds to a significant result, for example, the first difference being positive implies that the mean response among AIANs is greater than the mean response among Ws. Then the null and alternative hypotheses corresponding to the ith difference, where i denotes AIAN, ASIAN, BAA, H, HPI, and MULT, can be stated as

$$H_{oi} : \mu_i - \mu_W \leq 0 \text{ and } H_{ai} : \mu_i - \mu_W > 0. \quad (2)$$

Finally, let $p_i (i = AIAN, ASIAN, BAA, H, HPI, MULT)$ denote a p-value corresponding to testing H_{oi} versus H_{ai}. As a result, we have six pairs of hypotheses and six p-values.

Suppose the overall error rate for assessing Goal 1 (Goal 2) is fixed at α-level. Then to assess Goal 1 (to detect all significant mean differences), we should first rescale each p-value p_i, for example, via Bonferroni, Holm, or Hochberg approaches [6, 12–14]. This rescaling is essential to control the overall error rate at the nominal α-level. For example, in our case with six null hypotheses, the p-values rescaled via Bonferroni method are given as $6p_i$ (i.e., we multiply each original p-value by six). Second, we compare each rescaled p-value with α. If $p_i \leq \alpha$, then we reject H_{oi} and conclude that the ith difference is significant (positive); if $p_i > \alpha$, then we accept H_{oi} and conclude that the ith difference is not significant. As a result, we draw an individual conclusion regarding significance of each mean difference. Alternatively to the above hypothesis testing, one could construct the lower $100(1-\alpha/6)\%$ confidence intervals for the mean differences in Eq. (1) and use the lower bounds to differentiate between the significant and insignificant mean differences; this approach was discussed elsewhere [15].

To assess Goal 2 (to demonstrate that all differences are significant), one can use the Min test that is an intersection-union test [16–21]. The p-value for the Min test, denoted by p, is simply the largest p-value among the six individual p-values:

$$p = \max\{p_{AIAN}, p_{ASIAN}, p_{BAA}, p_H, p_{HPI}, p_{MULT}\}. \quad (3)$$

If $p \le \alpha$, then we reject all H_{0i}s and conclude that all mean differences are significant (positive); if $p > \alpha$, then we fail to reject all H_{0i}s and conclude that there is at least one insignificant mean difference. Note that we cannot comment on the significance of an individual mean difference, because we tested whether all mean differences are significant (i.e., whether the smallest mean difference is significant). Nonetheless, the Min test is more suitable for assessing Goal 2 than Bonferroni approach or another approach proposed for assessing Goal 1. A statistical method is usually proposed for a specific problem and thus, the methods should be used accordingly: the union-intersection hypothesis (Goal 1) should be tested via Bonferroni or another union-intersection test, while the intersection-union hypothesis (Goal 2) should be tested via the Min or another intersection-union test [12, 22].

Alternatively to the Min test, we can use the Strassburger-Bretz-Hochberg (SBH) confidence interval approach as follows [23, 24]. First, we compute the lower $100(1-\alpha)\%$ confidence intervals for the mean differences in Eq. (1). Let these bounds be denoted as L_{AIAN}, L_{ASIAN}, L_{BAA}, L_H, L_{HPI}, and L_{MULT}. Second, let L denote the smallest bound among these bounds, that is,

$$L = \min\{L_{AIAN}, L_{ASIAN}, L_{BAA}, L_H, L_{HPI}, L_{MULT}\}. \quad (4)$$

Then the SBH lower $100(1-\alpha)\%$ confidence interval for the smallest mean difference is given by $(L, +\infty)$. If $L > 0$, we conclude that all mean differences are significant, and if $L \le 0$, then we conclude that there is at least one insignificant mean difference.

We note that one needs to identify the appropriate statistical method to compute the individual p-values and confidence bounds. The choice depends on the study design, probability distributions, and other statistical considerations. The Min test and the SBH interval were discussed for parallel and factorial designs, where sample mean responses followed normal distributions with known variances or unknown (common) variance, as well as Binomial and several other distributions [20, 21, 23–27]. In addition, one needs to decide whether the analyses should adjust for explanatory factors, for example, sociodemographic characteristics [28–30]. Such adjustments may help reduce the effect of confounding factors and therefore, improve estimation [31, 32]. For example, Golden et al. examined how much smokers pay for a pack of cigarettes, on average, in the United States using data from the 2010–2011 Tobacco Use Supplement to the Current Population Survey (TUS-CPS) [1]. Among several design-based multiple linear regression models for the mean purchase price per pack (PPP) used in the study, one model adjusted for smokers' sociodemographic and smoking-related characteristics, cigarette purchase attributes, and the survey wave [1].

Despite availability and benefits of the Min test and SBH interval, these methods have not received much attention in behavioral sciences. We illustrate benefits of using these methods over Bonferroni method and simplicity of applications of these methods. We consider a study of racial and ethnic disparities in cigarette purchase prices conducted to demonstrate that W daily smokers, on average, purchase cigarettes at lower prices than do AIAN, ASIAN, BAA, H, HPI, and MULT daily smokers in the United States. This goal was motivated by results of a prior study revealing that BAA, H, and ASIAN/HPI (ASIAN and HPI combined) smokers paid higher PPP, on average, relative to W smokers, in the United States in the period from 2010 to 2011 [1].

2. Methods and results

2.1 Using data to derive the p-values and lower confidence interval bounds

We used the pooled 2010–2011 and 2014–2015 TUS-CPS data for adult daily smokers (n = 34,728) who reported the price of the last self-purchased pack or carton of cigarettes. The reported prices were used to compute the (average) PPP. The overall cohort was representative of about 23,370,261 adult daily smokers, where 12% were 18–24 years old, 38% were 25–44 years old, and 50% were 45+ years old, and 54% were men and 47% were women. The racial/ethnic representation was as follows: 76% were W, 11% were BAA, 8% were H, 2% were MULT, 2% were ASIAN, 1% were AIAN, and less than 1% were HPI. All racial/ethnic groups were well represented in the sample: the smallest number of respondents (96) corresponded to HPI daily smokers. Additional sample characteristics have been described in a prior study of purchasing cigarettes on Indian reservations [33].

We fixed the overall error rate at α = 5% and fitted a design-based multiple linear regression ($R^2 \approx 30\%$, $F(25, 160) \approx 257$, $p < 0.0001$) to model the mean PPP as a function of daily smokers' characteristics, location of the purchase (on/off Indian reservation), survey mode (phone, in-person), and survey period (2010–2011, 2014–2015). The daily smokers' characteristics included race/ethnicity, age, sex, marital status, education, employment record, region of residency (West, South, Midwest, and Northeast), metropolitan area of residency (metro, nonmetro), and heavy smoking indicator. The analysis incorporated statistical methods recommended in the methodological guidelines for analysis of the CPS and CPS supplements [34, 35]. Specifically, because the CPS incorporates complex sampling, we estimated variance using balanced repeated replications [36]. The main and 160 replicate weights for this approach have been made available for public use by the U.S. Census Bureau [34, 35]. The analysis was performed using SAS®9.4 software [37]; the SAS®9.4 Survey Package procedures suitable for analysis of TUS-CPS have been discussed elsewhere [38]. **Table 1** depicts the estimated model coefficients and their standard errors for all covariates. As is shown in **Table 1**, smokers' sex and survey mode (phone, in-person) were not significant.

Table 1 presents the individual p-values for comparisons of racial/ethnic populations of daily smokers versus W daily smokers (based on the model): $p_{AIAN} < 0.0001$, $p_{ASIAN} < 0.0001$, $p_{BAA} < 0.0001$, $p_H < 0.0001$, $p_{HPI} = 0.0002$, and $p_{MULT} = 0.0087$.The individual lower 95% confidence interval bounds for the mean PPP difference for each racial/ethnic population of daily smokers relative to W daily smokers were computed using the formula:

$$L_i = \hat{d}_i - t_{0.95,df=160} SE\left(\hat{d}_i\right), i = AIAN, ASIAN, BAA, H, HPI, MULT, \tag{5}$$

where $\hat{d}_i$ denotes the estimated mean PPP difference relative to W daily smokers (the point estimate for the ith mean difference), $SE\left(\hat{d}_i\right)$ is the standard error of the estimate (computed using the balanced repeated replications), and $t_{0.05,df=160} = 1.6544$ is the 95th percentile of the central t-distribution with 160 degrees of freedom (the number of degrees of freedom matches the number of the replicate weights) [34–36]. We note that there are alternative methods to construct the lower bounds, for example, using the standard normal distribution instead of the central t-distribution [34, 36].

Figure 1 depicts the lower bounds L_is and the estimated mean differences $\hat{d}_i$s for all racial/ethnic populations (relative to the W population). These bounds were computed using *proc surveyreg* procedure with *lsmestimate* statements (with "cl," "e,"

Factor	Estimated coefficient	Standard error	p-Value*
Intercept	3.64	0.12	*
Race/ethnicity (reference group is W)			
AIAN versus W	0.61	0.13	*
ASIAN versus W	0.62	0.09	*
BAA versus W	0.51	0.04	*
H versus W	0.61	0.06	*
HPI versus W	0.83	0.22	0.0002
MULT versus W	0.22	0.08	0.0087
Age (reference group is 45+ years old)			
18–24 years old versus 45+ years old	0.19	0.04	*
25–44 years old versus 45+ years old	0.20	0.02	*
Sex			
Female versus male	0.00	0.02	0.9052
Marital status (reference group is widowed/divorced/separated)			
Married (living with a spouse)	0.02	0.02	0.4878
Never married	0.22	0.03	*
Highest level of education (reference group is some college/Bachelor's degree)			
Graduate degree	0.11	0.08	0.1632
High school/equivalent	−0.14	0.02	*
Less than high school	−0.15	0.03	*
Employment status (reference group is unemployed)			
Employed (at work or absent) versus unemployed	0.16	0.03	*
Not in labor force versus unemployed	−0.15	0.04	*
Place where cigarettes were purchased (reference group is "on Indian reservation") (reference group is "yes)			
No versus yes	1.57	0.12	*
U.S. region of residency			
Midwest versus West	0.09	0.03	0.0091
Northeast versus West	1.75	0.05	*
South versus West	−0.62	0.03	*
Metropolitan area of residency			
Metropolitan area versus nonmetropolitan area	0.32	0.03	*
Heavy smoking indicator			
Heavy (20+ cigarettes per day) versus non-heavy smoker	−0.20	0.02	*
Survey mode			
Personal interview versus phone interview	−0.01	0.02	0.5205
Survey period			
2010–2011 versus 2014–2015	−0.38	0.02	*

*p-value < 0.0001.

Table 1.
Design-based multiple linear regression for the mean cigarette price per pack.

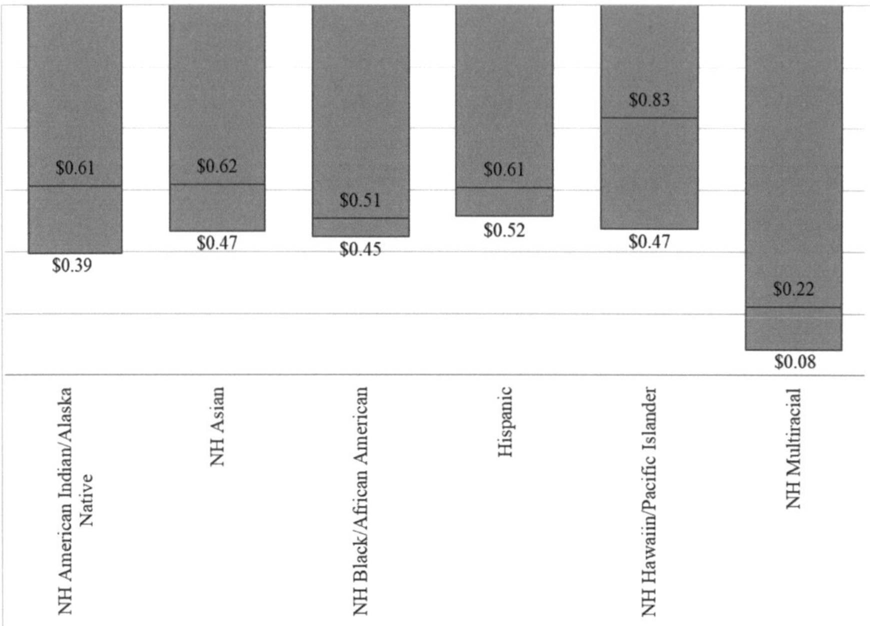

Figure 1.
Individual lower 95% confidence intervals for the mean price per pack differences relative to non-Hispanic (NH) White daily smokers; the lower number corresponds to the lower bound and the upper number corresponds to the point estimate for the mean difference. For example, AIAN daily smokers, on average, pay at least $0.39 more per pack of cigarettes than do NH White daily smokers, and the point estimate for the difference is $0.61.

"upper," and "alpha = 0.05" options) when fitting the model using SAS software. Alternatively, we could use the *lsmeans* statement (with "adj = bon," "cl," and "alpha = 0.1" options), and select the comparisons of interest out of all 21 pair-wise comparisons reported and note the lower bound of the two-sided 90% confidence interval reported in the output.

2.2 Demonstrating the study goal via the min test and SBH confidence interval

The p-value for the Min test is $p = 0.0087$, indicating that at 5% significance level we reject the null hypothesis in favor of the alternative. The corresponding SBH lower 95% confidence interval bound for the mean PPP difference is $0.08 (see **Figure 1**). Therefore, all six racial/ethnic groups of daily smokers paid, on average, higher PPP relative to W daily smokers in the United States in the periods from 2010–2011 to 2014–2015.

If instead of the Min test we used the Bonferroni approach, then the adjusted p-values would be less than 0.0006 for four comparisons (AIAN versus W, ASIAN versus W, BAA versus W, and H versus W), 0.0012 for one comparison (HPI versus W), and 0.0522 for one comparison (MULT versus W). Therefore, we would conclude that only AIAN, ASIAN, BAA, H, and HPI daily smokers pay higher PPP, on average, than do W daily smokers; and would fail to demonstrate that all six considered racial/ethnic groups of daily smokers pay higher PPP, on average, relative to W daily smokers.

3. Discussion

The choice of the reference group as "W daily smokers" was based on the study goal and prior studies of cigarette purchasing behaviors of smokers [1, 33]. The choice of the reference group as well as the statistical methods should always align with the study goal and should be made prior to the data analysis. Specifically, when examining racial/ethnic disparities, using "W" as the reference group could be logical in some studies but not logical in the other studies. For example, if the study goal is to show that purchasing cigarettes on Indian reservations is most prevalent among AIAN smokers, then "AIAN smokers" should be chosen as the reference group. In addition, while both Bonferroni method and the Min test are simple to use, in practice, only Bonferroni method results in individual conclusions regarding each comparison. However, Bonferroni method is less powerful than the Min test when applied to an intersection-union problem (to assess Goal 2) [6, 12].

The study indicated that W daily smokers paid significantly less for cigarettes, on average, than the other six racial/ethnic groups of daily smokers in the United States in the period from 2010–2011 to 2014–2015. The earlier reported finding (see model 6 in [1]) was that non-Hispanic White smokers, on average, paid significantly less for cigarettes than did BAA, AIAN, ASIAN/HPI (combined), and H smokers, and paid similar prices to the prices paid by "other non-Hispanic" smokers [1]. While the results might seem to disagree, the direct comparisons between these two findings are problematic, because the studies concerned different populations of smokers (daily smokers in our study, and daily and occasional smokers in the prior study) and time periods (overall 2010–2011 and 2014–2015 in our study, and 2010–2011 in the prior study). Moreover (though, the authors did not mention the method they used to adjust for multiple comparisons, if any), the authors considered the union-intersection problem that is conceptually different from the intersection-union problem addressed in our study [1].

Our study has several potential limitations. First, we considered the population of daily smokers, and thus, results should not be generalized to other populations of smokers such as occasional smokers. Indeed, daily and occasional smokers have very different cigarette purchasing behaviors, for example, daily smokers are more likely to purchase cigarettes in cartons rather than packs and travel to another state or Indian reservations to purchase cigarettes at lower prices [1, 39, 40]. Second, the analysis was based on a certain regression model where the mean PPP was modeled as a function of smokers' characteristics, location of the purchase, survey mode, and survey period. Another model could potentially lead to a different conclusion, for example, only two out of six models indicated significantly higher mean PPP for AIAN smokers relative to W smokers [1]. Another potential limitation is a lack of a theoretical proof that the SBH interval for the smallest mean PPP difference has indeed confidence level of $100(1-\alpha)\%$. The probability coverage of the SBH confidence interval depends on the probability coverage of the individual confidence intervals for the mean differences [23]. Because we used the statistical methods outlined in the CPS methodological guidelines for constructing the individual intervals, we believe that the resulting SBH interval has the probability coverage close to $100(1-\alpha)\%$ level.

Future research may target development and implementation of procedures for the Min test and SBH interval. Specifically, the software packages developed for analysis of complex survey data currently offer just a few multiple comparison methods. For example, the SAS Survey Package offers a built-in procedure for Bonferroni adjustments but lacks procedures for the multiple testing (interval estimation) such as the Min test (SBH interval). Availability of the "Min test" and "SBH interval" procedures would enable researchers to incorporate these methods directly in their analyses of complex survey data.

4. Conclusion

In our study, results of the Min test (and SBH interval) were different from the results of the Bonferroni method. Specifically, using the Min test (and SBH interval), we demonstrated that all six racial/ethnic groups of daily smokers paid, on average, higher PPP relative to W daily smokers in the United States in the periods from 2010–2011 to 2014–2015. However, using the Bonferroni method, we failed to demonstrate this claim. This discrepancy highlights the importance of choosing the appropriate statistical method for assessing the minimum among multiple mean differences (relative to one reference population). Availability of the "Min test" and "SBH interval" procedures in survey packages would help facilitate application of these methods in behavioral research.

Acknowledgements

The authors are thankful to Richard Pack, B.S., for providing editing comments.

Conflict of interest

There is no conflict of interest.

Funding

Research reported in this publication was supported by the National Institute on Minority Health and Health Disparities of the National Institutes of Health under Award Number R01MD009718. The content is solely the responsibility of the authors and does not necessarily represent the official views of the National Institutes of Health.

Author details

Julia N. Soulakova* and Trung Ha
College of Medicine, University of Central Florida, Orlando, FL, USA

*Address all correspondence to: julia.soulakova@ucf.edu

References

[1] Golden SD, Kong AY, Ribisl KM. Racial and ethnic differences in what smokers report paying for their cigarettes. Nicotine & Tobacco Research. 2016;**18**(7):1649-1655

[2] Soulakova JN, Li J, Crockett LJ. Race/ethnicity and intention to quit cigarette smoking. Preventive Medicine Reports. 2017;**5**:160-165. Available from: http://linkinghub.elsevier.com/retrieve/pii/S2211335516301632

[3] Cokkinides VE, Halpern MT, Barbeau EM, Ward E, Thun MJ. Racial and ethnic disparities in smoking-cessation interventions: Analysis of the 2005 National Health Interview Survey. American Journal of Preventive Medicine. 2008;**34**(5):404-412

[4] Tran S-TT, Rosenberg KD, Carlson NE. Racial/ethnic disparities in the receipt of smoking cessation interventions during prenatal care. Maternal and Child Health Journal. 2010;**14**(6):901-909. Available from: http://link.springer.com/10.1007/s10995-009-0522-x [Accessed: 10 December 2019]

[5] Dmitrienko A, Soulakova JN, Millen BA. Three methods for constructing parallel gatekeeping procedures in clinical trials. Journal of Biopharmaceutical Statistics. 2011;**21**(4):768-786

[6] Hochberg Y, Tamhane A. Multiple Comparison Procedures. New York: Wiley; 1987

[7] Dmitrienko A, Offen W, Wang O, Xiao D. Gatekeeping procedures in dose–response clinical trials based on the Dunnett test. Pharmaceutical Statistics. 2006;**5**(1):19-28. Available from: http://doi.wiley.com/10.1002/pst.190 [Accessed: 22 February 2018]

[8] Dmitrienko A, Offen WW, Westfall PH. Gatekeeping strategies for clinical trials that do not require all primary effects to be significant. Statistics in Medicine. 2003;**22**(15):2387-2400. Available from: http://doi.wiley.com/10.1002/sim.1526 [Accessed: 22 February 2018]

[9] Soulakova JN. Comparison of gatekeeping and other testing methods for identifying superior drug combinations in bifactorial designs with isotonic parameters. Journal of Biopharmaceutical Statistics. 2011;**21**(4):635-649

[10] Chen X, Luo X, Capizzi T. The application of enhanced parallel gatekeeping strategies. Statistics in Medicine. 2005;**24**(9):1385-1397. Available from: http://doi.wiley.com/10.1002/sim.2005 [Accessed: 22 February 2018]

[11] Sugitani T, Morikawa T. Gatekeeping strategies and graphical approaches in clinical trials with hierarchically structured study objectives: A review. Japanese Journal of Biometrics. 2017;**38**(1):41-78

[12] Hsu JC. Multiple Comparisons: Theory and Methods. New York: CRC Press; 1996

[13] Soulakova JN. On identifying effective and superior drug combinations via Holm's procedure based on the Min tests. Journal of Biopharmaceutical Statistics. 2009;**19**(2):280-291

[14] Holm S. A simple sequentially rejective multiple test procedure on JSTOR. Scandinavian Journal of Statistics. 1979;**6**(2):65-70. Available from: https://www.jstor.org/stable/4615733?seq=1#page_scan_tab_contents [Accessed: 25 July 2018]

[15] Johnson RA, Wichern DW. Applied Multivariate Statistical Analysis. Upper Saddle River, New Jersey: Prentice Hall; 2002. pp. 232-234

[16] Lehmann EL. Testing multiparameter hypotheses. Annals of Mathematical Statistics. 1952;**23**:541-552

[17] Berger RL. Multiparameter hypothesis testing and acceptance sampling. Technometrics. 1982;**24**(4):295-300. Available from: http://www.jstor.org/stable/1267823?origin=crossref [Accessed: 20 February 2018]

[18] Berger RL, Hsu JC. Bioequivalence trials, intersection-union tests and equivalence confidence sets. Statistical Science. 1996;**11**(4):283-319. Available from: https://projecteuclid.org/download/pdf_1/euclid.ss/1032280304 [Accessed: 20 February 2018]

[19] Berger RL. Likelihood Ratio Tests and Intersection-Union Tests. Raleigh, North Carolina: Institute of Statistics Mimeo Series Number 2288; 1996. Available from: http://citeseerx.ist.psu.edu/viewdoc/download?doi=10.1.1.17.4041&rep=rep1&type=pdf [Accessed: 20 February 2018]

[20] Laska EM, Meisner MJ. Testing whether an identified treatment is best. Biometrics. 1989;**45**(4):1139. Available from: http://www.jstor.org/stable/2531766?origin=crossref [Accessed: 20 February 2018]

[21] Laska EM, Tang D-I, Meisner MJ. Testing hypotheses about an identified treatment when there are multiple endpoints. Journal of the American Statistical Association. 1992;**87**(419):825. Available from: http://www.jstor.org/stable/2290221?origin=crossref [Accessed: 20 February 2018]

[22] Casella G, Berger RL. Statistical Inference. 2nd ed. Pacific Grove, CA: Duxbury Press; 2001

[23] Strassburger K, Bretz F, Hochberg Y. Compatible Confidence Intervals for Intersection Union Tests Involving Two Hypotheses. Institute of Mathematical Statistics; 2004. Vol. 47. pp. 129-142. Available from: http://projecteuclid.org/euclid.lnms/1196285631 [Accessed: 20 February 2018]

[24] Soulakova JN. Generalized confidence intervals compatible with the Min test for simultaneous comparisons of one subpopulation to several other subpopulations. Communications in Statistics - Theory and Methods. 2017;**46**(19):9441-9449

[25] Hung HMJ. Global tests for combination drug studies in factorial trials. Statistics in Medicine. 1996;**15**(3):233-247 Available from: http://doi.wiley.com/10.1002/%28SICI%291097-0258%2819960215%2915%3A3%3C233%3A%3AAID-SIM167%3E3.0.CO%3B2-H [Accessed: 22 February 2018]

[26] Hung HMJ, Chi GYH, Lipicky RJ. Testing for the existence of a desirable dose combination. Biometrics. 1993;**49**(1):85. Available from: http://www.jstor.org/stable/2532604?origin=crossref [Accessed: 22 February 2018]

[27] Westfall P, Ho S-Y, Prillaman B. Properties of multiple intersection-union tests for multiple endpoints in combination therapy trials. Journal of Biopharmaceutical Statistics. 2001;**11**(3):125-138

[28] Cartmell KB, Miner C, Carpenter MJ, Vitoc CS, Biggers S, Onicescu G, et al. Secondhand smoke exposure in young people and parental rules against smoking at home and in the car. Public Health Reports. 2011;**126**(4):575-582

[29] Johnson T, Mott J. The reliability of self-reported age of onset of tobacco, alcohol and illicit drug use. Addiction. 2001;**96**(8):1187-1198. Available from: http://onlinelibrary.wiley.com/doi/10.1046/j.1360-0443.2001.968118711.x/full [Accessed: 15 December 2016]

[30] Soulakova JN, Crockett LJ. Unassisted quitting and smoking cessation methods used in the United States: Analyses of 2010-2011 tobacco use supplement to the current population survey data. Nicotine & Tobacco Research. 2017;**20**(1):30-39

[31] Jiang J. Linear and Generalized Linear Mixed Models and their Applications. New York, USA: Springer Science & Business Media, LLC New York; 2007

[32] Charles EM, Shayle RS, John MN. Generalized, Linear, and Mixed Models. 2nd ed. Hoboken, New Jersey, USA: John Wiley and Sons Inc.; 2008

[33] Soulakova JN, Pack R, Ha T. Patterns and correlates of purchasing cigarettes on Indian reservations among daily smokers in the United States. Drug and Alcohol Dependence. 2018;**192**:88-93. Available from: http://www.ncbi.nlm.nih.gov/pubmed/30243144 [Accessed: 12 October 2018]

[34] U.S. Bureau of Labor Statistics, U.S. Census Bureau. Design and Methodology: Current Population Survey, Technical Paper 66 [Internet]. 2006. Available from: https://www.census.gov/prod/2006pubs/tp-66.pdf

[35] U.S. Department of Commerce, U.S. Census Bureau. National Cancer Institute and Food and Drug Administration co-sponsored Tobacco Use Supplement to the Current Population Survey. 2014-15. Technical Documentation [Internet]. 2016. Available from: https://www.census.gov/programs-surveys/cps/technical-documentation/complete.html

[36] Wolter KM. Introduction to Variance Estimation. New York, USA: Springer; 2007. p. 447

[37] SAS Institute Inc. SAS® 9.4 Product Documentation. Cary, NC, USA: SAS Institute Inc; 2013

[38] Ha T, Soulakova JN. Statistical Analyses of Public Health Surveys Using SAS® Survey Package. In: SESUG Paper 189 [Internet]. 2017. Available from: http://analytics.ncsu.edu/sesug/2017/SESUG2017_Paper-189_Final_PDF.pdf

[39] Cornelius ME, Driezen P, Hyland A, Fong GT, Chaloupka FJ, Cummings KM. Trends in cigarette pricing and purchasing patterns in a sample of US smokers: Findings from the ITC US surveys (2002-2011). Tobacco Control. 2015;**24**(Supplement 3):iii4-iii10. Available from: http://www.ncbi.nlm.nih.gov/pubmed/24917617 [Accessed: 22 February 2018]

[40] Pesko MF, Kruger J, Hyland A. Cigarette price-minimization strategies by U.S. smokers. American Journal of Public Health. 2012;**102**(9):19-21

Chapter 11

Intensive Computational Method Applied for Assessing Specialty Coffees by Trained and Untrained Consumers

Gilberto Rodrigues Liska, Luiz Alberto Beijo, Marcelo Ângelo Cirillo, Flávio Meira Borém and Fortunato Silva de Menezes

Abstract

The sensory analysis of coffees assumes that a sensory panel is formed by tasters trained according to the recommendations of the American Specialty Coffee Association. However, the choice that routinely determines the preference of a coffee is made through experimentation with consumers, in which, for the most part, they have no specific ability in relation to sensory characteristics. Considering that untrained consumers or those with basic knowledge regarding the quality of specialty coffees have little ability to discriminate between different sensory attributes, it is reasonable to admit the highest score given by a taster. Given this fact, probabilistic studies considering appropriate probability distributions are necessary. To access the uncertainty inherent in the notes given by the tasters, resampling methods such as Monte Carlo's can be considered and when there is no knowledge about the distribution of a given statistic, p-Bootstrap confidence intervals become a viable alternative. This text will bring considerations about the use of the non-parametric resampling method by Bootstrap with application in sensory analysis, using probability distributions related to the maximum scores of tasters and accessing the most frequent region (mode) through computational resampling methods.

Keywords: probability, Monte Carlo, bootstrap, GEV distribution, Mantiqueira Serra, height, consumers

1. Introduction

Basically the methodology involved in the analysis of sensory data is summarized in a set of experimental and statistical techniques applied with the purpose of verifying the quality or the degree of acceptance of a given product, without, however, disregarding the characteristics of the individuals, with respect to your sensory skills. In this context, two distinct groups of consumers can be inserted, that is, consumers who have some enhanced sensory ability (s), resulting from product training or knowledge and totally lay consumers.

Faced with this situation, it becomes plausible to admit that a sensory analysis, applied to a group of trained consumers, being able to discriminate small differences between the samples, the results provided by the evaluations will show little variation [1]. Therefore, a sensory experiment carried out with this group shows a greater agreement with the procedures standardized by [2], since the objective assessments would be more homogeneous for the perception of uniformity, sweetness, defects, among others, mentioned by [3, 4].

In an opposite situation, considering a group of untrained consumers, it is more likely that the evaluations will present heterogeneous results, in such a way that the statistical treatment to be given in the analysis of these results may include the atypical observations, classified as outliers arising from the evaluation. Individual to each consumer [5, 6].

It is worth mentioning that the heterogeneity between the observations may be the result of uncontrollable factors, such as, for example, genetics, fatigue, unwillingness to carry out all tests and differences between the abilities of consumers, as well as external causes such as, for example, the geographical origin of a particular product whose qualities or characteristics are due exclusively or essentially to the geographical environment, including natural and chemical factors, which, among others, mention variations in chemical composition due to the genetic variability between cultivars that influence the sensory quality of coffees [7–12].

Given countless causes that are supposed to be the sources that cause outliers in a sensory analysis and reporting the analysis of the quality of coffees, special coffees can be highlighted. Following the definition given by [2], in summary, a coffee is said to be special, as it presents superior quality to its competitors in relation to its origin, absence of defects, processing and/or sensory expressions such as aroma, flavor.

The results of the sensory evaluation are established on a scale ranging from 0 to 10 in which these values represent the increasing levels of coffee quality. According to the analysis protocol [2] the results of the sensory evaluation vary according to a scale where the grades 6, 7, 8, 9 correspond respectively to: good, very good, excellent and exceptional. When the grades are less than 6, the coffees are declared to be of a quality below the Specialty Grade.

Respecting these characteristics, Coffee arabica cultivars are potential coffees worthy of being classified as special [13, 14]. However, studies related to the interference of the environment and geographic origin can influence the quality of the drink. [14], in a study interacting quality with environmental factors, concluded that the coffees with the highest scores in a contest held in the state of Minas Gerais, were produced in colder regions with milder temperatures and annual precipitation index around 1600 mm [15]. In this context, in humid regions it is recommend that processing be performed prioritizing peeled and desmucilated coffees. Thus, the quality of the coffee would be inferred without the interference of defects.

In the case of statistical methodology, it is highlighted that the usual methods of analysis, in general, are sensitive to outlier observations, these being plausible to have arisen in a sensory analysis carried out by untrained consumers [5, 15].

Due to this fact and assuming that the assignment of maximum sensory scores can be understood as random phenomena, in the sense that there are variations in the judgment of different consumers, this work aims to propose the use of some distributions belonging to the generalized extreme value distribution class in sensory analysis. For this purpose, this work analyzes a sensory experiment to evaluate four special coffees produced in the Serra da Mantiqueira Region of Minas Gerais, differentiated in preparation and geographical identification classified by different altitudes.

Bootstrap, developed by Efron in the 70s, can be used in many situations. It is based on a simple, yet powerful idea that the sample represents the population, so analogous characteristics of the sample should give us information about the characteristics of the population. Bootstrap helps to learn about these sample characteristics by taking resamples (samples with replacement of the original sample) and we use this information to infer about the population [16].

In this sense, to detect a difference in the judgment of special coffees by trained and untrained tasters, a test built via non-parametric Bootstrap will be proposed for the mode of distribution of extreme values that best fits the data set.

2. Modeling maximum sensory scores and numerical procedure

In accordance with the opinion of the Ethics and Research Council, registered with the CAAE: 14959413.1.0000.5148, the preparation of the Samples of 100% Arabica coffee was done by removing all defective beans and toast, respecting the maximum period of 24 hours for tasting.

The roasting point was determined visually, using the color classification system by means of standardized discs (SCAA/Agtron Roast Color Classification System). Regarding the preparation of the drink, the concentration of 7% w/v was maintained using filtered water ready for consumption, free of any contaminants and without added sugar. With these specifications, four types of specialty coffees, coded in the samples by A, B, C and D given the description in **Table 1**.

For each type of coffee, the following sensory characteristics were assessed in the acceptance test: aroma, body, hardness, and final score, in four sessions, with the participation of a volunteer group of consumers with basic knowledge in regard to sensory analysis of coffees and another group without basic knowledge. **Table 2** provides a list of the tasters, as well as the sensory characteristics assessed by each taster, in which a_{ij} represents the score given by taster i $(i = 1, 2, \ldots, n_1, n_1 + 1, n_1 + 2, \ldots, n_2)$, such that $n_1 + n_2 = n$, for the sensory characteristic × coffee j $(j = 1, 2, \ldots, 16)$ combination.

In the test, four different types of coffee were evaluated in terms of their sensory characteristics, flavor, acidity, body and note. In different sessions, voluntary consumers were grouped into two classes: (a) people with the habit of consuming coffee, but who do not have basic knowledge about specialty coffees and (b) people with the habit of consuming coffee and trained with information basic information about specialty coffees.

The fit of the probability distributions was carried out, considering the random variable X representing the maximum consumers' sensory scores for the each type of coffee (**Table 1**), totaling in a sample of 696 observations.

Bearing in mind that the highest score provided by a tester will be considered, this being considered as a block, the distribution of the maximums, according to the Fisher-Tippet theorem, is the generalized extreme values distribution (GEV). Its probability density function is defined by:

Type	Genotype	Altitude	Processing
A	Bourbon	Above 1200 m	Natural
B	Acaia	Below 1100 m	Pulped natural
C	Acaia	Below 1100 m	Natural
D	Bourbon	Above 1200 m	Pulped natural

Table 1.
Description of specialty coffees evaluated in the sensory analysis with untrained consumers.

Condition	Taster	Sensory characteristic 1				...	Sensory characteristic 4			
		A	B	C	D	...	A	B	C	D
Trained	1	a_{11}	a_{12}	a_{13}	a_{14}	...	a_{113}	a_{114}	a_{115}	a_{116}
	2	a_{21}	a_{22}	a_{23}	a_{24}	...	a_{213}	a_{214}	a_{215}	a_{216}
	. . .	. . .	. . .	. . .	. . .	...	. . .	. . .	. . .	. . .
	n_1	a_{n11}	a_{n12}	a_{n13}	a_{n14}	...	a_{n113}	a_{n114}	a_{n115}	a_{n116}
Untrained	1	$a(n1+1)1$	$a(n1+1)2$	$a(n1+1)3$	$a(n1+1)4$	...	$a(n1+1)13$	$a(n1+1)14$	$a(n1+1)15$	$a(n1+1)16$
	. . .	. . .	. . .	. . .	. . .	...	. . .	. . .	. . .	. . .
	n_2	a_{n21}	a_{n22}	a_{n23}	a_{n24}	...	a_{n213}	a_{n214}	a_{n215}	a_{n216}

Table 2.
Tabulated representation of the sensory characteristics of the specialty coffees assessed.

$$f(x;\mu,\sigma,\xi,)=\frac{1}{\sigma}\left\{\left[1+\xi\left(\frac{x-\mu}{\sigma}\right)\right]^{\left(-\frac{1}{\xi}\right)-1}\exp\left\{-\left[1+\xi\left(\frac{x-\mu}{\sigma}\right)^{-\left(\frac{1}{\xi}\right)}\right]\right\}\right\}, \quad (1)$$

where $-\infty < x < \mu - \sigma/\xi$ when $\xi < 0$ resulting in the Weibull (μ, σ,ξ). When $\mu - \sigma/\xi < x < \infty$ for $\xi > 0$ results in Fréchet (μ, σ,ξ). When $\lim_{\xi\to 0} f(x;\mu,\sigma,\xi)$ leads to the Gumbel distribution. The parameters μ, σ and ξ are the location, scale and shape parameters.

The probability that a maximum score will be greater than realization of a score, represented by x is defined as

$$P[X > x] = 1 - P[X \leq x] = 1 - F\left(x;\hat{\theta}\right), \quad (2)$$

where $\hat{\theta}$ corresponds to the vector of maximum likelihood estimates [17, 18]. This method requires that the maximum scores are independent and identically distributed [19], which was assessed by the Ljung Box test, explained follow. $F\left(x;\hat{\theta}\right)$ is the cumulative distribution function (cdf) of GEV probability density function. Its cdf is given by

$$F(x;\mu,\sigma,\xi) = \exp\left[-\left(1+\xi\left(\frac{x-\mu}{\sigma}\right)\right)^{\frac{-1}{\xi}}\right]. \quad (3)$$

The mode (Mo) of the pdf in Eq. (1) is given by

$$Mo(X) = \mu + \sigma\frac{(1+\xi)^{-\xi}-1}{\xi}, \quad (4)$$

and in the $\lim_{\xi\to 0} f(x;\mu,\sigma,\xi)$ case, the mode is simplified to $Mo(X) = \mu$.

The goodness of fit for each distribution was validated using the Kolmogorov Smirnov (KS) adherence test in conjunction with the Q-Q plots [20, 21]. The Q-Q plot consists of the points

$$\left\{\left(F^{-1}\left(p_i\right), x_i\right), i = 1, \ldots, n\right\}, \quad (5)$$

where $F^{-1}(p_i)$ is the inverse function of the cumulative distribution function of a given probability distribution, p_i are the percentiles and x_i are the data used to fit the model, ordered in ascending and n the sample size.

According to [22], the Kolmogorov–Smirnov (KS) test is used to assess the fit of a probability distribution to the original data. It is based on the analysis of the proximity or adjustment between the sample distribution function $\hat{F}(x_i)$ and the population distribution function under the null hypothesis,$F_0(x_i)$. The test statistic (D) is given by,

$$D = \max\left|F_0(x_i) - \hat{F}(x_i)\right| \tag{6}$$

In the KS test, the hypothesis of interest are given by H_0: The distribution function from which the sample is derived follows the distribution function that is assumed to be known; that is, $F(x) = F_0(x_i)$ and H_1: $F(x) \neq F_0(x_i)$ [23]. Then, the value (Eq. (6)) must be compared with the critical value (using tables), for the significance level of the test. According on the result, the null hypothesis is rejected or not. The null hypothesis is also rejected if the p-value is lower than the significance level adopted.

Regarding the verification of the assumption of independence of the observations, such that is required by the maximum likelihood method for estimating parameters, the Ljung-Box (LB) test was used. According to [24], it is a statistical test used to find out if there are non-zero autocorrelation groups. To do this, it tests total randomness based on the number of deviations. The test hypotheses are H_0: all autocorrelation coefficients are equal to zero and H_1: not all autocorrelation coefficients are equal to zero. The test statistic is

$$Q = n(n+2)\sum_{k=1}^{s}\frac{r_j^2}{(n-j)}, \tag{7}$$

where n is the number of observations, s is the number of coefficients in testing autocorrelation, r_j is the autocorrelation coefficient (for the deviation) and Q the test statistic. If the sample values of Eq. (7) exceed the critical value of a Chi-Squared distribution with s degrees of freedom, then at least one deviation r is statistically different from zero at the specified significance level, that is, H_0 is rejected. H_0 is also rejected if p-value is lower than the adopted significance level. It should be noted that if H_0 is rejected, it can be said that the data are independent. In both tests, the significance level of 1% was adopted [25].

In order to make an inference about the most frequent score among the tasters, it is necessary to know the sample distribution of the quantity in Eq. (4). For that, an alternative would be to use resampling methods, which one of them will be presented below.

The Bootstrap resampling process consists of resampling B samples $P^{*(1)}, P^{*(2)}, \ldots, P^{*(B)}$, with replacement, independent and identically distributed of the n highest marks awarded by trained and untrained tasters. Estimates of the parameter of interest can be obtained, denoted by $\hat{\theta}^*_{(i)}$, for each sample, which is $\theta = Mo(X)$. With that we will obtain the vector $\hat{\boldsymbol{\theta}}^* = \left(\hat{\theta}^*_{(1)}, \hat{\theta}^*_{(2)}, \ldots, \hat{\theta}^*_{(B)}\right)$ and from the vector $\hat{\boldsymbol{\theta}}^*$ it is possible to obtain the Bootstrap distribution of the $\hat{\theta}$ estimator.

Once the empirical distribution of the $\hat{\theta}$ estimator is obtained, confidence intervals for θ can be estimated. The Bootstrap confidence interval based on the Bootstrap distribution percentiles of θ, described in [16, 26], is known as the p-Bootstrap confidence interval. In a more formal way, the confidence interval can be constructed by following the following steps:

(Step 1) Draw, with replacement, of P, one Bootstrap sample P^*;
(Step 2) From Bootstrap sample P^*, obtain $\hat{\theta} = Mo(X)$;
(Step 3) Repeat the steps 1 and 2 B times;
(Step 4) From the vector $\hat{\boldsymbol{\theta}}^* = \left(\hat{\theta}^*_{(1)} \leq \hat{\theta}^*_{(2)} \leq \ldots \leq \hat{\theta}^*_{(B)}\right)$, for α significance level $(0 < \alpha < 1)$, the p-Bootstrap confidence interval with $100 \times (1-\alpha)\%$ level of confidence is given by $IC_{(1-\alpha)}(\theta) : \left[\hat{\theta}^*_{(k_1)}; \hat{\theta}^*_{(k_2)}\right]$, where $k_1 = (B+1)(\alpha/2)$ and $k_2 = (B+1)(1-\alpha/2)$ are the highest integers that are not greater than$(B+1)(\alpha/2)$ and $(B+1)(1-\alpha/2)$, respectively; and $\hat{\theta}^*_{(k_1)}$ is the $100(\alpha/2)\%$-percentile of the Bootstrap empirical distribution; and $\hat{\theta}^*_{(k_2)}$ is the $100(1-\alpha/2)\%$-percentile of the Bootstrap empirical distribution [16, 26].

Finishing the proposed methodology, the computational resources available in the R software [27, 28] were used through the *boot* and *evd* [29] packages to fitting the probability distributions for sensory scores, hypothesis tests and construction of Bootstrap confidence intervals.

3. Experimental results

The following results correspond to the parameter estimates for the probability distributions fitted for the two classes of tasters, as well as the p-values referring to the validation of the probabilistic model fitted for the sensory scores.

With these specifications, given a level of significance of 1%, it is noted the confirmation of the fit in the sensory scores for each coffee, therefore, there is statistical evidence to assume that GEV distribution is adequate to model the maximum sensory grades of the evaluated coffees (**Table 3**). It should be noted that the fact that we have p-values greater than 1% for the KS test indicates that there is statistical evidence for the acceptance of the test's null hypothesis, as can be seen in Section 2. The test used, however, according to [30], should only be used for completely specified distributions, that is, when there are no unknown parameters that need to be estimated from the sample. Otherwise, the test is very conservative. One solution would be to obtain, via simulation, the theoretical quantiles of the Kolmogorov Smirnov test to compare them with the quantiles obtained from the sample. A similar procedure for the Gumbel distribution was carried out by [31].

Coffee	**Group**	**Parameter estimates**			**KS (p-value)**	**LB (p-value)**
		$\hat{\mu}$	$\hat{\sigma}$	$\hat{\xi}$		
A	Untrained	5.9471	2.4105	−0.6569	0.9077	0.0803
	Trained	6.9345	1.6259	−0.6156	0.8908	0.2359
B	Untrained	5.9326	2.4624	−0.5721	0.9466	0.3306
	Trained	6.6031	1.9455	−0.5736	0.8255	0.0110
C	Untrained	6.4290	2.2108	−0.6348	0.9485	0.6084
	Trained	7.0595	1.3382	−0.5485	0.9998	0.9823
D	Untrained	7.8676	2.0437	−0.9582	0.9962	0.9625
	Trained	7.8113	1.8183	−0.8221	0.6543	0.6924

Table 3.
Parameters estimates and results of the Kolmogorov–Smirnov and Ljung-box tests for the maximum scores given by consumers in the sensory evaluation of the special coffees named in A, B, C and D.

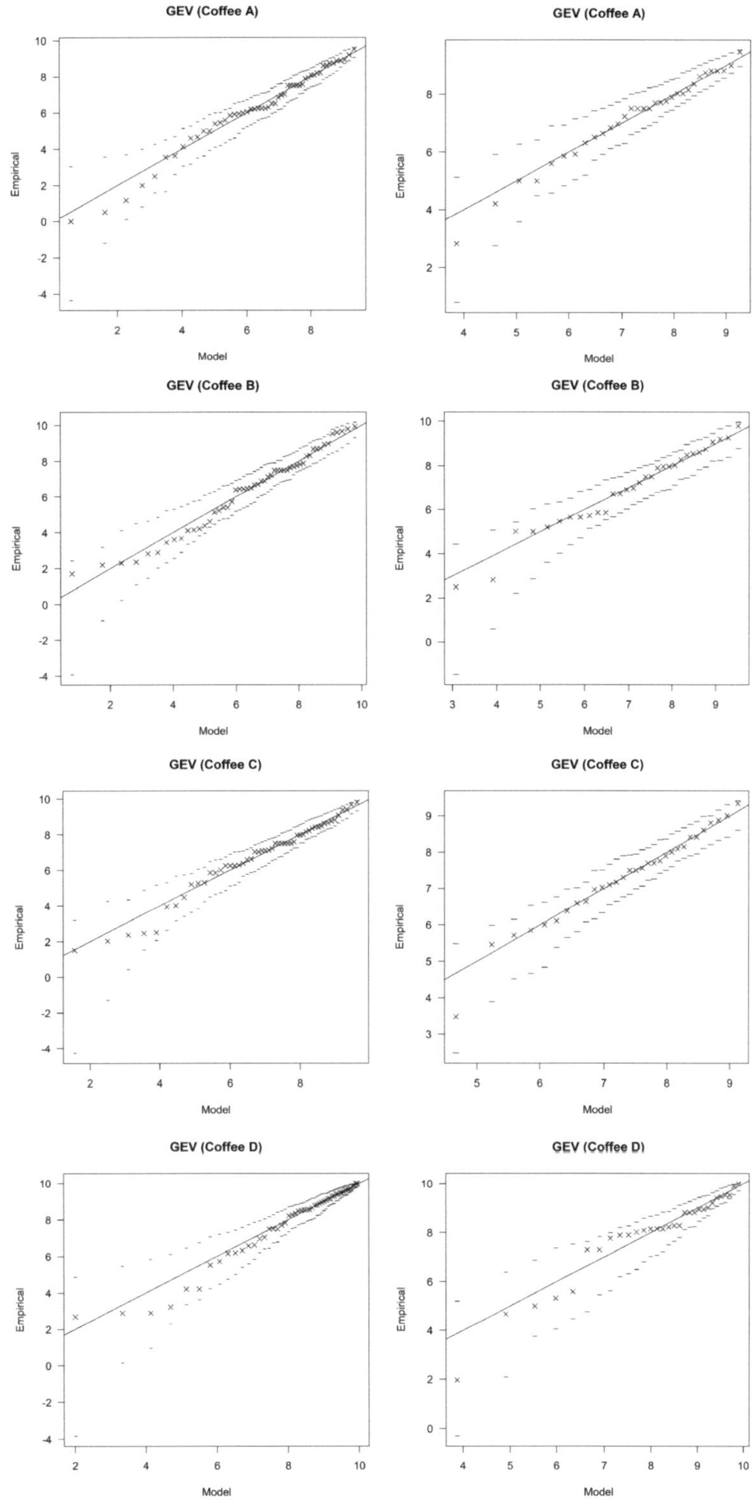

Figure 1.
Q-Q plot referring to the fitted GEV distributions for the maximum sensory scores obtained in the evaluation of each special coffee for the group of untrained (left) and trained (right) tasters.

Alternatively, inspection of fit quality can be assessed via Q-Q plots graphs. They are shown in **Figure 1**.

In this sense, the validation of the GEV distribution is corroborated in the Q-Q plots shown in **Figure 1**, because for all the specialty coffees evaluated, the theoretical quantiles showed a linear behavior and close to the straight identity with the observed quantiles and the points being, in their mostly, contained in the 95% confidence interval. It should also be noted that the quantiles have a trend to converge to a region located as an upper tail. All p-values of the Ljung Box test are higher than 1%, thus showing the acceptance of the null hypothesis of the test, as described in Section 2. It can be concluded, therefore, that the maximum scores given by trained and untrained tasters they are independent. We should highlight that we have used these tests to verify the assumptions of the Extreme Value Theory models, but that they could be used for other interests, such as in the trend analysis of hydro-climatic series [32–34]. Failure to observe these assumptions can lead to fitted models parameter estimates, as well return levels estimates, biased and/or under/overestimated. For these situations, Bayesian methods, regression or time series based on the Box-Jenkins methodology could be considered [35, 36].

In function of the confirmatory results related to the GEV distribution goodness of fit, given the estimates of the parameters for this distribution applied in the maximum sensory scores given by consumers in the evaluations carried out for each coffee, we proceeded with the calculations of the probabilities for an individual to supply a grade higher than a given grade. The results are described in **Table 4**.

Before that, the distribution modes were calculated as shown in **Table 4**, in order to verify the similarity between the grades provided by trained and untrained tasters. It is observed that occasionally they can be considered very close. For specialty coffees A and B, trained tasters provided higher grades more frequently than untrained tasters and for specialty coffees C and D the opposite occurred.

Although the similarity between the modes of the grades attributed by the tasters is evident, this similarity is not associated with any level of confidence, since the similarity is only punctual. To circumvent this situation, confidence intervals were constructed using the non-parametric Bootstrap method, as shown in Section 2. Thus, it can be stated with 95% confidence that the grades most frequently attributed to coffee A by trained tasters and not trained do not differ statistically, since the point estimate for the fashion of the notes is contained in the respective confidence intervals and they are overlapping.

Coffee	Group	Mode	$q_{0.025}$	$q_{0.975}$	$P[X > q_{0.025}]$ (%)	$P[X > q_{0.975}]$ (%)	Difference (%)
A	Untrained	7.8	6.9	8.9	47.2	7.7	39.5
	Trained	8.1	7.3	9.0	54.1	8.2	45.8
B	Untrained	7.6	6.2	9.2	59.2	8.2	51.0
	Trained	7.9	6.6	9.2	62.4	7.1	55.4
C	Untrained	8.1	7.2	9.0	47.6	9.5	38.1
	Trained	7.9	7.1	8.9	62.2	8.1	54.1
D	Untrained	9.9	9.1	10.0	32.8	0.2	32.6
	Trained	9.5	8.6	10.0	44.5	0.7	43.9

The probabilities P[X > q0.025], P[X > q0.975] and Difference are given in percentages.

Table 4.
Maximum scores modes given by consumers in the sensory panel of specialty coffees named in A, B, C and D and their respective 95% confidence intervals ($q_{0.025}$ and $q_{0.975}$).

More specifically speaking, for coffee A, the initial mode estimate is 7.8 points for untrained tasters, i.e., $\hat{\theta}_0 = Mo(X)$ is 7.8 points, that is contained in the 95% confidence interval for Bootstrap mode ($\hat{\theta}^*_{(2.5\%)} = 6.9$ points and $\hat{\theta}^*_{(97.5\%)} = 8.9$ points). Likewise for trained tasters, the initial estimate for mode is 8.1 points for trained tasters, that is, $\hat{\theta}_0 = Mo(X)$, is 8.1 points, that is contained in the 95% confidence interval for Bootstrap mode ($\hat{\theta}^*_{(2.5\%)} = 7.3$ points and $\hat{\theta}^*_{(97.5\%)} = 9.0$ points), indicating, therefore, that the scores attributed to coffee A by trained and untrained tasters are similar with 95% confidence.

According to the results described in **Table 4**, it is clear that given a sensory panel made up of untrained consumers, there is a probability that all consumers will have a sensory score higher than 6.0, indicating that whatever the taster is, among the types of specialty coffees studied, no coffee will be classified with quality below the Specialty Grade, since all the coffees analyzed showed a high probability that the most frequent grade is higher than 6. On a 9-point verbal hedonic scale, it can be concluded that, in general, consumers have a trend to be indifferent to the agradability of specialty coffees.

Figure 2 presents the histogram and the Q-Q plot for the mode of the fitted distribution for the grades given by the untrained tasters for coffee A in **Table 1**. The histogram suggests that the empirical distribution of $\theta = Mo(X)$ is a normal and this fact is corroborated by the Q-Q plot, since the one-to-one proportionality is maintained considering the quantiles of the standard normal versus the observed quantiles. Similar results were observed for all other specialty coffees, however these results will not be shown.

When considering an expressive score worthy of international competitions, having a reference higher than 8, the probability of a consumer providing an occurrence of a note being higher than 8 or the coffee being classified as excellent is relatively low for all evaluated coffees (**Table 4**). It is also noted that the probability of a consumer assigning a grade between 9.1 and 10.0 is 32.8%, that is, it can be interpreted that coffee D to be considered exceptional by a consumer is 32.8%. In addition, coffee D is the one with the least amplitude in probability, corresponding to the column "Difference" in **Table 4**, which indicates that it is a type of coffee that provided low variability between the grades attributed by the tasters. On the other hand, coffee B showed greater variability between the grades attributed by

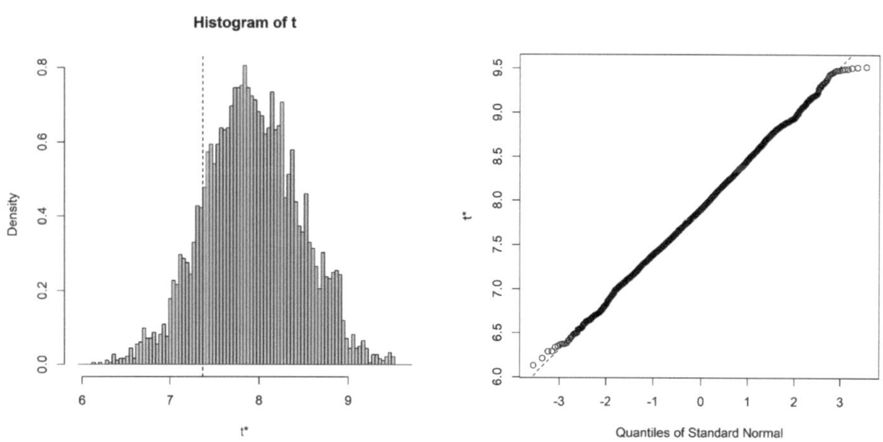

Figure 2.
Histogram for the 5000 values of the bootstrap modes for the scores of untrained tasters and the respective normal Q-Q plot.

the tasters, since the difference $P[X > q_{0.025}] - P[X > q_{0.975}]$ is the largest among the analyzed coffees.

Therefore, in the evaluation of the four specialty coffees, given the low probabilities, it can be said that a sensory experiment carried out with the objective of discriminating the specialty coffees, is done with consumers who present more improved training.

Figure 3 shows graphically the agreement between the scores given by trained (blue hatched) and untrained (black hatched) tasters, according to the results shown in **Table 3**.

The importance of using bootstrap procedures in the analysis of responses that corroborate with these scores is relevant for statistically validating the scores obtained in international competitions, since it assumes that subjective and / or unknown factors, related to the different sensory perceptions of the tasters may suggest violations in the sample distribution, and as a consequence, the estimates of the probabilistic model are distorted. Thus, through successive resampling, an empirical distribution for each parameter is generated in connection with the assumed probabilistic model, and inferences will be made with better precision and accuracy. The amplitude of the confidence interval in **Figure 3** reflects the precision

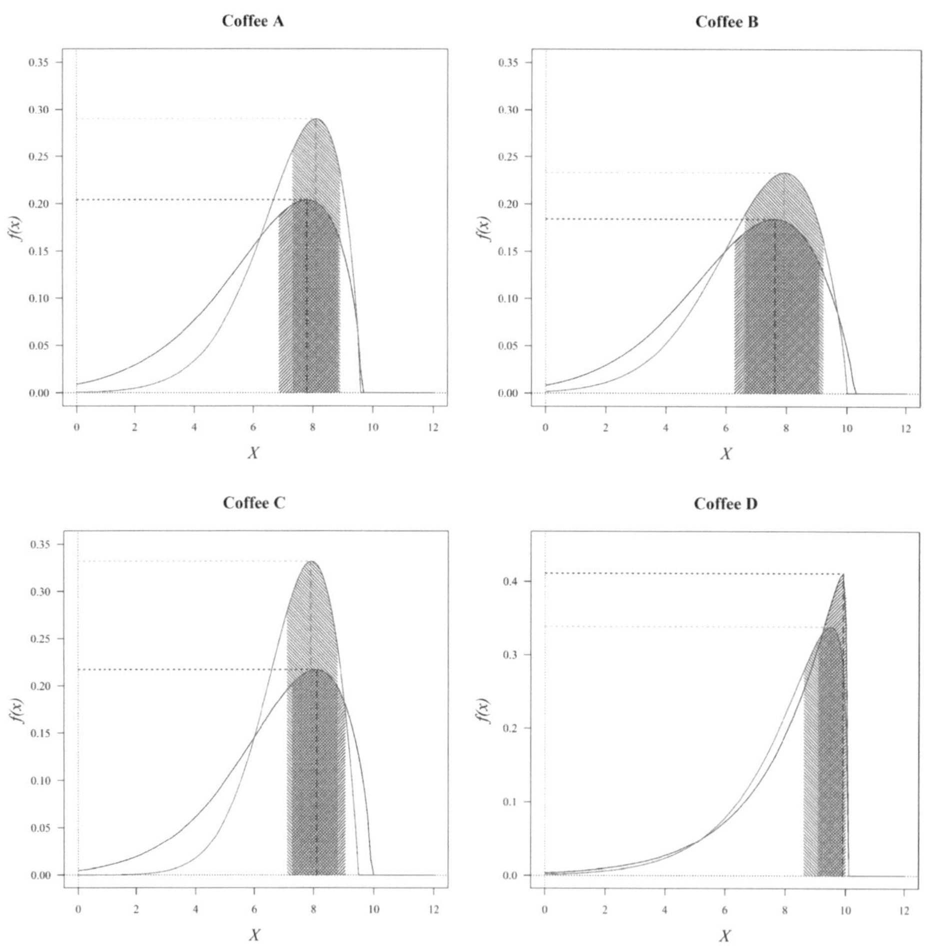

Figure 3.
Graphical representation of the probability distributions adjusted for the notes attributed by untrained (black) and trained (blue) tasters and the respective 95% confidence intervals for the maximum scores modes attributed to each special coffee.

of the estimates for the maximum notes mode, given the GEV distribution. Other confidence intervals via bootstrap could be considered, such as bootstrap-*t* and BCa [37, 38]. We emphasize that the strategy adopted is innovative in the context of sensory notes and the comparison of confidence intervals can be done as future work.

4. Conclusions and final remarks

The GEV distribution can be applied to the sensory analysis of specialty coffees, whose sensorial panel presents an heterogeneity among consumers.

The probabilities obtained by this distribution show that the sensory analysis of specialty coffees performed by untrained consumers indicates that they are able to differentiate specialty coffees and provide similar scores to the sensory analysis performed by consumers with prior training.

The proposed inference made it possible to attribute some degree of uncertainty regarding the occurrence of sensory scores in the different types of specialty coffees studied and to indicate which group each coffee belongs to with high probability according to the Specialty Grade.

It can be recommended that more intensive training with tasters or the application of the proposed methodology with tasters with international certification should be considered with a view to assessing specialty coffees against a reference score of 9 points, since for the present study, only coffee D has a high probability of presenting this note. It should be noted that according to the analysis protocol provided by Specialty Coffee Association of America, the results of the sensory evaluation vary according to a scale where the grades upper to 9 correspond to exceptional coffee.

The study has some limitations that provide directions for future research, although the GEV distribution is specific for analyzing maximum values, the data generating mechanism truncates the maximum score at 10. This characteristic could be taken into account, fitting the model to truncated data. Some proposals have appeared in the literature to consider truncation in the estimation process by maximum likelihood, but there is no consolidated methodology yet. Therefore, it is a possibility for further studies that may be the subject of future research.

Acknowledgements

The authors are grateful to the National Council for Scientific and Technological Development (CNPq—Conselho Nacional de Desenvolvimento Científico e Tecnológico), the Minas Gerais State Research Support Foundation (FAPEMIG—Fundação de Amparo para Pesquisa do Estado de Minas Gerais), the Coordination for the Improvement of Higher Education Personnel (CAPES—Coordenação de Aperfeiçoamento de Pessoal de Nível Superior), and the National Coffee Science and Technology Institute (INCT/Café—Instituto Nacional de Ciência e Tecnologia do Café).

Conflict of interest

The authors declare no conflict of interest.

Author details

Gilberto Rodrigues Liska[1*], Luiz Alberto Beijo[2], Marcelo Ângelo Cirillo[3], Flávio Meira Borém[4] and Fortunato Silva de Menezes[5]

1 Department of Agroindustrial Technology and Rural Socioeconomics, Federal University of São Carlos, Araras, São Paulo State, Brazil

2 Department of Statistics, Federal University of Alfenas, Alfenas, Minas Gerais State, Brazil

3 Department of Statistics, Federal University of Lavras, Lavras, Minas Gerais State, Brazil

4 Department of Agricultural Engineering, Federal University of Lavras, Lavras, Minas Gerais State, Brazil

5 Department of Physics, Federal University of Lavras, Lavras, Minas Gerais State, Brazil

*Address all correspondence to: gilbertoliska@ufscar.br

References

[1] de Oliveira Fassio L, Malta M, Carvalho G, Liska G, de Lima P, Pimenta C. Sensory Description of Cultivars (*Coffea Arabica* L.) Resistant to Rust and Its Correlation with Caffeine, Trigonelline, and Chlorogenic Acid Compounds. Beverages. 2016 Jan 18;2(1):1.

[2] Specialty Coffee Association of America. SCAA Protocols - Cupping Specialty Coffee [Internet]. 2015. Available from: www.scaa.org

[3] Alves HMRA, Volpato MML, Vieira TGC, Borém FM, Barbosa JN. Características ambientais e qualidade da bebida dos cafés do estado de Minas Gerais. Inf Agropecuário. 2011;32(261): 1–12.

[4] Lingle T. The coffee cupper's handbook : systematic guide to the sensory evaluation of coffee's flavor. Fourth edi. Long Beach California: Specialty Coffee Association of America; 2011.

[5] Liska GR, De Menezes FS, Cirillo MA, Borém FM, Cortez RM, Ribeiro DE. Evaluation of sensory panels of consumers of specialty coffee beverages using the boosting method in discriminant analysis. Semin Agrar. 2015;36(6).

[6] Ferreira HA, Liska GR, Cirillo MA, Borém FM, Ribeiro DE, Cortez RM, et al. Selecting A Probabilistic Model Applied to the Sensory Analysis of Specialty Coffees Performed with Consumer. IEEE Lat Am Trans. 2016;14(3).

[7] Malta MR, Chagas SJ de R. Avaliação de compostos não-voláteis em diferentes cultivares de cafeeiro produzidas na região sul de Minas Gerais. Acta Sci - Agron [Internet]. 2009 [cited 2020 Oct 23];31(1):57–61. Available from: http://www.scielo.br/scielo.php?script=sci_arttext&pid=S1807-86212009000100010&lng=en&nrm=iso&tlng=pt

[8] Chagas E do N, Morais AR de, Cirillo MA, Figueiredo LP, Borém FM. Selection of robust estimator usedin analysis of sensory characteristics and identification of environments conducive to specialty coffee production. Adv Crop Sci [Internet]. 2013 [cited 2020 Oct 23];3(8):515–24. Available from: http://repositorio.ufla.br/jspui/handle/1/13024

[9] Silva FLF, Nascimento GO, Lopes GS, Matos WO, Cunha RL, Malta MR, et al. The concentration of polyphenolic compounds and trace elements in the *Coffea arabica* leaves: Potential chemometric pattern recognition of coffee leaf rust resistance. Food Res Int [Internet]. 2020 Aug;134: 109221. Available from: https://linkinghub.elsevier.com/retrieve/pii/S0963996920302465

[10] Malta MR, Fassio L de O, Liska GR, Carvalho GR, Pereira AA, Botelho CE, et al. Discrimination of genotypes coffee by chemical composition of the beans: Potential markers in natural coffees. Food Res Int [Internet]. 2020 Aug;134: 109219. Available from: https://linkinghub.elsevier.com/retrieve/pii/S0963996920302441

[11] Fassio L de O, Malta MR, Liska GR, Carvalho GR, Botelho CE, Pereira AA, et al. Performance of arabica coffee accessions from the active germplasm bank of Minas Gerais, Brazil as a function of dry and wet processing: a sensory approach. Aust J Crop Sci [Internet]. 2020 Jun 20;14(6):1011–8. Available from: https://www.cropj.com/fassio_14_6_2020_1011_1018.pdf

[12] Fassio LO, Malta MR, Carvalho GR, Pereira AA, Silva AD, Liska GR, et al. Discrimination of Genealogical Groups of Arabica Coffee by the Chemical

Composition of the Beans. J Agric Sci. 2019 Sep 30;11(16):141.

[13] Figueiredo LP, Borém FM, Cirillo MÂ, Ribeiro FC, Giomo GS, Salva TDJG. The Potential for High Quality Bourbon Coffees From Different Environments. J Agric Sci [Internet]. 2013 Sep 15 [cited 2020 Oct 23];5(10):87–98. Available from: http://www.ccsenet.org/journal/index.php/jas/article/view/27842

[14] Barbosa JN, Borem FM, Cirillo MA, Malta MR, Alvarenga AA, Alves HMR. Coffee Quality and Its Interactions with Environmental Factors in Minas Gerais, Brazil. J Agric Sci [Internet]. 2012 Mar 31 [cited 2020 Oct 23];4(5):181–90. Available from: http://www.ccsenet.org/journal/index.php/jas/article/view/13784

[15] Borém FM, Cirillo M, de Carvalho Alves AP, dos Santos CM, Liska GR, Ramos MF, et al. Coffee sensory quality study based on spatial distribution in the Mantiqueira mountain region of Brazil. J Sens Stud. 2020 Apr 1;35(2).

[16] Efron B, Tibshirani RJ. An Introduction to the Bootstrap [Internet]. CRC Press; 1994 [cited 2020 Oct 23]. 456 p. Available from: https://books.google.com.br/books/about/An_Introduction_to_the_Bootstrap.html?id=gLlpIUxRntoC&redir_esc=y

[17] Casella G, Berger RR. Statistical Inference. 2nd ed. Thomson Learning; 2001. 688 p.

[18] Mendes BV de M. Introdução à análise de eventos extremos. Rio de Janeiro: E-papers Serviços Editoriais Ltda; 2004. 232 p.

[19] Coles S. An Introduction to Statistical Modeling of Extreme Values [Internet]. London: Springer London; 2001. 221 p. (Springer Series in Statistics). Available from: http://link.springer.com/10.1007/978-1-4471-3675-0

[20] Blain GC. Dry months in the agricultural region of Ribeirão Preto, state of São Paulo-Brazil: an study based on the extreme value theory. Eng Agrícola [Internet]. 2014 Oct;34(5): 992–1000. Available from: http://www.scielo.br/scielo.php?script=sci_arttext&pid=S0100-69162014000500018&lng=en&tlng=en

[21] Moral RA, Hinde J, Demétrio CGB. Half-Normal Plots and Overdispersed Models in R : The hnp Package. J Stat Softw [Internet]. 2017;81(10). Available from: https://www.jstatsoft.org/v081/i10

[22] Hartmann M, Moala FA, Mendonça MA. Estudo das precipitações máximas anuais em Presidente Prudente. Rev Bras Meteorol [Internet]. 2011 Dec;26 (4):561–8. Available from: http://www.scielo.br/scielo.php?script=sci_arttext&pid=S0102-77862011000400006&lng=pt&tlng=pt

[23] Ferreira RV de C, Liska GR. Análise probabilística da temperatura máxima em Uruguaiana, RS. Rev Bras Agric Irrig. 2019 Jul 25;13(3):3390–401.

[24] Ljung GM, Box GEP. On a Measure of Lack of Fit in Time Series Models. Biometrika [Internet]. 1978 Aug;65(2): 297. Available from: https://www.jstor.org/stable/2335207?origin=crossref

[25] Martins ALA, Liska GR, Beijo LA, Menezes FS de, Cirillo MÂ. Generalized Pareto distribution applied to the analysis of maximum rainfall events in Uruguaiana, RS, Brazil. SN Appl Sci [Internet]. 2020 Sep 5;2(9):1479. Available from: http://link.springer.com/10.1007/s42452-020-03199-8

[26] Rizzo ML. Statistical Computing with R [Internet]. Chapman and Hall/CRC; 2007. 416 p. Available from: https://www.crcpress.com/Statistical-Computing-with-R/Rizzo/p/book/9781584885450

[27] R Core Team. R: A language and environment for statistical computing. Vienna, Austria; 2018.

[28] RStudio Team. RStudio: Integrated Development for R. [Internet]. Boston: RStudio, Inc; 2015. Available from: http://www.rstudio.com/

[29] Stephenson AG. evd: Extreme Value Distributions. R News [Internet]. 2002;2 (2):31–2. Available from: https://cran.r-project.org/doc/Rnews/Rnews_2002-2.pdf

[30] Crutcher HL. A Note on the Possible Misuse of the Kolmogorov-Smirnov Test. J Appl Meteorol [Internet]. 1975 Dec 1 [cited 2020 Oct 23];14(8):1600–3. Available from: http://journals.ametsoc.org/jamc/article-pdf/14/8/1600/4968526/1520-0450

[31] Bautista EAL, Zocchi SS, Angelocci LR. Fitting the generalized extreme value distribution (GEV) to the maximum wind speed data in Piracicaba, São Paulo, Brazil. Rev Matemática e Estatística. 2004;22(1):95–111.

[32] Tan ML, Samat N, Chan NW, Lee AJ, Li C. Analysis of Precipitation and Temperature Extremes over the Muda River Basin, Malaysia. Water [Internet]. 2019 Feb 6;11(2):283. Available from: http://www.mdpi.com/2073-4441/11/2/283

[33] Sá EAS, Moura CN de, Padilha VL, Campos CGC. Trends in daily precipitation in highlands region of Santa Catarina, southern Brazil. Ambient e Agua - An Interdiscip J Appl Sci [Internet]. 2018 Feb 16;13(1):1–13. Available from: http://www.scielo.br/scielo.php?script=sci_arttext&pid=S1980-993X2018000100312&lng=en&nrm=iso&tlng=en

[34] Salviano MF, Groppo JD, Pellegrino GQ. Análise de Tendências em Dados de Precipitação e Temperatura no Brasil. Rev Bras Meteorol [Internet]. 2016 Mar;31(1):64–73. Available from: http://www.scielo.br/scielo.php?script=sci_arttext&pid=S0102-77862016000100064&lng=pt&tlng=pt

[35] Aguirre AFL, Nogueira DA, Beijo LA. Análise da temperatura máxima de Piracicaba (SP) via distribuição GEV não estacionária: uma abordagem bayesiana. Rev Bras Climatol [Internet]. 2020 Sep 21 [cited 2020 Nov 25];27:496–517. Available from: http://dx.doi.org/10.5380/abclima.v27i0.73763

[36] Liska GR, Sáfadi T, Bortolini J, Beijo LA. Estimativas de velocidade máxima de vento em Piracicaba-SP via Séries Temporais e Teoria de Valores Extremos. Rev Bras Biometria. 2013;31 (2):295–309.

[37] DiCiccio TJ, Efron B. Bootstrap confidence intervals. Stat Sci [Internet]. 1996 [cited 2020 Nov 12];11(3):189–212. Available from: https://projecteuclid.org/euclid.ss/1032280214

[38] Jung K, Lee J, Gupta V, Cho G. Comparison of Bootstrap Confidence Interval Methods for GSCA Using a Monte Carlo Simulation. Front Psychol. 2019 Oct 11;10:1–11.

Section 5

Predictability

Chapter 12

The Periodic Restricted EXPAR(1) Model

Mouna Merzougui

Abstract

In this chapter, we discuss the nonlinear periodic restricted EXPAR(1) model. The parameters are estimated by the quasi maximum likelihood (QML) method and we give their asymptotic properties which lead to the construction of confidence intervals of the parameters. Then we consider the problem of testing the nullity of coefficients by using the standard Likelihood Ratio (LR) test, simulation studies are given to assess the performance of this QML and LR test.

Keywords: nonlinear time series, periodic restricted exponential autoregressive model, quasi maximum likelihood estimation, confidence interval, LR test

1. Introduction

Since the 1920*s*, linear models with Gaussian noise have occupied a prominent place, they have played an important role in the specification, prevision and general analysis of time series and many specific problems were solved by them. Nevertheless, many physical and natural processes exhibit nonlinear characteristics that are not taken into account with linear representation and are better explicated and fitted by nonlinear models. For example, ecological and environmental fields present phenomena close to the theory of nonlinear oscillations, such as limit cycle behavior remarked in the famous lynx or sunspot series, leading to the consideration of more complex models from the 1980*s* onwards. A first nonlinear model possible is the Volterra series which plays the same role as the Wold representation, for linear series. The interest of this representation is rather theoretical than practical, for this reason, specific parametric nonlinear models were presented as the *ARCH* and Bilinear models suitable for financial and economic data, threshold AutoRegressif (*TAR*) and exponential *AR* (*EXPAR*) models suitable for ecological and meteorological data. These nonlinear models have been applied with great success in many important real-life problems. Basics of nonlinear time series analysis can be found in [1–3] and references therein.

Amplitude dependent frequency, jump phenomena and limit cycle behavior are familiar features of nonlinear vibration theory and to reproduce them [4, 5] introduced the exponential autoregressive (*EXPAR*) models. The start was by taking an autoregressive (*AR*) model Y_t, say, and then make the coefficients dependent in an exponential way of Y_{t-1}^2.

Several papers treated the probabilistic and statistic aspects of *EXPAR* models. A direct method of estimation is proposed by [5], it consists to fix the nonlinear coefficient in the exponential term at one of a grid of values and then estimate the other parameters by linear least squares and use the AIC criterion to select the final parameters, necessary and sufficient conditions of stationarity and geometric ergodicity for the *EXPAR*(1) model are given by [6], the problem of estimation of nonlinear time series in a general framework by conditional least squares CLS and maximum likelihood ML methods is treated by [7] with application in *EXPAR* models, a forecasting method is proposed by [8], the *LAN* property was shown in [9] and asymptotically efficient estimates was constructed there for the restricted *EXPAR*(1), a genetic algorithm for estimation is used in [10], Bayesian analysis of these models is introduced in [11], a parametric and nonparamtric test for the detection of exponential component in *AR*(1) is constructed by [12], sup-tests are constructed by [13] with the trilogy Likelihood Ratio (LR), Wald and Lagrange Multiplier (LM) for linearity in a general nonlinear *AR*(1) model with *EXPAR*(1) as special cases, the extended Kalman filter (*EKF*) is used in [14]. Given that nonlinear estimation is time consuming [15] proposed to estimate heuristically the nonlinear parameter from the data and this is a very interesting remark because when the nonlinear parameter is known we get the Restricted *EXPAR* model. The applications of the *EXPAR* model are multiple: ecology, hydrology, speech signal, macroeconomic and others see, for example, [16–21].

On the other hand, fitted seasonal time series exhibiting nonlinear behavior such cited before and having a periodic autocovariance structure by *SARIMA* models will be inadequate. These models are linear and the seasonally adjusted data may still show seasonal variations because the structure of the correlations depends on the season. The solution is the use of a periodic version of *EXPAR* models. The notion of periodicity, introduced by [22], was used to fit hydrological and financial series and allowed the emergence of new classes of time series models such as Periodic *GARCH*, Periodic Bilinear, *MPAR* model. Motivated by all this, we introduced recently the Periodic restricted *EXPAR*(1) model see [23], which consists of having different restricted *EXPAR*(1) for each cycle and we established a most stringent test of periodicity since a periodic model is more complicated than a nonperiodic one and its consideration must be justified. We studied the problem of estimation by the least squares (*LS*) method in [24] and the test of Student was used for testing the nullity of the coefficients in the application. Traditionally, the step of estimation must be followed by tests of nullity of coefficients and the major tests used are Wald, LR and LM tests. We used a Wald test for testing the nullity of one coefficient and consequently testing linearity in [25].

In this chapter, we will present the quasi maximum likelihood (*QML*) estimation of the parameters, which are the *LS* estimators in [24] under the assumption that the density is Gaussian, these estimators are asymptotically normal under quite general conditions. This will play a role in the construction of the confidence interval for the parameters and then we treat the problem of testing the nullity of parameters which lead us to a linearity test using the standard and well known LR test. This test is based on the comparison between the maximum of the constrained and unconstrained quasi log likelihood, see for example [26] or [27], the null hypothesis is accepted, if the difference is small enough or equivalently H_0 ought to be rejected for large values of the difference. The problem is standard because the periodic model is restricted, i.e. the nonlinear parameter is known and for the other parameters 0 is an interior point of the parameter space, then the LR statistic asymptotically follows the χ^2 distribution under H_0 just like the Wald test, but we chose the former because it does not require estimation of the information matrix. It is known that the two tests are asymptotically equivalent and may be identical see [26] for more details.

The chapter is organized as follows. In Section 2, we introduce the Restricted *PEXPAR* model and we present the asymptotic normality of the QML estimators and we construct confidence intervals of the parameters. Section 3 provides the LR test for nullity of one coefficient and a test for linearity, a small simulation shows the efficiency of these tests.

2. The Periodic Restricted $EXPAR(1)$ model and QML estimation

2.1 Restricted $PEXPAR(1)$ model

Let $\{Y_t\}_{t\geq 1}$ be a seasonal stochastic process with period S ($S\geq 2$).

Definition 1

The process $\{Y_t\}_{t\geq 1}$ is a Periodic Restricted EXPonential AutoRegressive model (restricted $PEXPAR(1)$) of order 1 if it is a solution of the nonlinear difference equation given by

$$Y_t = \left(\varphi_{t,1} + \varphi_{t,2}\exp\left(-\gamma Y_{t-1}^2\right)\right)Y_{t-1} + \varepsilon_t, \quad t\in\mathbb{N}, \tag{1}$$

where $\{\varepsilon_t\}_{t\geq 1}$ is $iid\left(0,\sigma_t^2\right)$, $\varphi_{t,1}$ and $\varphi_{t,2}$ are the autoregressive parameters and $\gamma > 0$ is the known nonlinear parameter. A heuristic determination of γ from data is

$$\hat{\gamma} = -\frac{\log\varepsilon}{\max\limits_{1\leq t\leq n} Y_t^2}, \tag{2}$$

where ε is a small number and n is the number of observations. (cf. [15]).

The autoregressive parameters and the innovation variance are periodic of period S, that is,

$$\varphi_{t+kS,1} = \varphi_{t,1}, \varphi_{t+kS,2} = \varphi_{t,2} \text{ and } \sigma_{t+kS}^2 = \sigma_t^2 \ , \forall k, \ \ t\in\mathbb{N}. \tag{3}$$

To point out the periodicity, let $t = i + S\tau, i = 1, \dots, S$ and $\tau\in\mathbb{N}$, then Eq. (1) becomes

$$Y_{i+S\tau} = \left(\varphi_{i,1} + \varphi_{i,2}\exp\left(-\gamma Y_{i+S\tau-1}^2\right)\right)Y_{i+S\tau-1} + \varepsilon_{i+S\tau}, i = 1, \dots, S, \ \ \tau\in\mathbb{N} \tag{4}$$

In Eq. (4), $Y_{i+S\tau}$ is the value of Y_t during the i-th season of the cycle τ and $\varphi_{i,1}$, $\varphi_{i,2}$ are the model parameters at the season i. It is clear that the parameters depend on $Y_{i+S\tau-1}$ in the sense that for large $|Y_{i+S\tau-1}|$ we have $\varphi_{i,1} + \varphi_{i,2}\exp\left(-\gamma Y_{i+S\tau-1}^2\right) \sim \varphi_{i,1}$ while for small $|Y_{i+S\tau-1}|$: $\varphi_{i,1} + \varphi_{i,2}\exp\left(-\gamma Y_{i+S\tau-1}^2\right) \sim \varphi_{i,1} + \varphi_{i,2}$ of course the change is done smoothly between these regimes. In application, the restricted $PEXPAR(1)$ model is fitted to seasonal time series displaying nonlinearity features like amplitude dependent frequency.

These forms of models are new in the literature of the time series it is interesting to make several simulations to see their characteristics. An important fact is their property of non normality as is shown by histogram in **Figure 1** and confirmed by the test of Shapiro Wilk where the $p - value = 0.008226$ is less than 0.05. The realization of the process (A) is given in **Figure 1** from it and from the correlogram we can see that the process is stationary in each season due to the fast decay to 0 as h increases. Another interesting fact, that these types of models can exhibit, is the limit cycle behavior which is a well known feature in nonlinear vibrations and is one of possible mode of oscillations. Such phenomena is shown in **Figure 2** from model (B).

$$\text{Model (A)}: \begin{cases} Y_{1+2\tau} = \left(-0.3 + 2\exp\left(-Y_{2\tau}^2\right)\right)Y_{2\tau} + \varepsilon_{1+2\tau} \\ Y_{2+2\tau} = \left(-0.8 + \exp\left(-Y_{1+2\tau}^2\right)\right)Y_{1+2\tau} + \varepsilon_{2+2\tau} \end{cases}. \tag{5}$$

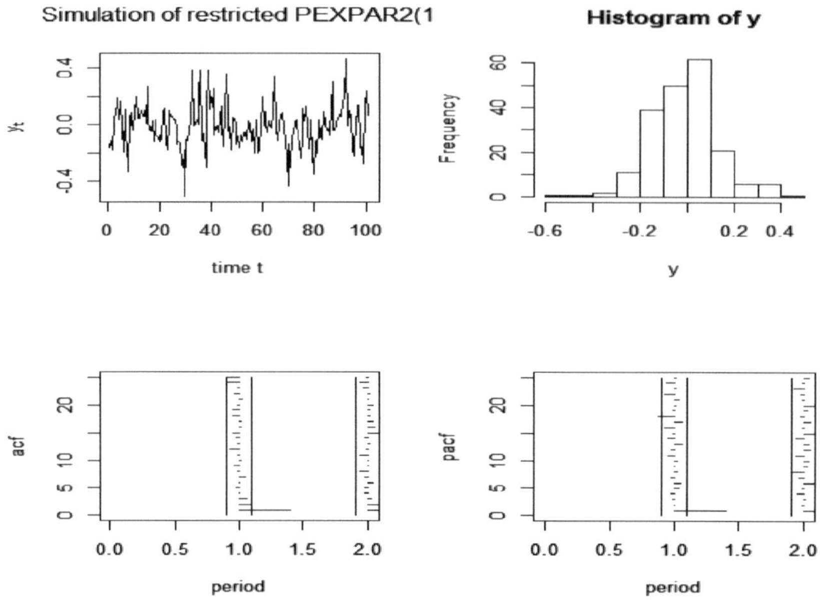

Figure 1.
Realization of (A) with corresponding histogram and correlogram.

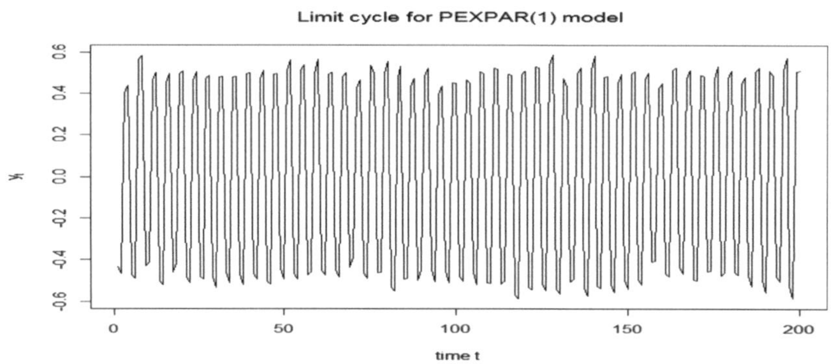

Figure 2.
Limit cycle from PEXPAR₂(1) model.

$$\text{Model (B)}: \begin{cases} Y_{1+2\tau} = \left(0.2 - 1.5\exp\left(-Y_{2\tau}^2\right)\right)Y_{2\tau} + \varepsilon_{1+2\tau} \\ Y_{2+2\tau} = \left(0.8 + 0.3\exp\left(-Y_{1+2\tau}^2\right)\right)Y_{1+2\tau} + \varepsilon_{2+2\tau} \end{cases}. \tag{6}$$

2.2 QML Estimation

Let $\underline{\varphi} = \left(\underline{\varphi}'_1, \dots, \underline{\varphi}'_S\right)' \in \mathbb{R}^{2S}$ the parameter vector where $\underline{\varphi}_i = \left(\varphi_{i,1}, \varphi_{i,2}\right)', i = 1, \dots, S$. We want to estimate the true parameter $\underline{\varphi}_0$ from observations $Y_1, \dots, Y_n$ where $n = mS$ which means that we have m full period of data. The problem is resolved by the *QML* method and under the conditions:

$A1$: The Periodical restricted exponential autoregressive parameters $\underline{\varphi}$ satisfy the stationary periodically condition of (1). A sufficient condition is given by $\left|\varphi_{i,1}\right| < 1, \varphi_{i,2} \in \mathbb{R}, i = 1, \dots, S$.

$A2$: The periodically ergodic process $\{Y_t; t \in \mathbb{N}\}$ is such that $E\left(Y_t^4\right) < \infty$, for any $t \in \mathbb{N}$.

Periodic stationarity has not been treated for this model so stationarity is required for each season hence $A1$. We can replace the assumption $A2$ by $E\left(\varepsilon_t^4\right) < \infty$, for any $t \in \mathbb{N}$, since $E(\varepsilon_t^4) < \infty \Rightarrow E(Y_t^4) < \infty$. Under this condition significant outliers are improbable and the existence of the information matrix is guaranteed.

Given initial value Y_0, the conditional log likelihood of the observations evaluated at $\underline{\varphi}$ depends on f. The *QML* estimator is obtained by replacing f by the $N(0, \sigma_t^2)$:

$$L_n\left(\underline{\varphi}, Y_1, \dots, Y_n\right) = -\frac{mS}{2}\log(2\pi) - \frac{m}{2}\sum_{i=1}^{S}\log\left(\sigma_i^2\right) - \sum_{i=1}^{S}\sum_{\tau=0}^{m-1}\frac{\left(Y_{i+S\tau} - \left(\varphi_{i,1} + \varphi_{i,2}\exp\left(-\gamma Y_{i+S\tau-1}^2\right)\right)Y_{i+S\tau-1}\right)^2}{2\sigma_i^2}, \tag{7}$$

(assuming) $\sigma_i \neq 0$.

Let $\underline{\hat{\varphi}}$ the *QML* estimator, one can see that maximizing L_n is equivalent to minimization of the quantity:

$$Q_n\left(\underline{\varphi}\right) = \frac{1}{n}\sum_{t=1}^{n}\left(Y_t - \left(\varphi_{t,1} + \varphi_{t,2}\exp\left(-\gamma Y_{t-1}^2\right)\right)Y_{t-1}\right)^2. \tag{8}$$

The initial value is unknown but its choice is not important for the asymptotic behavior of the *QML* estimator so we put $Y_0 = 0$, which defines the operational criterion

$$\tilde{Q}_n\left(\underline{\varphi}\right) = \frac{1}{S}\sum_{i=1}^{S}\tilde{Q}_{i,m}\left(\underline{\varphi}_i\right) \tag{9}$$

and

$$\tilde{Q}_{i,m}\left(\underline{\varphi}_i\right) = \frac{1}{m}\sum_{\tau=0}^{m-1}\left(Y_{i+S\tau} - \left(\varphi_{i,1} + \varphi_{i,2}\exp\left(-\gamma Y_{i+S\tau-1}^2\right)\right)Y_{i+S\tau-1}\right)^2. \tag{10}$$

The first order condition of the *QML* minimization problem is a system of 2*S* linear equations with 2*S* unknowns. The solution is

$$\begin{bmatrix}\hat{\varphi}_{i,1}\\ \hat{\varphi}_{i,2}\end{bmatrix} = \begin{bmatrix}\sum_{\tau=0}^{m-1} Y_{S\tau+i-1}^2 & \sum_{\tau=0}^{m-1} Y_{S\tau+i-1}^2\exp\left(-\gamma Y_{S\tau+i-1}^2\right)\\ \sum_{\tau=0}^{m-1} Y_{S\tau+i-1}^2\exp\left(-\gamma Y_{S\tau+i-1}^2\right) & \sum_{\tau=0}^{m-1} Y_{S\tau+i-1}^2\exp\left(-2\gamma Y_{S\tau+i-1}^2\right)\end{bmatrix}^{-1} \times \begin{bmatrix}\sum_{\tau=0}^{m-1} Y_{S\tau+i-1}Y_{S\tau+i}\\ \sum_{\tau=0}^{m-1} Y_{S\tau+i-1}Y_{S\tau+i}\exp\left(-\gamma Y_{S\tau+i-1}^2\right)\end{bmatrix} \tag{11}$$

$$\hat{\sigma}_i^2 = \frac{1}{m}\sum_{\tau=0}^{m-1}\left(Y_{S\tau+i} - \left(\hat{\varphi}_{i,1} + \hat{\varphi}_{i,2}\exp\left(-\gamma Y_{S\tau+i-1}^2\right)\right)Y_{S\tau+i-1}\right)^2.$$

We remark that the *QML* estimator is the *LS* estimator and we can proof the next theorem in the same way.

Theorem

The *QML* estimator is strongly consistent and we have for $i = 1, \dots, S$

$$\sqrt{m}\begin{bmatrix}\hat{\varphi}_{i,1}-\varphi_{i,1}\\ \hat{\varphi}_{i,2}-\varphi_{i,2}\end{bmatrix}\underset{m\to\infty}{\overset{\mathcal{L}}{\to}} N\left(0_2,\sigma_i^2\begin{pmatrix}E\left(Y_{i-1}^2\right) & E\left(X_{i-1}^2\exp\left(-\gamma Y_{i-1}^2\right)\right)\\ E\left(Y_{i-1}^2\exp\left(-\gamma Y_{i-1}^2\right)\right) & E\left(Y_{i-1}^2\exp\left(-2\gamma Y_{i-1}^2\right)\right)\end{pmatrix}^{-1}\right). \tag{12}$$

Furthermore, $\underline{\hat{\varphi}}_{i,m}$ and $\underline{\hat{\varphi}}_{j,m}$ are asymptotically independant, $i \neq j,\ i,j = 1, \dots, S$.

Proof

The proof is very standard in the literature of time series. The consistency is based on an ergodicity argument and for the normality a central limit version for martingale differences is used. The detail is similar to the *LSE* (see [24]) hence it is omitted. The independence of the $\varepsilon_{i+S\tau}$ implies that all the terms for $i \neq j$ are zero, this implies that $\sqrt{m}\left(\underline{\hat{\varphi}}_{i,m}-\underline{\varphi}_i\right)$ and $\sqrt{m}\left(\underline{\hat{\varphi}}_{j,m}-\underline{\varphi}_j\right)$, $i \neq j$, are asymptotically uncorrelated.

The *QML* estimators (Eq. (4)) yields a point estimator, a confidence interval (*CI*) gives a region where the parameters fall in with a given probability (usually 95% or 90%). Based on the asymptotic normality of the *QML* estimators, with asymptotic probability $1-\alpha$, $\varphi_{i,j}$ is in the interval

$$\left(\hat{\varphi}_{i,j} \pm \Phi_{1-\alpha/2}\frac{\hat{\sigma}}{\sqrt{m_i}}\sqrt{(\Gamma_i)_{jj}}\right),\ j = 1,2, i = 1, \dots, S, \tag{13}$$

where

$$\Gamma_i = \begin{pmatrix}E\left(Y_{i-1}^2\right) & E\left(Y_{i-1}^2\exp\left(-\gamma Y_{i-1}^2\right)\right)\\ E\left(Y_{i-1}^2\exp\left(-\gamma Y_{i-1}^2\right)\right) & E\left(Y_{i-1}^2\exp\left(-2\gamma Y_{i-1}^2\right)\right)\end{pmatrix}^{-1}, \tag{14}$$

and $\Phi_{1-\alpha/2}$ is the $1-\alpha/2$ quantile of the standard normal distribution. That is, the *CI* contains the true parameters in $100(1-\alpha)\%$ of all repeated samples.

To examine the performance of the *QML* estimators, we construct *CI* of the parameters from the simulation of restricted $PEXPAR_2(1)$ model with parameters: $\underline{\varphi}_1 = (-0.8, 1.2)'$ and $\underline{\varphi}_2 = (0.4, -0.9)'$ with sizes $n = 200, 500$ and 1000 and for the significance levels: $\alpha = 10\%$ and 5% and 1000 replications. From the **Tables 1–3** we deduce that the parameters are well estimated and when n increases the length of *CI* decreases showing that the estimates are consistent. Obviously, a higher confidence level produces wider *CI*.

$n = 200$	$CI(\varphi_{1,1})$	$CI(\varphi_{1,2})$	$CI(\varphi_{2,1})$	$CI(\varphi_{2,2})$
$\alpha = 10\%$	$[-0.8206, -0.7796]$	$[1.1136, 1.2576]$	$[0.3801, 0.4119]$	$[-0.9559, -0.8156]$
$\alpha = 5\%$	$[-0.8126, -0.7661]$	$[1.0893, 1.2618]$	$[0.3808, 0.4194]$	$[-0.9967, -0.8260]$

Table 1.
CI of parameters for n = 200.

$n = 500$	$CI(\varphi_{1,1})$	$CI(\varphi_{1,2})$	$CI(\varphi_{2,1})$	$CI(\varphi_{2,2})$
$\alpha = 10\%$	$[-0.8038, -0.7879]$	$[1.1551, 1.2113]$	$[0.3874, 0.4001]$	$[-0.9015, -0.8470]$
$\alpha = 5\%$	$[-0.8097, -0.7912]$	$[1.1783, 1.2453]$	$[0.3873, 0.4020]$	$[-0.9104, -0.8448]$

Table 2.
CI of parameters for n = 500.

$n = 1000$	$CI(\varphi_{1,1})$	$CI(\varphi_{1,2})$	$CI(\varphi_{2,1})$	$CI(\varphi_{2,2})$
$\alpha = 10\%$	$[-0.8027, -0.7952]$	$[1.1909, 1.2191]$	$[0.3978, 0.4040]$	$[-0.9072, -0.8793]$
$\alpha = 5\%$	$[-0.8030, -0.7938]$	$[1.1797, 1.2139]$	$[0.3958, 0.4034]$	$[-0.9119, -0.8791]$

Table 3.
CI of parameters for n = 1000.

3. Likelihood Ratio tests

3.1 Test for the Nullity of One Coefficient

The asymptotic normality of the *QML* in Eq. (12) can be exploited to perform tests on the parameters. This problem is very standard, especially when 0 is an interior point of the parameter space and can be done with the trilogy: Wald, LR and LM tests. We treated the former in [25] and in this chapter, we will use the LR test which is based upon the difference between the maximum of the likelihood under the null and under the alternative hypotheses and has the advantage of not estimating information matrix. In this section, we are interested in testing assumptions of the form

$$H_0 : \varphi_{i,2} = 0 \;\left(\text{or } H_0 : \varphi_{i,1} = 0\right) \text{ vs } H_1 : \varphi_{i,2} \neq 0 \;\left(\text{or } H_1 : \varphi_{i,1} \neq 0\right), \tag{15}$$

for some given i. Under H_1, we have the *QML* estimator $\hat{\underline{\varphi}}_i$ given by Eq. (11) and mean square error $\tilde{Q}_{i,m}\left(\hat{\underline{\varphi}}_i\right)$ given by Eq. (10) and $\tilde{\underline{\varphi}}_i = \begin{pmatrix} \tilde{\varphi}_{i,1} \\ 0 \end{pmatrix}$, is the *QML* estimator given under H_0 where

$$\tilde{\varphi}_{i,1} = \frac{\sum_{\tau=0}^{m-1} Y_{S\tau+i-1} Y_{S\tau+i}}{\sum_{\tau=0}^{m-1} Y_{S\tau+i-1}^2} \tag{16}$$

and the corresponding mean square error under the null

$$\tilde{Q}_{i,m}\left(\tilde{\underline{\varphi}}_i\right) = \frac{1}{m}\sum_{\tau=0}^{m-1}\left(Y_{i+S\tau} - \tilde{\varphi}_{i,1} Y_{i+S\tau-1}\right)^2. \tag{17}$$

The usual LR statistic is

$$\lambda_{i,m} = \frac{L\left(\tilde{\underline{\varphi}}_i, \tilde{\sigma}_i^2\right)}{L\left(\hat{\underline{\varphi}}_i, \hat{\sigma}_i^2\right)} = \left(\frac{\tilde{Q}_{i,m}\left(\hat{\underline{\varphi}}_i\right)}{\tilde{Q}_{i,m}\left(\tilde{\underline{\varphi}}_i\right)}\right)^{\frac{m}{2}} \tag{18}$$

then the test rejects H_0 at the asymptotic level α when

$$\begin{aligned} LR_{i,m} &= -2\log\lambda_{i,m} \\ &= m\log\frac{\tilde{Q}_{i,m}\left(\tilde{\underline{\varphi}}_i\right)}{\tilde{Q}_{i,m}\left(\hat{\underline{\varphi}}_i\right)} > \chi_1^2(1-\alpha), \end{aligned} \tag{19}$$

where $\chi_1^2(1-\alpha)$ is the $(1-\alpha)-$quantile of the χ^2 distribution with 1 degree of freedom.

In the same manner we can test the nullity of $\varphi_{i,1}$ by taken $\underline{\tilde{\varphi}_i} = \begin{pmatrix} 0 \\ \tilde{\varphi}_{i,2} \end{pmatrix}$ and

$$\tilde{Q}_{i,m}\left(\underline{\tilde{\varphi}_i}\right) = \frac{1}{m}\sum_{\tau=0}^{m-1}\left(Y_{i+S\tau} - \tilde{\varphi}_{i,2}\exp\left(-\gamma Y_{i+S\tau-1}^2\right)Y_{i+S\tau-1}\right)^2. \tag{20}$$

Example 1

In the simulation we focused on testing the nullity of $\varphi_{i,2}$ only. We simulated 1000 independent samples of length $n = 200$ and 500 of 3 models.

Model I: Periodic autoregressive ($PAR_2(1)$) with the parameters $\underline{\varphi} = (-0.7, 0.4)'$.

Model II: Restricted $PEXPAR_2(1)$ with the parameters $\underline{\varphi} = (-0.7, 0, 0.4, -2)'$ and $\gamma = 1$

Model III: Restricted $PEXPAR_2(1)$ with the parameters $\underline{\varphi} = (-0.7, 1.5, 0.4, -2)'$ and $\gamma = 1$.

The model I is chosen to calculate the level, the model III is chosen to calculate the power, the choice of model II is to show that the test is efficient since in the first cycle we have an $AR(1)$ and in the second cycle a restricted $EXPAR(1)$. On each realisation we fitted a restricted $PEXPAR_2(1)$ model by QML and carried out tests of $H_0 : \varphi_{i,2} = 0$ against $H_1 : \varphi_{i,2} \neq 0$. The rejection frequencies at significance level 5% and 10% are reported in **Tables 4** and **5**. **Figure 3** shows the asymptotic distribution of $LR_{i,m}$ under the null hypothesis. From the tables we see that the levels of the LR test are pretty well controlled since for $n = 500$, we note a relative rejection frequency of 5.5% for $\varphi_{1,2}$ and 5.1% for $\varphi_{2,2}$, which are not meaningfully different from the nominal 5%, the same remark is made for $\alpha = 10\%$ where the relative rejection frequency is of 9.5% and 10.3%. From model III, the rejection frequencies which represent the empirical power increase with the length n indicating the good performance and the consistency of the test. To illustrate that the asymptotic distribution of $LR_{i,m}$ under the null hypothesis is the standard χ_1^2 we have the

Model	α	$\varphi_{1,2}$	$\varphi_{2,2}$
I	5%	0.052	0.066
	10%	0.105	0.103
II	5%	0.054	0.998
	10%	0.117	1
III	5%	0.967	0.930
	10%	1	0.997

Table 4.
The rejection frequency computed on 1000 replications of simulations of length $n = 200$.

Model	α	$\varphi_{1,2}$	$\varphi_{2,2}$
I	5%	0.055	0.051
	10%	0.095	0.103
II	5%	0.053	1
	10%	0.098	1
III	5%	0.991	1
	10%	1	1

Table 5.
The rejection frequency computed on 1000 replications of simulations of length $n = 500$.

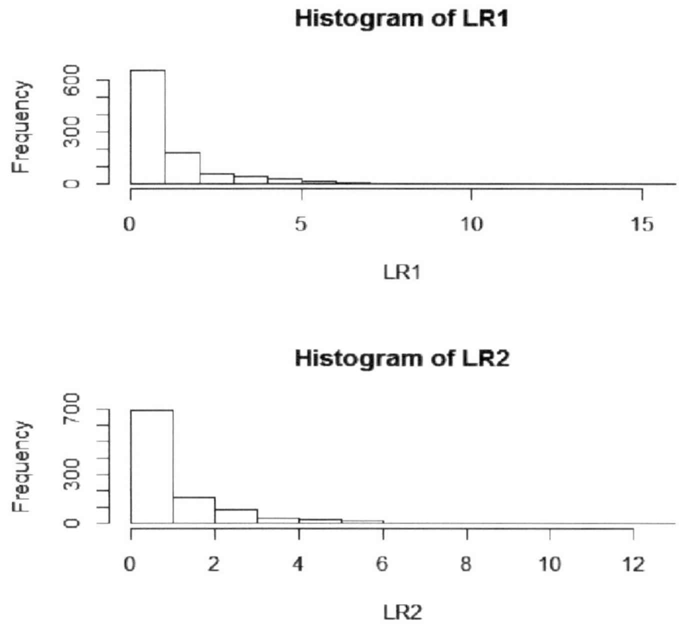

Figure 3.
Asymptotic distribution of LR.

histograms in **Figure 3** where we see that the distribution of $LR_{i,m}$ has the well known shape of χ_1^2.

3.2 Test for linearity in Restricted PEXPAR(1) model

The most important case to test is when $\varphi_{i,2} = 0, \forall i,$ which correspond to the linear periodic autoregressive model ($PAR_S(1)$) of period S. The null hypothesis is then

$$H_0 : \varphi_{i,2} = 0, \forall i \;\text{ vs }\; H_1 : \exists i / \varphi_{i,2} \neq 0. \tag{21}$$

H_1 correspond to the restricted $PEXPAR_S(1)$ model, that is, the linear $PAR_S(1)$ model is nested within the nonlinear restricted model and it can be obtained by limiting the parameters $\varphi_{i,2}$ to be zero $\forall i$, hence we have a problem of testing the linearity hypothesis.

The standard LR test statistic is

$$\lambda_m = \left(\sum_{i=1}^{S} \frac{\tilde{Q}_{i,m}\left(\hat{\underline{\varphi}}_i\right)}{\tilde{Q}_{i,m}\left(\tilde{\underline{\varphi}}_i\right)} \right)^{\frac{m}{2}}. \tag{22}$$

The test rejects H_0 at the asymptotic level α when

$$\begin{aligned} LR_m &= -2 \log \lambda_m \\ &= m \sum_{i=1}^{S} \log \frac{\tilde{Q}_{i,m}\left(\tilde{\underline{\varphi}}_i\right)}{\tilde{Q}_{i,m}\left(\hat{\underline{\varphi}}_i\right)} > \chi_S^2(1-\alpha), \end{aligned} \tag{23}$$

where $\chi_S^2(1-\alpha)$ is the $(1-\alpha)$−quantile of the χ^2 distribution with S degrees of freedom which is simply the number of supplementary parameters in H_1.

Model	α	LR test ($n = 200$)	LR test ($n = 500$)
I	5%	0.0615	0.0581
	10%	0.1225	0.1075
II	5%	0.9999	1
	10%	0.9999	1

Table 6.
The rejection frequency computed on 10000 replications.

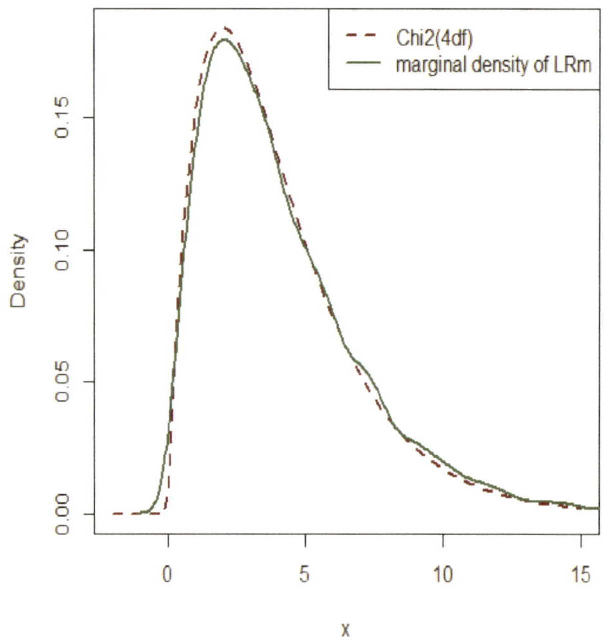

Figure 4.
Asymptotic distribution of LR_m.

Example 2

Table 6 shows the rejection frequency computed on 10000 replications of simulations of length $n = 200$ and 500 from the 2 models.

Model I: $PAR_4(1)$ with the parameters $\underline{\varphi} = (-0.8, 0.5, 0.9, -0.4)'$.

Model II: Restricted $PEXPAR_4(1)$ with $\underline{\varphi} = (-0.8, 2, 0.5, -1.5, 0.9, 1.1, -0.4, 0.6)'$ and $\gamma = 1.$, **Figure 4** shows the asymptotic distribution of LR_m under the null hypothesis. The results show that the empirical levels are acceptable, for $n = 500$, we have a relative rejection frequency of 5.81% (resp. 10.75%) which is very close to 5% (resp. 10%), the empirical power increases with the size n which means that the test is consistent. The rejection region is $\{LR_m > \chi^2_4(1-\alpha)\}$, where $\chi^2_4(1-\alpha)$ is the $(1-\alpha)-$ quantile of the χ^2 distribution with 4 degrees of freedom. From **Figure 4**, we see that the asymptotic distribution of LR_m (in full line) is close to the χ^2_4 (in dashed lines), this confirm the above theoretical result.

4. Conclusion

The periodic restricted *EXPAR* model is added to the family of nonlinear and periodic models. Interest is focused on estimation and testing problems. The periodic stationarity allows to calculate the QML estimators and derived tests of coefficients, cycle by cycle, and therefore use standard techniques. From this point of view, we can extend several results concerning the classical *EXPAR* to the periodic case.

Author details

Mouna Merzougui
LaPS Laboratory, University Badji Mokhtar Annaba, Algeria

*Address all correspondence to: merzouguimouna@yahoo.fr

References

[1] Priestley MB. *Non-linear and Non-stationary Time Series Analysis*. New York: Academic Press; 1988

[2] Tong H. *Nonlinear Time Series: a Dynamical System Approach*. Oxford: Oxford University Press; 1990

[3] Douc R, Moulines E, Stoffer D. *Nonlinear Time Series: Theory, Methods, and Applications with R Examples.* Texts in Statistical Science. CRC Press; 2013

[4] Ozaki T. Non-linear time series models for non-linear random vibrations. Journal of Applied Probability. 1980;**17**:84-93

[5] Haggan, V. and Ozaki,T. (1981), Modeling nonlinear random vibrations using an amplitude-dependent autoregressive time series model. *Biometrik* Vol.68, No.1, PP(96–189).

[6] Chan KS, Tong H. On the use of the deterministic Lyapunov function for the ergodicity of stochastic difference equations. Advances in Applied Probability. 1985;**17**(3):666-678

[7] Tjøstheim D. Estimation in Nonlinear Time Series Models. *Elsevier Science Publishers B.V. North-Holland*. Stochastic Processes and their Applications. 1986;***21****:251-273*

[8] Al-Kassam MS, Lane JA. Forecasting Exponential Autoregressive Models Of Order 1. Journal Of Time Series Analysis. 1989;**10**(2):95-113

[9] Koul HL, Schick A. Efficient estimation in nonlinear autoregressive time series models. Bernoulli. 1997;**3**: 247-277

[10] Shi Z, Aoyama H. Estimation of exponential autoregressive time series model by using genetic algorithm. Journal of Sound and Vibration. 1997; ***205***(*3*)*:309-321*

[11] Ismail, M. A. (2001). Bayesian Analysis of Exponential AR Models. *The Far East Journal of Statistics*, Vol 5, pp 1–15. (India)

[12] Allal, J. and El Melhaoui, S.(2006). Optimal Detection Of Exponential Component. *Journal Of Time Series Analysis* Vol. 27, No. 6, PP(793–810).

[13] Francq, C., Horvath, L. and Zakoian, J.M. (2008). Sup-tests for linearity in a general nonlinear AR(1) model when the supremum is taken over the full parameter space. *MPRA Paper* No. 16669.

[14] Ghosh H, Gurung B, Gupta P. Fitting EXPAR Models Through the Extended Kalman Filter. Sankhyā : The Indian Journal of Statistics. 2015;*77*(*1*): *27-44*

[15] Shi Z, Tamura Y, Ozaki T. Monitoring the stability of BWR oscillation by nonlinear time series modeling. Annals of Nuclear Energy. 2001;**28**:953-966

[16] Terui, N. and Van Dijk, H. K. (1999). Combined Forecasts from Linear and Nonlinear Time Series Models. *Econometric Institute Report* EI-9949/A.

[17] Ishizuka K, Kato H, Nakatani T. Speech signal analysis with exponential autoregressive model, *Proc. the 30th International Conference of Acoustics, Speech*. Signal Processing. 2005;**1**: 225-228

[18] Amiri, E., (2012). Forecasting GDP Growth rate with Nonlinear Models. *1st International Conference on Econometrics Methods and Applications*. ICEKU2012 (25–27).

[19] Gurung, B. (2013) An Application of Exponential Autoregressive (EXPAR) Nonlinear Time-series Model.

International Journal of Information and Computation Technology. ISSN 0974–2239, 3(4), 261–266.

[20] Katsiampa P. *A new approach to modelling nonlinear time series: Introducing the ExpAR-ARCH and ExpAR-GARCH models and applications*. 4th Student Conference on Operational Research. In: 34–51. 2014

[21] Azouagh N, El Melhaoui S. An Exponential Autoregressive model for the forecasting of annual sunspots number. Electronic Journal of Mathematical Analysis and Applications. 2019;**7**(3):17-23

[22] Gladyshev EG. Periodically correlated random sequences. Soviet. Math. 1961;**2**:385-388

[23] Merzougui M, Dridi H, Chadli A. Test for periodicity in restrictive EXPAR models. Communication in Statistics-Theory and Methods. 2016;**45**(9): 2770-2783

[24] Merzougui, M. (2017) Estimation in periodic restricted EXPAR(1) models. *Communications in Statistics - Simulation and Computation, D.O.I. 10.1080/03610918.2017.1361975.*

[25] Merzougui, M. (2020) Wald Tests in the restricted Periodic *EXPAR*(1) model. Biostatistics and Biometrics Open Access Journal. V 10(1). DOI: 10.19080/BBOAJ.2020.10.555780.

[26] Engle RF. *Handbook of Econometrics*. Vol. II: Elsevier Science Publishers BV; 1984

[27] Bierens HJ. *Estimation, testing, and specification of cross-section and time series models*. Cambridge University Press; 1994

Chapter 13

Severe Testing and Characterization of Change Points in Climate Time Series

James Ricketts and Roger Jones

Abstract

This paper applies misspecification (M-S) testing to the detection of abrupt changes in climate regimes as part of undertaking severe testing of climate shifts versus trends. Severe testing, proposed by Mayo and Spanos, provides severity criteria for evaluating statistical inference using probative criteria, requiring tests that would find any flaws present. Applying M-S testing increases the severity of hypothesis testing. We utilize a systematic approach, based on well-founded principles that combines the development of probative criteria with error statistical testing. Given the widespread acceptance of trend-like change in climate, especially temperature, tests that produce counter-examples need proper specification. Reasoning about abrupt shifts embedded within a complex times series requires detection methods sensitive to level changes, accurate in timing, and tolerant of simultaneous changes of trend, variance, autocorrelation, and red-drift, given that many of these measures may shift together. Our preference is to analyse the raw data to avoid pre-emptive assumptions and test the results for robustness. We use a simple detection method, based on the Maronna-Yohai (MY) test, then re-assess nominated shift-points using tests with varied null hypotheses guided by M-S testing. Doing so sharpens conclusions while avoiding an over-reliance on data manipulation, which carries its own assumptions.

Keywords: severe testing, misspecification testing, abrupt shifts, unit-roots, change-points

1. Introduction

Anybody examining sudden changes in data needs to ask, "Does this mean what I think it means? Are there other explanations?" Further, the evidence needed to overturn the acceptance of a generally held position requires high probative value, supporting the proposed position and addressing the accepted one; and should be convincing to the investigator and others.

This paper addresses this issue by presenting a systematic approach that combines the development of probative criteria with error statistical testing, illustrating it with a specific investigation of climate. The approach is developed from previous work on the philosophy of statistics, which is relatively new to climate work [1–6].

Climate, like many areas of natural science, depends heavily on statistical induction for the interpretation of physically-based behavior. Many popular statistical tools are generalized tests, framed against broad statistical assumptions that may be challenged by complex physical processes. The implicit assumptions of tests must be considered, as must the linkage between those processes, the accessible data, and statistical models. Where competing alternatives cannot be correctly distinguished by the tests and specific data chosen for that purpose, the data is misspecified with respect to the statistical models or the model selection processes.

The particular aspect addressed here is model specification with respect to the data. Probative criteria drawing from theory and interpretations of physical behavior cannot be applied correctly, if the tests to not adequately represent those criteria, or distinguish between them.

1.1 Illustrative example: abrupt shifts in climate signals

A number of publications now address an area of some controversy – the hypothesis that under greenhouse gas-induced radiative forcing, climate changes in a step-like manner [7–12]. The controversy arises because it is almost universally accepted that the forced response of climate change, especially global mean surface temperature (GMST), responds rapidly to forcing and hence is trend-like; albeit embedded in a very complex "error" process which yields highly structured residuals.

Our paper from 2017 (JR2017) [12] and the PhD thesis of Ricketts (R2019) [13], in addressing this controversy, required the development of automated, reliable and unbiased detection of shifts, and importantly various means of ensuring that presumptive shifts were not artefacts of the detection method and the structured residuals.

We built on the concepts of severe testing [3] and misspecification testing [2], and we adapted a framework of models to connect theory and data [14, 15]. Thus we could severely test two propositions: (*H1*) forced warming and natural variability proceed gradually and independently, with the response to forced warming best represented as trend-like; and (*H2*) forced warming and natural variability interact so that patterns of response may project onto modes of climate variability – either one-way as proposed by Corti et al. [16] or two-way as proposed by Branstator [17] – in either case giving rise to abrupt state-like transitions in the signal.

JR2017 showed that *H2* was preferred to *H1* in all of six tests of a severe testing regime; R2019 also showed that abrupt shifts relate directly to warming; in their extent, frequency and intensity; and more so at finer scale.

1.2 Structure of the rest of the paper

Firstly we very briefly introduce severe testing.

Then we introduce our version of a framework that connects hypotheses about the physical world to statistically based tests that license inductions about models of the world.

Next we spend more time on misspecification testing (M-S), which was proposed as an approach to determining whether the assumptions needed to reliably model the statistical variables are met [2].

2. Severe testing

Severe testing, proposed by Mayo and Spanos, is based on the intuition that "Data x0 in test T provide good evidence for inferring H (just) to the extent that H passes severely with x0, i.e., to the extent that H would (very probably) not have survived the test so well were H false." [3]. They propose that a severity criterion supplies a meta-statistical principle for evaluating statistical inferences (their page 328), where the severity of testing is not assigned to hypothesis H, but to the testing procedure.

In the preface of [6] we read the following, "If little or nothing has been done to rule out flaws in inferring a claim, then it has not passed a severe test. In the severe testing view, probability arises in scientific contexts to assess and control how capable methods are at uncovering and avoiding erroneous interpretations of data. … A claim is severely tested to the extent that it has been subjected to and passes a test that probably would have found flaws, were they present."

Severe testing is beginning to be picked up by the climate community (e.g. [18–20]). It was applied to an analysis of optimal fingerprint methods in climatology [18] and to address issues of model tuning in climate projections [20]. Severe testing forms a core methodology of JR2017, R2019, and a conference paper [21] (RJ2017).

3. The theoretical mechanistic/statistical inductive (TM/SI) framework

The TM/SI framework borrows from a strong body of earlier work (e.g. [4, 14, 15, 22, 23]) and was outlined in Section 2 of JR2017 to provide support for reasoning about climate where the scientific debate had been muddied by competing claims from outside the science community.

The approach follows Haig [15], and employs the concept of severe testing [3], and in keeping with it, error-statistical methods [4, 23]. It requires a carefully reasoned matching between scientific hypotheses about the physical world, with statistical hypotheses about the observed data.

The theoretical-mechanistic part consists of the physical aspects, components, relationships and measurable quantities.

The statistical-inductive part consists of the process of drawing conclusions about specified hypotheses concerning the system given the physical model, real-world data and statistical tests.

The goal is to construct a chain of reasoning that ties physical hypotheses, *H1* .. *Hn*, to statistical hypotheses *h1* .. *hn*. That is, features of the world map to defined outcomes of statistical tests (preferably one to one). One to one mapping meets a requirement of severe testing. Misspecification testing assists this mapping.

3.1 The TM/SI structure

Suppes [14] suggested that science employs a values hierarchy of models that ranges from experimental experience to theory, claiming that theoretical models, high on the hierarchy, are not compared directly with empirical data, which are low on the hierarchy. Rather, he said, they are compared with models of the data, which are higher than data on the hierarchy. Following on, Haig describes an egalitarian framework in which three different types of models are interconnected and serve to structure error-statistical inquiry [15].

He describes: Primary models which break a research question into a set of local hypotheses; Experimental models which "structure the particular models", and link Primary models to Data models; which in turn generate and model raw data, and check that the data satisfies the assumptions of experimental models. Although Haig does not fully explain experimental models which "structure the particular models" it seems implicit that they map hypotheses to model components and processes. He leaves to his data models the role of checking that data meets the assumptions of experimental models.

To summarize, the TM/SI was constructed with physically grounded work in mind, and adapts Haig's approach. Physical entities and their relationships about which we propose hypotheses guided by Physical models are the Primary models. These link to the Statistical models which support reasoning with an inductive framework, via Data models which includes Sampling procedures. Sampling procedures guide the accumulation of data on which we reason. All data sampling procedures and statistical tests are framed against *ruling assumptions*. Violation of the ruling assumptions weakens statistical inference.

Physical Model: Concerns the system of physical entities and their interactions. Entities have measurable properties, which are accessed through Sampling procedures.

Statistical model. A mapping between a sampled set of observations and a set of parameterized probability distributions. This is informed by an error model – the theoretical behavior and characteristic distribution of sampling error, generally assumed to be random. If properly specified, the statistical model(s) license(s) valid statistical inductions about hypotheses, generally via statistical model selection from a specified statistical family.

Sampling procedures: Data models which cover the collection of measurements. Measurements are made, and treated (e.g. homogenized), and output to become sample data input to statistical models. The choice of sampling model (random sampling, averaging) influences subsequent induction since sampling error subsumes both random processes and statistical misspecification.

Severe Testing requires that these issues be accounted for so that to the extent possible, when features are present they are detected, and when they are not present they are not erroneously identified.

3.2 Applying the TM/SI to climate

3.2.1 Physical model: surface temperatures

To guide investigation we propose in JR2017 (a) physical model *M1* – a world in which average surface temperatures closely track forced warming, and natural variability is independent, and reflective of the indices of variability and by contrast (b) a physical model *M2* – which mirrors *M1* but in which there is interaction between forcing and natural variability. The *M2* world requires that Earth's surface temperature is additionally reflective of, and tracks, internal physical states of variability modes which may change abruptly, thus imprinting step-like shifts into the temperature records. These shifts mark state changes in the climate system, and represent the major response at decadal time scales of the climate system to the gradually increasing greenhouse forcing. Earth's surface temperature is sampled, but it is understood that this also reflects the overall state of heat transport in the fluid layers.

3.2.2 Sampling considerations

Observed climate data is derived over time using evolving and fallible instrumentation. This dictates the use of a wide variety of strategies to enable inter-comparisons. In our analyses we are concerned with annual or monthly averages which, in the case of gridded data, have been further averaged and re-interpolated spatially. We must consider the effects of these procedure.

Averaging implicitly assumes a signal/noise model where mean noise converges on zero at all time points to enhance the signal which is assumed to be represented equally in all samples. An influence travelling in space and time when averaged will appear as some form of non-stationarity in the time series of the mean.

Temperature records increase in spatial density over time, they are records of opportunity. Conditions over land and ocean differ. To enable inter-comparison with models they are re-interpolated onto regular grids. They are also homogenized prior to gridding to deal with instrumentation changes [24].

Both *M1* and *M2* worlds have time varying temperature records but because forced change in M1 propagates rapidly, averaging does not induce troublesome artefacts. This is not the case for *M2*. A step-like change occurring serially across regions may give rise to trends and auto-regression, and or may obscures more regional signals.

3.2.3 Statistical model(s)

Different statistical models are involved in detection of changes, and in the assessment of the relative merits of *M1* and *M2*. It is important that the probabilities from statistical *feature detection* not be also used for *model selection*.

In break-point analysis the family of segmented linear regression models is used. The choice of specific parameters from within a specified family is termed model selection, and would in our work include the serial selection of specific change-points. The MSBV differs from other approaches in that it does not terminate the search for change-points (feature detection) on the basis of an all of model information criterion such AIC (a model selection criterion), but usually earlier, when no segment can be sub-divided.

In our work, detection of such steps, supported by evidence that they are not artefacts provided by M-S testing supports constitutes support for *M2*, and thus support for *H2*.

4. Misspecification testing

Mayo and Spanos differentiate between model specification and model selection. An adequate model specification licenses primary statistical inference, and with it statistical model selection from the specified family. Serial feature detection in any time series is a form of model selection from a family of related models, reliant on model specification. It must be noted that for our work a series of tests are performed, a single detection test and multiple probative tests, but that as each is against an independent null, this does not involve a multiple-testing issue, instead increasing the overall power of the testing regime.

Chapter 2 of [25] defines experimental error as all extraneous variation outside experimental treatments, and states "Neither the presence of experimental errors or their causes need concern the investigator, provided his [sic] results are sufficiently accurate to permit definite conclusions to be reached". This definition still dominates statistical climatology. Climate data are not generally experimental, but often a feature of interest in climate data is investigated by treating natural variability as extraneous variation. Experimental design requires that statistical models are properly specified, however complex systems being observed may align to many different statistical models and have multiple features of interest, leading to the possibility of misspecification.

Mayo and Spanos [2] (MS2004) introduce a methodology for testing misspecifications in statistical models (M-S testing). Taking this as a point of departure we then propose that a full understanding of the assumptions of statistical models allows one to probe data for features even when available tests are misspecified.

Model specification delineates families of statistical models. For physical problems, the family would be misspecified if the available parameters do not properly reflect the physical processes [13].

In MS2004 the authors use an example of a linear regression model to address a problem of validation in regression models. Three general forms of M-S are recognized:

- Functional form misspecification in which a statistical model includes the correct parameters or variables but inside an incorrect function. For example, as x^2 instead of x^3 or sin(x).
- Missing parameter misspecification in which a parameter/variable is omitted.
- Irrelevant parameter misspecification in which unnecessary parameters are introduced.

4.1 Summary of MS2014

MS2004 says, "A full methodology of M-S testing, as we see it, would tell us how to specify and validate statistical models, and how to proceed when statistical assumptions are violated."

Statistical model specification (goals and assumptions) is different from statistical model selection (from an assumed family of models, with a heuristic (e.g. AIC))). We consider a statistical model M, selected from a family of models.

MS2004 considers firstly the primary questions of statistical inference, whether the assumptions needed to reliably model the data are met; and secondary questions, whether there are influential gaps between variables in a statistical model and primary questions. Primary questions are addressed within the selected statistical model M. Formally, the hypothesis H_{M}:

$$H_{\mathrm{M}} : \text{data } \mathbf{z} \text{ supports the probabilistic assumptions of statistical model } M. \quad (1)$$

Secondary questions are essentially meta-questions conducted outside the model M, they address the suitability of the test, given the data and require auxiliary models and put M's assumptions to the test. Formally this would test

$$\begin{aligned} &H_0 : \text{the assumption(s) of statistical model } M \text{ hold for data } \mathbf{z}, \\ &\text{against all possible assumptions by which } H_0 \text{ could fail } (H_{1.} \ldots H_{\mathrm{n}}). \end{aligned} \quad (2)$$

It is critically important to recognise that this use of multiple tests does not constitute multiple tests of HM. It augments and increases (not diminishes) the confidence in the conclusions.

They present a case study of an empirical relationship between the USA population (y_i) and a secret variable (x_i). They commence with a proposed explanatory model, with $R^2 = 0.995$ and p-value nearly zero. Here, $\hat{u}_i$ represents the estimated error process.

$$M0 : y_i = 167.115 + 1.907 x_i + \hat{u}_i \quad (3)$$

Concerning H_{M}. An assumption of the regression is that errors $\hat{u}_i$ are normally distributed, independent and identically distributed (NIID). Is this met? A runs test suggests not, and a parametric Durbin-Watson test suggests autocorrelation.

$$M1 : y_i = \beta_0 + \beta_1 x_i + u_i, u_i = \rho u_{i-1} + \varepsilon_i. \tag{4}$$

However an alternative AR(2) model is then shown to explain more variance *without* the $\beta_1 x_i$ term.

$$M2 : y_i = \beta_0 + \beta_1 y_{i-1} + \beta_2 y_{i-2} + \hat{u}_i \tag{5}$$

Probing the model M0 shows it to be misspecified due to an irrelevant variable. The secret variable x_i is the number of shoes owned by Spanos's grandmother!

4.2 Application to climate data

4.2.1 Abrupt changes in previous literature

In some papers step-like changes are introduced *en passant*, on the way to revealing or locating in time various phenomena. For instance the delineation of the Pacific Decadal Oscillation [26–28], or reduction in South-Western Western Australian rainfall [29]. In the last decade an astonishing number of papers addressed the so-called hiatus, many purporting to show that it never happened [30] or was simply routine variability [31, 32], or a methodological/statistical error [33], or suggesting that natural variability, internal variability and extrinsic factors combined with forced warming [34]. However others, one way or another, simply incorporate it as fact [35, 36].

From these and other papers and some personal communication, the objections/challenges to the existence of abrupt changes (including but not limited to the so-called hiatus) appear to be

1. Physical implausibility of step like changes in average temperatures.

2. Overcooking. In general, that warming is in fact more or less constant and positive, and more or less smoothly changing natural variability is imposed on it, with the result that a test for shifts is deceived by increases and decreases in the derivative of the sum.

3. Overcooking worsened by autocorrelation. As above but with at least some natural components following an autocorrelation model.

4. Model misspecification by virtue [sic] of step methods applied to trending data.

5. Non-determinism. Red noise/unit root processes masquerading as natural variability and/or as one off deterministic events. Non-determinism implies that detected events cannot be attributed to a deterministic physical model.

6. Presence of one or more sub-detection threshold deterministic events. This is a particularly nasty issue because (a) it affects detection of many phenomena, (b) it may deceive autocorrelation tests and unit-root tests as well as trend tests.

7. Conflict with objectors favoured model/approach.

Not all of these concern statistical M-S. Objection 1, physically implausibility of discontinuities in surface temperatures [37] can only result from an underlying assumption of a physical model where heat is dispersed rapidly and uniformly. Objection 7 is regrettable but not uncommon and not further considered.

4.2.2 Approach

Bearing in mind *M1* and *M2*, an important step is determining precisely what information is of primary importance. What variables are of interest and what features are important?

Step 1: As argued in JR2017, step-like shifts in temperatures are a feature of the abrupt state transitions of *M2* rather than the smooth transitions of *M1*.

Step 2: Matching alternative hypotheses, including those represented by the objections, to appropriate statistical tests. Consideration of the implicit choices made.

1. Physical implausibility is not considered further here.

2. Overcooking alone and ...

3. Overcooking with autocorrelation. The challenge here is to the meaning of abrupt shifts. If detection tests are finding the point of maximum (or minimum) derivatives of quasi-sinusoidal variability then the residuals of a segmented model will be heteroskedastic whether or not autocorrelation is present.

4. Step-change methods applied to (and deceived by) constantly trending data. In general segmenting such a process will yield segments which testing against a step and trend model will reveal to be co-linear.

5. Non-determinism relates to the interpretability of change-points and their relationship to any physical model.

6. Undetected deterministic events (events below detection thresholds or misspecification). The primary issue with these is that the error series is not random and tests assuming such are ill founded. This includes autocorrelation and trend tests.

4.3 Types of tests

In what follows, three classes of testing useful during analysis of step-like change-points have been identified. The first two involve testing the segment of data within which a change-point is found. The third asks if a multiple change-point model is adequate.

1. Does the requirements of the detection test (specifically that it should be sensitive to small shifts while precise in the timing) open the door to deception? Do individual change-points remain if more parameters are allowed?

2. Are changes reasonably regarded as deterministic or is there evidence of non-determinism which would support objection 5?

3. Does the full set of change-points explain the necessary degree of variation? Are the residuals homoskedastic?

4.3.1 Detection test

Complex climate time-series data is almost certainly misspecified for *any* change-point detection test – thus the goal is adequate applicability to questions of interest. In testing for multiple change-points, many methods, including the MSBV, examine data only between presumptive lower and upper bounding points and restart estimation of the distribution parameters. The assumptions of the basic detection methods used must be considered.

Issues potentially arising include false detection, timing errors, and false negatives. Timing error includes misplacement and imprecision. The MSBV incorporates a resampling strategy [38] which reduces imprecision. False positives (deterministic and non-deterministic) can be uncovered by post-detection assessments, but false negatives introduce down-stream non-stationarities that interfere with detection of later change-points. Combining tests with differing assumptions and different nulls probes for both non-deterministic and deterministic causes of false results including sub-detection threshold events.

Analysis of covariance (ANCOVA) is used post-detection of a change-point by MSBV (which does not consider trend) to ensure that the presence of the change-point provides explanatory power in an unconstrained disjoint linear statistical model which allows trend. It does not attempt to locate an alternative change-point – the Zivot-Andrews test however, see below, does this in passing. ANOVA tests for change of trend and change of level are obtained in passing but in R2019, final p-values for change-points are obtained from only from ANCOVA.

4.3.2 Tests for heteroskedasticity for segmentation of data with change-points

The full set of change-points in an entire sequence is tested here by the studentized Breusch-Pagan test (hereafter SBP test) for homoskedasticity of the residuals of the disjoint multi-segment model (JR2017 utilised the equivalent White's test [39]). An adequate model explanation of a time series, under the assumption of i.i.d. error, should have a featureless residual. This test has a null of homoskedasticity, rejected in favour of heteroskedasticity at low probabilities.

4.3.3 Tests for stationarity in a segment

Our detection test and the subsequent probability assignments by ANCOVA or ANOVA, and the further misspecification testing all assume serial independence either in the null or contrast hypothesis.

In these tests the segment containing a provisional change-point is tested for features that may deceive tests for shifts and trends. The MY test, ANCOVA, and where used ANOVA tests, have ruling assumptions of serial independence. The MSBV, and other multiple break tests assume some form of censorship between provisional data segments (determination of change-points within provisional bounds includes only the data within the bounds); but tests of the overall model assume homogeneity of error, thus of variance (e.g. the Akaike Information Criterion or AIC). The SBP also assumes this. All of these above tests are formalised as null hypothesis statistical tests (NHST) and as such they each are subject to their own ruling assumptions. The ruling assumptions are incorporated in the interpretation of the tests.

Autocorrelation in climate time-series is variously treated; some propose its estimation and removal [40], some warn against this idea [41]. Some treat it as a short term process and a cause of deception in change-point analyses [42], others have treated it as a persistent signal [43]. In climate signals, autocorrelation often appears to be time varying. Therefore we apply the MSBV without adjustment for autocorrelation and perform post-detection analysis to determine whether the detection test is likely to have been interfered with. In general, regression based statistical tests assume the absence of deterministic step-like changes and of unit-root, or red-noise progression.

The term unit-root refers to processes with a characteristic equation that has a value of one. If a unit characteristic is moving average, the error is integrated order zero or I(0), if it is auto-regressive, it is integrated order one, I(1). I(0) processes tend to revert to a mean, I(1) processes follow a martingale [44], and is dominated by red-noise. The integration order defines the number of successive differencing operations required to produce a trend-stationary series.

4.3.4 Residuals compared to initial data

In our work, both the raw data, and the residuals after removal of internal steps and trends, are tested. The rationale for testing both derives from the formulation of the tests themselves, since in these tests, the deterministic and non-deterministic components are separately parameterised. The set of tests chosen are from the econometric literature, and each is framed as a null hypothesis significance test (NHST) with its own specific assumptions. Each test poses either H_0 or H_1 as presence of an assumed non-deterministic unit root progression (see Chapter 2) in data, and the alternatives are chosen from a small range of deterministic features. Crucially, each must be interpreted in the light of its own ruling assumptions.

4.3.5 The full process applied to a single time-series

a. The MSBV is applied to delineate provisional change-points. The resulting statistical model would be accepted as the best estimate (i.e. further testing of change-points not warranted) if the time-series of the residuals was *known* to be i.i.d., *and* underlying physical processes were fully deterministic, and fully reflected in the time series. However this should not be simply assumed.

b. The segment containing each provisional change-point is tested to ensure that to a feasible extent, physically plausible types of deception are not present, and that change-points are deterministic, not stochastic quirks.

c. The set of detected change-points is treated as a disjoint segmented model and the residuals examined for evidence of a misfit of model to data.

The program of tests thus sharpens the error-statistical reasoning component of the TM/SI framework.

4.3.6 Deceptive features detectable with unit root tests

The application of deterministic methods such as OLS to non-deterministic data progression such as a random walk is a misspecification; the results may be deceptive with meaningless shifts and/or trends. Unit root (UR) tests probe the data for features that can superficially imitate deterministic structural changes by cumulative random walks, a red or near red progression. It has been shown by Monte Carlo methods that a test for deterministic trends will find deterministic trends in about 85% of realizations that contain only a stochastic (UR) trend [45]. However combinations of UR tests may also be used to detect both stochastic and deterministic non-stationary sequences, due their varied ruling assumptions and constructions.

Because multiple UR tests are performed, and because they each have differing ruling assumptions, the tests are interpreted in terms of evidence for and against stationarity in the underlying processes. In the econometric literature an exogenous change is one imposed upon a model from outside the model. We elected to retain this word where concepts were derived from economic papers as meaning an abrupt and deterministic change in a deterministic time-series.

4.3.7 Unit roots, non-stationarity, and climate

Transient unit root behaviour, if it occurred, could indicate some sort of regime change, temporarily decoupled from normal forcings. If, in addition, measured noise was not persistent this would show I(0) behaviour; or, if it were fully persistent, as I(1) behaviour. In regional signals in which this occurs, the region may also have become coupled to other sub-systems [46]. This could indicate that the underlying physical model is incomplete and that a missing variable misspecification has resulted. On the other hand, persistent unit root behaviour means that a deterministic change-point analysis is suspect.

The Earth system is constrained so that the overall temperature cannot solely follow a pure random walk – at worst it would follow a Brownian bridge (i.e. sequences where the end-points are meaningful and accepted as deterministic but the path is apparently a random walk [47]). However the composition of summary deterministic signals, such as the GMST, involves manipulations that can produce data that existing unit root tests will identify as containing unit roots, and furthermore deceive deterministic tests in much the same way as random walk data. This issue was extensively examined in R2019 and is addressed later.

4.3.8 Detecting unit root presence

Random walk progression may be present in climate data because of transient physical conditions, or because the data is unrelated to the physical processes assumed (M-S due to irrelevant variables). Additionally there may be features in the data that do not correspond to any of a shift, a trend change, or unit root behaviour (M-S due to missing variables), and UR tests are potentially sensitive to this. This source of deception must also be dealt with. Other features may be present in the data but not detected. For instance, a step-like shift well above a detectability threshold may be present together with a number of small, deterministic shifts below detectability, and this latter may be taken to be evidence of stochastic drift by a UR test.

The unit root based tests used here all inherit in one form or another the Dickey Fuller (DF) model [48].

$$Y_t = \mu + \beta t + \rho Y_{t-1} + e_t \tag{6}$$

ρ represents the portion of the signal (Y_{t-1}) carried forward by autocorrelation, β represents the (deterministic) linear trend, μ represents the intercept, and e_t is the i.i.d. error with zero mean and a constant variance σ^2. If $\rho = 0$ this describes a deterministic trend with no autocorrelation, if $0 > \rho < 1$ there is a deterministic trend with a degree of autocorrelation, and if $\rho = 1$, regardless of other parameters it contains a unit root. If all other parameters are zero and ρ equals one, then there is no deterministic trend, no offset, and Y_i is a random walk. This formulation is modified and sometimes rearranged in different ways by the three UR tests used here.

It is important to note that time-series of successive differences of a step-change in an otherwise stationary time series will contain only one out of range difference. Hence the DF model is intrinsically insensitive to deterministic step changes. Another important property of a unit root process is that the variance of the process increases over time, whereas the variance of a stationary process is constant. This gives a second strategy for determining unit root like behaviour – testing for diverging variance. The Kwiatkowski-Phillips-Schmidt-Shin test (KPSS), [49] examines the properties of the variance rather than the fitted parameters, and it primarily focussed on determination of stationarity. As a result it is more sensitive to exogenous changes.

5. Proposed tests and strategies

The unit root methods used are all coded in R and are, (a) a development of the DF test, the Augmented Dickey-Fuller test (ADF), which takes H_0 of a I(1) unit root against an alternative H_1 of a presumption of no unit root (in this implementation trend and multiply lagged autocorrelation is allowed for), (b) two variants of the KPSS, which takes a H_0 of stationarity (or trend-stationarity) rejecting it in favour of an alternative H_1 of a presumption of unit root, and (c) the Zivot-Andrews test (ZA) [50], which takes a H_0 of I(1) unit root behaviour with a possible endogenous drift against an alternative H_1 of trend-stationarity with exogenous structural change. A trend change or a step change would constitute an exogenous structural change.

Use of a combination of UR tests is not new. The combination of ADF and of KPSS testing has been used before in order to add precision to an analysis of monthly inflation expectations (e.g. [51] Appendix B).

The tests are being applied to data within which a single presumptive deterministic, exogenous, step-like changes was detected. No such change is allowed for in the KPSS and ADF tests, the presumption of unit-root in H_0 or H_1 of the above tests is reinterpreted as evidence of non-stationarity. Evidence of unit-root like behaviour is then sought by examination of the residuals after the removal of the deterministic internal trends and shifts detected in the data.

In general, where evidence of a unit-root is detected, it may be due to undetected deterministic features, and hence will be initially treated as evidence of either deterministic non-stationarity or stochastic non-stationarity.

For all of the above tests, the R implementations take published critical values of the test statistic at the 0.01, 0.05, and 0.1 levels. The KPSS implementation interpolates the test statistic against these values to give probabilities between 0.01 and 0.1, the ADF and ZA implementations simply give the critical values and the test statistic.

None of the tests proposed consider unit root presence or absence when possible structural breaks (such as shifts or trend changes) exist under both the null and alternate hypotheses. The problem is under active consideration [52–54].

5.1 ADF

The ADF test is a variation of the Dickey Fuller test for trend stationarity in the possible presence of unit root. It has a null hypothesis of unit root against an alternative of stationarity after compensation for auto-correlation [48, 55]. The ADF test has relatively low power, and in this type of application a finding of a UR may be because of a single deterministic permanent shift or trend-change [56], as noted above.

Eq. (6) is expanded to allow for multiple lags in the case of the Augmented Dickey Fuller (ADF) test, taking advantage of the recursive nature of the formula. This is more explicit below where k multiple lags are included as $\sum_{j=2}^{k} \rho_j \Delta y_{t-j+1}$. The difference series is then computed,

$$\Delta Y_i = b_0 + b_1 t + (\rho_1 - 1) Y_{i-1} + \sum_{j=2}^{k} \rho_j \Delta y_{t-j+1} + e_t \tag{7}$$

A unit root exists if $\rho_1 = 1$. The number of lags can be specified by the user or, as here, selected by using an information criterion.

The ADF test implementation used is programmed in R, available in the package 'urca' [57], and estimate and removes auto-correlation then applies a DF test. The code allows for three variants, are available, (a) a unit root, (b) a unit root with drift, and (c) a unit root with drift and a deterministic time trend – which corresponds to the model of Eq. (7) (above) and which we use. We select suitable autocorrelation lags on the basis of an information criterion, using the call "ur.df (ys, type = "trend", lags = 7, selectlags = "AIC")" following Hacker [58]. The resulting possible reduction in power in the test (inability to distinguish unit root from near unit root) is compensated by other tests in the suite. The test assumes no exogenous change, and H_0 may be accepted in the presence of one ([59], page 76).

5.2 KPSS

There are two variant of the KPSS test used here to test for level and trend stationarity. These tests invert the sense of the testing with respect to the ADF test, rejecting an H_0 of stationarity in favour of H_1, a presumption of a unit root. In this case a regime shift may well appear as H_1, with a step change being non level stationary and a trend change being non trend stationary. We use the R package 'tseries' [60] and invoke the two tests as kpss.test(ys), to test for level stationarity (henceforth KPSS-L) and kpss.test(ys,null = "Trend") to test for trend stationarity (henceforth KPSS-T).

KPSS tests are designed to give weight to stationarity. Assuming that the time-series can be decomposed into the sum of a deterministic trend, a random walk and a stationary error, the model of Eq. (6) is re-parameterised as follows with r_t representing the random walk

$$\begin{aligned} Y_t &= r_t + \beta t + u_{1t} \\ r_t &= r_{t-1} + u_{2t} \end{aligned} \tag{8}$$

Where u_{1t} is a stationary process, and u_{2t} is an i.i.d. process with zero mean and a variance σ^2.

If $\sigma^2 = 0$ then r_t is constant and the stationary processu_{1t} dominates. If not, then a unit root enters via u_{2t} and r_t is a random walk. Under a random walk, variance increases with time. Therefore this expectation is tested by estimating the variance using the Newey-West estimator [61] s^2 . To test for trend stationarity, a residual series ($\{e_1..e_n\}$) is given by residuals of an OLS linear regression ($\{e_1..e_n\}$).). To test for level stationarity the residual series is replaced by $e_t = y_t - \bar{y}$. Then for both cases, partial sums of residuals are defined as $S_t = \sum_{i=1}^{i} e_i$ and for T samples, the test statistic is given as

$$LM = \frac{\sum_{i=1}^{T} S_i^2}{s^2 T^2} \tag{9}$$

Both the ADF test and the ZA test below, perform by estimating an auto-regression parameter by OLS, whereas the KPSS tests examine the properties of the variance of the time series (KPSS-L) or of the difference series (KPSS-T).

5.3 Zivot-Andrews test

The previous tests are confounded by deterministic/exogenous change (steps or shifts), and additionally a combination non-deterministic and deterministic change must be detected.

The Zivot-Andrews test (ZA) [50] tests for the presence of a unit root (with a possible deterministic/exogenous change) against an alternative of stationarity with at most one exogenous change. An advantage is that the test also returns a time of a possible exogenous change [62] – but note that an exogenous change can be any of step, transient or trend change.

The code is in the R package “urca”, called as “ur.za(ys, model = “both”)”, which allows for changes in trend or steps. H_0 is UR without exogenous change. H_1 is trend-stationary with a possible exogenous change at an unknown time.

The ruling assumptions are (a) that there is at most one exogenous structural change (b) in a multivariable model, that only one exhibits unit root. In either of these cases other tests are preferred [54]. Here, we are testing a single variable with intervals bounded by breaks within which we have already detected exactly one break, whilst others may be below a detectability threshold. It has been previously shown that rejection of the null of a unit root could be due to a structural break even in the presence of unit root [63], whilst the presence of more than one break in the absence of a unit root may lead to the acceptance of the H_0 of UR [64].

Acceptance of H_0 does not imply merely UR, but rather, UR without exactly one deterministic break, [56], and thus H_1 means not UR or not a single break. Given we know there is a break (detected by MSBV, confirmed by ANCOVA), H_1 is reinterpreted as not UR, or more than one break.

The model used here is that documented by Zivot and Andrews (50) as Model (C). The model follows the ADF approach and its equation contains more complex parameters for: intercept and change of intercept (a step-like change), $\hat{\mu} + \theta DU_t(\hat{\lambda})$; and trend and change of trend, $\hat{\beta}t + \hat{\gamma}DT_t^*(\hat{\lambda})$. The remaining parameters are similar to the ADF; autocorrelation with lags, $\hat{\alpha}y_{t-1} + \sum_{j=1}^{k}\hat{c}_j\Delta y_{t-j}$ and the presumed i.i.d. error …

$$y_t = \hat{\mu} + \theta DU_t(\hat{\lambda}) + \hat{\beta}t + \hat{\gamma}DT_t^*(\hat{\lambda}) + \hat{\alpha}y_{t-1} + \sum_{j=1}^{k}\hat{c}_j\Delta y_{t-j} + e_t \tag{10}$$

Circumflexes above represent estimates of parameters. $\hat{\lambda}$ is a value that is minimised during the search for the most likely time of a break, $DU_t(\hat{\lambda}) = 1$ if $t > T\lambda$, the time of change, 0 otherwise, and $DT_t^*(\hat{\lambda}) = t - T\lambda$ if $t > T\lambda$, 0 otherwise. Parameters estimated include the time of change and each of the parameters of the above model. $\hat{\lambda}$ is estimated so as to minimise the one side t-statistic for $\alpha = 1$, which in turn leads to rejection of the null. One should note that in the absence of any deterministic change-point the test functioned as a stationarity test when empirically assessed.

5.4 Empirical quantification of false determination rates

All of these tests are posed as null hypothesis tests. As such they only reject the null hypothesis at a particular level once sufficient evidence is found against it, and when the data size is limited, the power (the probability of correctly rejecting the null hypothesis) is similarly reduced. Therefore, in R2019 (page 100) the four tests were each tested separately for their false positive and false negative rates using a Monte Carlo method.

This aids interpretation since data segments vary in length. Before proceeding further, one may ask how meaningful the nominal p-values are, or as in R2019, one can determine the minimum data length required to allow acceptance of a finding of both UR and non-UR separately, for each test.

5.5 Applying these UR tests

Assuming an objective change-point method has been used bounded between two objectively determined change-points. Do the assumptions of the detection method hold for the segment of data and to what extent?

These tests are all applied to the segments of data within which a single change-point has already been provisionally identified. The change-point itself is not otherwise considered. However, since the climate data being tested provisionally contains a deterministic change and only the ZA test is formulated with this as a ruling assumption, findings of non-stationarity may be caused by the presence of additional deterministic change-points below detection thresholds.

Level stationarity is not simply a zero trend, since data with zero trend may be either deterministically or stochastically level, and even if deterministic may not be linear. A deterministic change-point detection method may return indeterminate change-points given non-linear trend. The residuals around stochastic trend will retain a UR characteristic. Trend stationary data has level stationary residuals, as do discontinuous trend stationary data fitted appropriately.

A segment of data with a valid change point should not be found to be level stationary, it should not be in a segment with unit root behaviour, and if it shows trending behaviour this should not be due to a drifting unit root. It should also have low p-values by ANCOVA.

5.5.1 Level stationarity

The KPSS-L test is used here with an expectation that segments of climate data in which a change-point occurs contain a step-like shift but may also contain a change of trend. Hence it is used as a cross check. Further, once the deterministic internal shift and trend components are removed the residual should be both level and trend stationary. Level non-stationarity in the segment and level stationarity in the residuals supports the existence of a change-point.

5.5.2 Trend stationarity

Data with a provisional change of trend is expected to be non-trend-stationary. Data with constant trend and a step-like change may show as trend stationary depending on the assumptions of the specific test. The KPSS-T test and the ADF test as formulated here may return different results in the presence of a step-like shift and no trend change, with the ADF test showing trend stationarity and the KPSS showing non-stationarity.

5.5.3 Unit root/non-stationarity in the absence of any deterministic change

The presence of a unit root may cause the data to mimic either a step-like change or a change of trend. In either case the MSBV can return a step-like change. All four unit-root tests are expected to detect this, with the ADF being less powerful, partly due to a potential to overfit autocorrelation lags. Since the detection method has provisionally detected a change-point, tests on the residuals would likely all show non-stationarity, and similarly testing of the segment itself. The ZA test would likely be the most powerful.

5.5.4 Unit root/non-stationarity in the presence of deterministic change

This is a complex issue. The combination of UR and deterministic trend is potentially explosive [58]. On the other hand the climate system is physically bounded and so at worst the combination may appear as step-like. If tests support unit root in both the segment data and its residuals, either a genuine unit root is present or multiple deterministic changes are. Data with apparent UR that disappears in the residuals is consistent with a single deterministic change. However data with multiple change points is misspecified for all tests.

5.5.5 Misspecification due to use of averaging

As discussed, the TM/SI identifies the sampling model as a point of consideration, and data conditioning may itself be a misspecification to a given investigation. Climate data is not homogenous. Averaging is often assumed to increase the signal to noise ratio (S/N) but more localised features may fall below detectability thresholds. For step-like changes occurring at different times in different components, the steps are diluted but potentially, autocorrelation is induced. Further, if the changes differ slightly in time over a number of components then the deterministic shift-like changes may be confused with either stochastic or deterministic trend. Similarly trend changes: Only if the step-like or trend change happens simultaneously across all processes will the S/N increase. Data conditioning methods that imposes or presume smoothing may turn steps into trends. If autocorrelation is present as part of the signal in the component's data together with trend, the situation is still more complex.

Table 1 summarises the conditions that can be diagnosed with these UR tests. **Tables 2** and **3** provide interpretations.

5.6 The averaging of multiple datasets with autocorrelation

The "order" of an AR process is the number of lags, and also the polynomial order required to fit the error terms. The Dickey-Fuller equation (Eq. (6)) describes an autoregressive single lag, i.e. an AR(1) process. The sum of two AR(1) processes is most compactly represented as an autoregressive-moving average (ARMA) process of greater order, ARMA(2,1) [65].

If p and q are the lag order of processes, then two AR processes combine into an ARMA process, where the first parameter of the ARMA is the order of the AR part, and the second is the order of the moving average (MA) part.

$$AR(p) + AR(q) = ARMA(p + q, \max(p, q)) \tag{11}$$

Properties of Test	ADF (trend and drift)	KPSS (level stationarity)	KPSS (trend stationarity)	ZA
Ruling assumptions	No exogenous change.	No exogenous change.	No exogenous change.	Not combined exogenous change and unit root. At most one exogenous change (shift or trend change).
Null hypothesis, H0	I(1) Unit Root after allowing for autocorrelation and trend.	Stationarity	Trend stationarity	I(1) Unit Root with drift and no exogenous change
Contrast hypothesis, H1	Presumption of trend stationarity	I(0) Unit Root	I(1) Unit Root	Deterministic with possible exogenous change at a date
When at most a single exogenous change is present				
If exogenous change and UR present	Accept H0, i.e. UR	Prefer H1, i.e. UR	Prefer H1, i.e. UR	May prefer H1, i.e. exogenous change
If exogenous change but UR not present	May accept H0, i.e. UR	If constant trend and step-change then will accept H0, stationarity. Otherwise prefer H1, i.e. UR	If step-change only then will accept H0, trend stationarity. A strong trend change will prefer H1, i.e. UR	Prefer H1, i.e. exogenous change
Unit root but no exogenous change	Accept H0, i.e. UR	Prefer H1, i.e. UR	Prefer H1, i.e. UR	Accept H0, i.e. UR
When multiple exogenous changes are present				
Plus Unit Root	Accept H0, i.e. UR	Prefer H1, i.e. UR	Prefer H1, i.e. UR	Accept H0, i.e. UR
No Unit Root	May accept H0, i.e. UR	Prefer H1, i.e. UR	May prefer H1, i.e. UR if exogenous trend changes present	May accept H0, i.e. UR
After removal of all exogenous change				
If Unit Root	Accept H0, i.e. UR	Prefer H1, i.e. UR	Prefer H1, i.e. UR	Accept H0, i.e. UR
No Unit Root	Prefer H1, trend stationarity	Accept H0, stationarity (unless residual trend remains)	Accept H0, trend stationarity (unless residual trend remains)	Prefer H1, exogenous change (even if there is none)
After removal of main exogenous change but with exogenous change still present				
If Unit Root	Accept H0, i.e. UR	Prefer H1, i.e. UR	Prefer H1, i.e. UR	May prefer H1 if exactly one exogenous change remains. H0 of more than one.

Properties of Test	ADF (trend and drift)	KPSS (level stationarity)	KPSS (trend stationarity)	ZA
No Unit Root	Prefer H1, trend stationarity (even if residual trend remains)	Accept H0, level stationarity	May prefer H1	Prefer H1, exogenous change

Table 1.
Unit root tests used and their main assumptions (reproduced from R2019, Table Ch4.1.2). Possibilities not formally considered may deceive these tests by supporting either the null or contrast hypotheses.

Initial data with a presumptive step change	Residual with internal step and trends removed	Interpretations
H_0 rejected, accept as Exogenous/Stationary	H_0 rejected, accept as Exogenous/Stationary	There is a deterministic change with stationary residual.
	H_0 not rejected, accept as Endogenous/Non stationary	There is a deterministic change with non-stationary residual
H_0 not rejected, accept as Endogenous/Non stationary	H_0 rejected, accept as Exogenous/Stationary	Residual is non-stationary with two deterministic changes
		Residual is stationary with two deterministic changes
	H_0 not rejected, accept as Endogenous/Non stationary	Residual is non-stationary with zero exogenous changes: step-change is false positive
		Residual is stationary apart from two or more undetected change-points
		Residual is non-stationary with more than two deterministic change-points

Table 2.
Reproduced from R2019 Table Ch4.1.3: Expected outcomes of the Zivot Andrews test, given data with a presumptive step-like change plus a variety of additional conditions. The first and second columns define results of the tests on the initial data segment and the residual with internal step and trend removed. The last column lists interpretations of the pairs of results.

Treating the result of $AR(1) + AR(1) = ARMA(2, 1)$ as an AR(1) process may be deceptive. And yet in many analyses, the issue of the composition of the data is at best brushed off, and autocorrelation is in general approximated as AR(1). In R2019 apparent unit root-like behaviour in some zonal ocean temperature data sets resolves to deterministic shifts at different times in sub-sectors of those zones, and this affects the determination of change-points.

5.7 Reasoning about change-points

It is possible to examine the data segment and its residual and to have greatly increased confidence that the change-point methods are adequate to the task, and to broadly classify change-points detected as potentially affected by (a) misspecification of the detection tests with data out of its applicability range, (b) random-walks, (c) presence of undetected change-points, (d) some forms of model family misspecification.

Initial data with a presumptive step change	Residual with internal step and trends removed	Interpretations
KPSS-T H_0 not rejected accept as Stationary. *ADF* H_0 rejected accept as Stationary.	*KPSS-T* H_0 not rejected accept as Stationary. *ADF* H_0 rejected accept as Stationary.	Residual is stationary, the single change-point did not have a trend change
	KPSS-T H_0 rejected accept as Non stationary. *ADF* H_0 not rejected accept as Non stationary.	Location of a single change-point misidentified so that the trend is also miscalculated
KPSS-T H_0 rejected accept as Non stationary. *ADF* H_0 not rejected accept as Non stationary.	*KPSS-T* H_0 not rejected accept as Stationary. *ADF* H_0 rejected accept as Stationary.	Residual is stationary and change-point included a trend change
	KPSS-T H_0 rejected accept as Non stationary. *ADF* H_0 not rejected accept as Non stationary.	The data segment is non-stationary and the provisional change-point may be a false positive.
		Residual is non-stationary. The initial segment contained a step and/or trend change.

Table 3.
Reproduced from R2019 Table Ch4.1.4: Expected outcomes of the KPSS-T and ADF tests, given data with a presumptive step-like change plus a variety of different conditions.

In Chapter 4 of R2019 the KPSS-T, ADF and ZA tests are combined to provide a classification scheme for change-points (**Table 4**). Using this classification scheme it became straight-forward to determine that regime changes over land and ocean

Classification	Reasoning and interpretation
Single, non-stationary	We accept the single exogenous change, but the residuals are not stationary, leaving open the possibility of undetected features. The ZA has reverted from Exogenous/stationary to Endogenous/non-stationary in the residuals, consistent with a single exogenous change plus a presumptive unit root. The presumptive unit root in the residuals is not reliably separable from multiple change-points below detectability.
Single, Stationary	We accept the step-change detected by the MSBV as the single exogenous change with no stochastic trend. The residuals are stationary supporting the single change-point. The ZA test does not change from exogenous/stationary
Single, N/A	We accept the step-change detected by the MSBV, without a valid ZA result, noting that there is insufficient data to probe further.
Non-stationary	We have evidence that the data segment contains sufficient non-stationarity as to cast doubt on the MSBV. The ZA test does not revert from endogenous/non-stationary and neither do the other tests. Hence the removal of a single change-point has had no apparent effect. Multiple change-points on top of a non-stationary background is too complex a situation to detect with these tests.
Multiple, Stationary	We may be dealing with a pair of exogenous changes. The ZA reverted from non-stationary to stationary with other tests consistent with this. Potentially a single additional undetected change-point, since two exogenous changes may be classified as an endogenous change in the ZA.
Stationary	Possible false positive or weak change in stationary data
N/A	Not classifiable/indeterminate

Table 4.
Extended from R2019 Table Ch4.1.5: classifications of data segments.

differ in complexity. Sharpening the testing, it also further supported the principal findings of R2019, that abrupt shifts relate directly to warming; in their extent, frequency and intensity; and more so at finer scale. For this paper the last two additional classes apply when ANCOVA does not support a change-point. An example is provided in the appendix.

6. Examples

Figure 1 below, illustrates the difference between analysis, commencing with the MSBV, of global mean temperature and the area averaged Northern mid-latitude (NML) temperature. While the step change after 1996 is very obvious in the zonal data, the change is sometimes disputed in the global signal. **Table 5**, adds strong support to the contention that the so-called hiatus was a significant event, but not on the basis of trends. The 1988 event in the NML corresponds to an atmospheric reorganisation and extensive biophysical changes regionally [10]. All of the change-points occur in data which is otherwise stationary.

Figure 2 below, illustrates the contribution to reasoning about the nature of decadal climate regimes which follows from a reasoned classification scheme. If the global temperatures are averaged over smaller areas, and then step-change points are calculated it becomes more likely that the data will present as stationary. This shows that the zonal data are not homogeneous with respect to regime shifts; that regime shifts are more regional. Note also the difference between land (almost always stationary) and ocean (less so), supporting the ocean as being more complex. There is also a tendency for land shifts to be a year or two delayed.

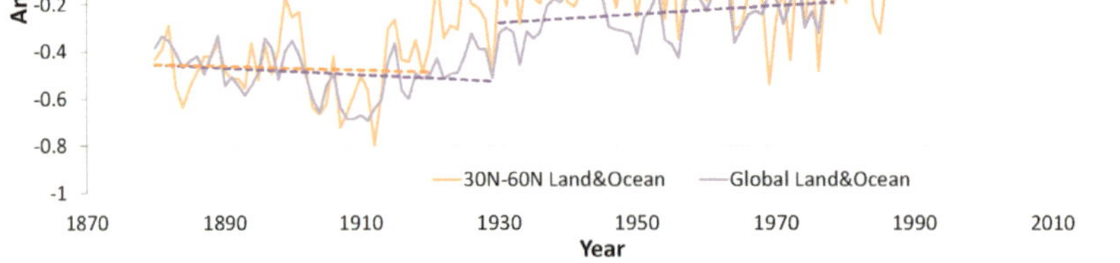

Figure 1.
Adapted from R2019, Figure Ch3.11. Step-like shifts in the Northern extra-tropics compared to those detected in global data.

7. Conclusions

The principal contribution of this paper is to expand on the use of misspecification testing to strengthen reasoning about abrupt shifts in time-series. We focus on climate data records and statistical model specification with respect to the data. Probing the misspecification of statistical models helps ensure that tests better represent probative criteria, and better distinguish between them.

DOI: http://dx.doi.org/10.5772/intechopen.98364

Zone	MSBV			KPSS-L		KPSS-T		ADF		Zivot Andrews			ANOVA/ANCOVA		
	First Changed Year	Internal Shift (°C)	Internal Trend Change (°C/Yr)	Data segment	Residuals	Data segment	Residuals	Data segment	Residuals	Data segment	Residuals	First Changed Year	ANOVA-Internal Shift(Pr)	ANOVA-Trend Change (Pr)	ANCOVA-Change-point(Pr)
30 N–60 N	1921	0.34	0.001	NS	S	NS	S	NS	S	S	S	1921	***	—	***
30 N–60 N	1988	0.37	−0.014	NS	S	NS	S	NS	S	S	S	1964	**	—	***
30 N–60 N	1997	0.43	0.019	NS	S	S	S	NS	S	NS	S	1997	***	—	**
Global	1930	0.25	0.003	NS	S	NS	S	NS	S	S	S	1914	***	*	***
Global	1979	0.12	0.009	NS	S	NS	S	NS	S	S	S	1946	**	*	***
Global	1997	0.16	−0.005	NS	S	S	S	S	S	S	S	1997	***	—	**

Table 5.
Adapted from R2019 Table A4.1.30: For the UR tests red text denotes results of tests where the data length may affect precision. NS = non-stationary, S = stationary. *** $p <= 0.001$, ** $0.001 > p <= 0.01$, * $0.01 < p >= 0.05$.

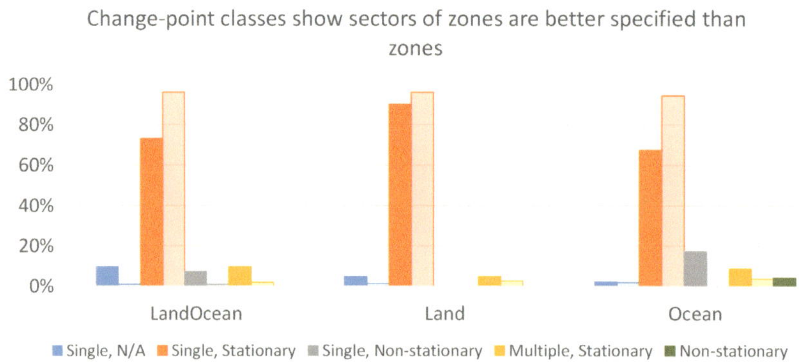

Figure 2.
Adapted from R2019, Figure Ch5.5, Data becomes less complex at finer scale, as evidenced by classification. The proportions of each class of change-point are shown for the same data averages over 30 degree zones (saturated colors) and 45 degree sectors of the same zones (unsaturated colors).

The TM/SI framework has been suggested as a variation on previously discussed inductive frameworks. While it assists adequate testing of physical hypotheses; none the less, climate is a complex system [66]. Thus the ongoing assessment of testing procedures, and with it model specification.

Misspecification testing supports severe testing. Severe testing is strengthened by improved one to one mapping between physical features and statistical test outcomes. The tests outlined here assist a probative analysis, firstly by adding nuance to the findings, and secondly by providing the basis of a change-point classification, they assist strong reasoning. They have been selected because they are individually automatable and complementary, and the utility of this has been indicated in the case study. The chain of reasoning involved in the use of multiple tests is complex but the final classification scheme is compact and as seen, informative.

A basis has also been established for potentially detecting signatures of a data composition misspecification whereby features emerge or submerge in composited data due to averaging of signals (especially ones moving in time and space). The signature is a reduction in non-stationarity when signals are decomposed or segmented using the MSBV as seen in **Figure 2**. The same issue also affects both autocorrelation and trend analysis simply because step-like dislocations in data are generally deceptive for the regressions embedded in many general methods.

Acknowledgements

R.N. Jones is a Professorial Fellow of Victoria Institute for Strategic Economic Studies in Melbourne. J. H. Ricketts was the holder of a Victoria University postgraduate research scholarship. The anonymous reviewers of a joint paper and a joint conference paper, and two thesis reviewers, all contributed substantial improvements and assisted development of the ideas expressed here. We would like to acknowledge with both gratitude and sorrow the lasting influence of our colleague and friend, Dr. Penny Whetton, who passed away too soon.

Notes

A number of tables and figure are adapted from the PhD thesis of JH Ricketts [13], mostly chapter 4. A peer reviewed joint paper [21] and a joint conference paper [21] are also sourced.

DOI: http://dx.doi.org/10.5772/intechopen.98364

Koninklijk Nederlands Meteorologisch Instituut (KNMI) make available the KNMI Climate Explorer and this was a valuable resource.

Other data has been sourced from, Met Office Hadley Centre, NASA, Goddard Institute for Space Studies and United States National Climatic Data Center.

A. Appendix

During sensitivity testing of the detection and characterization tests in R2019 simulations were run, including assessments of (a) the effects of shifts single and multiple shifts below detection thresholds, (b) multiple shifts close in time, (c) high levels of autocorrelation, (d) state switching between deterministic and stochastic data, and (e) curvilinear trends. This illustrative example is an extension of one part of that work.

A.1 Synthetic climate-like data

Following R2019, a suite of four artificial multi-step time series ('A' to 'D') was constructed and analyzed by MSBV then validation tests were run against both the shifts as detected by MSBV and as originally defined.

A is an artificial 200 year annual temperature consisting of random data (and a standard deviation, σ, of 0.44) with lag 1 autocorrelation of 25%, lag 7 autocorrelation of 10%, centered about zero, plus a quadratic trend curve rising 2.1 degrees. The degree of autocorrelation is consistent with the findings of [67].

Eight shifts random shift level (mean 1.5 σ) are added at defined times (Shifts of 1.5σ are less than MSBV reliability threshold) (**Table 6**).

To assess the suite presence of UR with deterministic trends plus shifts, shifts without trends, and UR alone, red-noise (summed white noise, $\mu = 0, \sigma = 0.44$) was added to A to produce set B, set C is the defined steps plus red-noise and D is red-noise only.

A.1.1 Studentized Breusch-Pagan test for heteroskedasticity

The studentized BP test was run for the disjoint regression of breaks detected (Break model), and also for the breaks as defined (**Table 7**). A linear model and a

Year	1954	1982	1998	2029	2035	2054	2070	2096	Total
Shifts in K (σ)	0.57 (1.30)	**0.34** **(0.77)**	0.72 (1.64)	0.85 (1.93)	1.00 (2.27)	0.61 (1.39)	0.94 (2.14)	**0.31** **(0.70)**	5.34 (12.14)

Table 6.
Adapted from R2019, Table Ch4.1.6 Synthetic Data Timing and extent of Shifts. Total Rise is shown both as anomaly and as standard deviations. Shifts of < 0.5 are not guaranteed to be found by MSBV and are bolded.

Dataset	Break model	Defined	Linear model	Quad model
A.	0.6802	0.6959	0.0241	0.0001
B.	0.0034	0.0270	0.0000	0.9108
C.	0.0000	0.0039	0.0000	0.5205
D.	0.2870	0.0024	0.0000	0.0457

Table 7.
Studentized Breusch-Pagan Test results. Green denotes 0.01 < p < 0.05, red 0.01 > p, black p > 0.05. A null hypothesis of homoscedasticity is rejected for low p-values.

quadratic model were also run for comparison. Data sets A and D appear to have homoskedastic residuals for their breaks given the detected shifts, and yet A is deterministic and D is non-deterministic. Datasets B and C, on the other hand appear homoskedastic given a quadratic model. Note that the SBP operates under an i.i.d assumption which is violated by sets B, C and D.

The breaks returned by the MSBV, and breaks defined, for A both form an adequate model. The breaks returned by the MSBV for D form an adequate model whereas the defined set – not present in D – does not. B contains a curvilinear trend, plus along with C, shifts which also induce an apparent curvilinearity. As can be seen from **Figure 3** the MSBV does quite well at locating the change-points.

Note: The data tested, residuals of detected change-points, will almost always appear to be deterministic when tested by the UR tests. This is because the residual

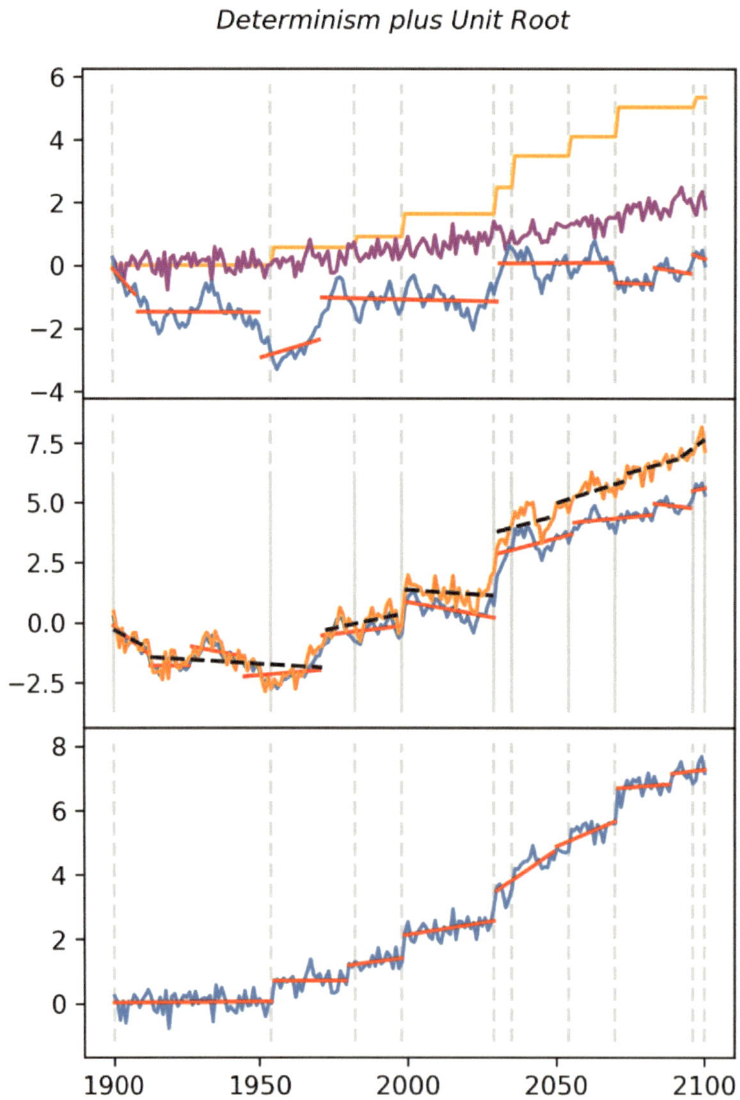

Figure 3.
Top: Blue is data set D, red is break-segments determined by MSBV. Magenta is auto-correlated noise plus quadratic trend. Orange is the defined shifts. Middle: Dataset C is shown as orange (black breaks), dataset B is blue (red breaks). Bottom: dataset A in blue, red is break-segments determined by MSBV. Vertical grey reference lines indicate defined shifts.

Defined	A	B	C	D
		1912(++S)	1912(NS)	1908(+NS)
1954	1954(+S)		1926(NS)	
			1944(+NS)	1950(+NS)
		1971(++S)	1970(+NS)	1970(NS)
1982	1979(+S)			
1998	1998(+S)	1998(+S)	1998(NS)	
2029	2029(+S)	2029(++S)	2029(+NS)	2030(NS)
2035				
2054	2049(S)	2049(S)	2055(+NS)	
2070	2070(+S)	2073(S)		2069(+NS)
	2088(S)		2082(++S)	2082(+)
2096		2091(−)	2095(+)	2095(+)

Table 8.
Changes defined and years detected in each dataset. Annotations denote segment classification, + is single change-point, ++ possible multiple changes, S is stationary, NS is non-stationary. Red denotes ANCOVA $p <= 0.05$.

of deterministic signal is expected to be deterministic, whereas the residual of a purely non-deterministic signal from which a deterministic components has been subtracted *acquires* a deterministic component and appears mean-reverting, i.e. I (0). There is no current method for dealing with multiple deterministic changes in a UR time series, and blended series such as B and C will not meet the criteria of a UR series. In fact looking at D only through the lens of the SBP and UR tests of the residuals does not distinguish it from a deterministic time series like A. The difference only becomes apparent when the individual change-points are tested (see **Tables 8** and **9**).

A.1.2 Analysis of individual change-points

The full analysis results are available on-line at https://cdn.intechopen.com/public/docs/230558_files.zip.

Set A. One pair of defined change-points violated an assumption of the MSBV that rejects shifts within a seven year refractory period (defined as 2029, 2035), selecting only 2029 which registers as a strong shift embedded in stationary data with an internal trend (notably the ZA test of the residuals locates 2035). When the data is broken up according to the defined shifts, 2035 registers as a strong shift in non-stationary data, and evidence for the internal trend weakens. The defined small shift in 2054 following 2035 was attributed to 2049 after 2029 but not supported by ANCOVA and the segment was classified as non-stationary. The ZA suggests a change in 2034 but non-stationarity in the residuals. All other change-points were detected as defined and classified as single change-points in trend-stationary data.

Datasets B though D represent increasingly UR dominated data. For B (combining deterministic trends and red noise), the only detected shift that is classified as a single shift in stationary data is 1998, all prior being classified as having possible multiple sub-detection shifts, and all following being rejected by ANCOVA although the segments are classed as stationary. Sets C (UR with shifts) and D (UR only), show that the MSBV by itself is vulnerable to non-determinism.

DataSet	Breaks	Single, Non-stationary	Non-stationary	Single, N/A	Stationary	Multiple, Stationary	Single, Stationary	N/A	Sum
A.	Found	0	1	0	1	0	5	0	7
A.	Defined	0	1	0	2	0	5	0	8
B.	Found	0	0	0	2	3	1	1	7
B.	Defined	0	2	0	2	1	3	0	8
C.	Found	4	3	1	0	1	0	0	9
C.	Defined	3	3	0	0	1	1	0	8
D.	Found	3	2	2	0	0	0	0	7
D.	Defined	4	2	1	1	0	0	0	8

Table 9.
Numbers of change-points assigned to each class. Note that C and D differ from A and B by having non-stationary residuals, where as B differs from A by displaying evidence of undetected multiple change-points.

DOI: http://dx.doi.org/10.5772/intechopen.98364

The principal indication that a change-point dominated time-series has an underlying difference stationarity (i.e. red, or brown noise) is given by examination of the segmentation and not the residuals.

Author details

James Ricketts* and Roger Jones
Victoria University, Melbourne, Australia

*Address all correspondence to: james.ricketts@live.vu.edu.au

References

[1] Mayo DG. An error-statistical philosophy of evidence. The nature of scientific evidence: Statistical, philosophical and empirical considerations. 2004:79-96.

[2] Mayo DG, Spanos A. Methodology in Practice: Statistical Misspecification Testing. Philosophy of Science. 2004;71 (5):1007-25. doi: 10.1086/425064.

[3] Mayo DG, Spanos A. Severe Testing as a Basic Concept in a Neyman–Pearson Philosophy of Induction. The British Journal for the Philosophy of Science. 2006;57(2):323-57.

[4] Mayo DG, Spanos A. Error statistics. Handbook of the philosophy of science. 2011;7:153-98.

[5] Spanos A, Mayo DG. Error statistical modeling and inference: Where methodology meets ontology. Synthese. 2015;192(11):3533-55.

[6] Mayo DG. Statistical Inference as Severe Testing: How to Get Beyond the Statistics Wars: Cambridge University Press; 2018.

[7] Jones RN. Detecting and attributing nonlinear anthropogenic regional warming in southeastern Australia. Journal of Geophysical Research: Atmospheres (1984–2012). 2012;117 (D4).

[8] Belolipetsky P. The Shifts Hypothesis-an alternative view of global climate change. arXiv preprint arXiv: 14065805. 2014.

[9] Belolipetsky P, Bartsev S, Ivanova Y, Saltykov M. Hidden staircase signal in recent climate dynamic. Asia-Pacific J Atmos Sci. 2015;51(4):323-30. doi: 10.1007/s13143-015-0081-6.

[10] Reid PC, Hari RE, Beaugrand G, Livingstone DM, Marty C, Straile D, et al. Global impacts of the 1980s regime shift. Global change biology. 2015.

[11] Bartsev S, Belolipetskii P, Degermendzhi A, editors. Multistable states in the biosphere-climate system: towards conceptual models. IOP Conference Series: Materials Science and Engineering; 2017: IOP Publishing.

[12] Jones RN, Ricketts JH. Reconciling the signal and noise of atmospheric warming on decadal timescales. Earth Syst Dynam. 2017;8(1):177-210. doi: 10.5194/esd-8-177-2017.

[13] Ricketts JH. Understanding the Nature of Abrupt Decadal Shifts in a Changing Climate. Melbourne: Victoria University; 2019.

[14] Suppes P. Models of data. In: Nagel E, Suppes P, Tarski A, editors. Logic, methodology, and philosophy of science: Proceedings of the 1960 International Congress; Stanford, CA: Stanford University Press.; 1962. p. 252-61.

[15] Haig BD. Tests of Statistical Significance Made Sound. Educational and Psychological Measurement. 2016: 0013164416667981.

[16] Corti S, Molteni F, Palmer T. Signature of recent climate change in frequencies of natural atmospheric circulation regimes. Nature. 1999;398 (6730):799-802.

[17] Branstator G, Selten F. “Modes of variability” and climate change. Journal of Climate. 2009;22(10):2639-58.

[18] Katzav J. Severe testing of climate change hypotheses. Studies in History and Philosophy of Science Part B: Studies in History and Philosophy of Modern Physics. 2013;44(4):433-41. doi: http://dx.doi.org/10.1016/j.shpsb .2013.09.003.

[19] Katzav J. Should we assess climate model predictions in light of severe tests? EOS, Transactions American Geophysical Union. 2011;92(23):195-.

[20] Katzav J, Dijkstra HA, de Laat ATJ. Assessing climate model projections: State of the art and philosophical reflections. Studies in History and Philosophy of Science Part B: Studies in History and Philosophy of Modern Physics. 2012;43(4):258-76. doi: http://dx.doi.org/10.1016/j.shpsb.2012.07.002.

[21] Ricketts JH, Jones RN. Characterizing change-points in climate series with a severe approach. In: Syme G, Hatton MacDonald D, Fulton B, Piantadosi J, editors. The 22nd International Congress on Modelling and Simulation (MODSIM2017); 3-8 December 2017; Hobart: The Modelling and Simulation Society of Australia and New Zealand Inc.; 2017.

[22] Salmon WC. Scientific explanation and the causal structure of the world: Princeton University Press; 2020.

[23] Mayo DG. Error and the growth of experimental knowledge: University of Chicago Press; 1996.

[24] Jeffrey SJ, Carter JO, Moodie KB, Beswick AR. Using spatial interpolation to construct a comprehensive archive of Australian climate data. Environmental Modelling and Software. 2001;16(4): 309-30.

[25] Cochrane W, Cox G. Experimental designs., 2nd edn (John Wiley & Sons: Sydney). 1957.

[26] Minobe S. A 50–70 year climatic oscillation over the North Pacific and North America. Geophysical Research Letters. 1997;24(6):683-6. doi: 10.1029/97GL00504.

[27] Mantua NJ, Hare SR, Zhang Y, Wallace JM, Francis RC. A Pacific interdecadal climate oscillation with impacts on salmon production. Bulletin of the american Meteorological Society. 1997;78(6):1069-79.

[28] Trenberth KE, Hurrell JW. Decadal atmosphere-ocean variations in the Pacific. Climate Dynamics. 1994;9(6): 303-19.

[29] Hope P, Drosdowsky W, Nicholls N. Shifts in the synoptic systems influencing southwest Western Australia. Climate Dynamics. 2006;26 (7-8):751-64. doi: 10.1007/s00382-006-0115-y.

[30] Rajaratnam B, Romano J, Tsiang M, Diffenbaugh N. Debunking the climate hiatus. Climatic Change. 2015:1-12. doi: 10.1007/s10584-015-1495-y.

[31] Lewandowsky S, Risbey JS, Oreskes N. The "Pause" in Global Warming: Turning a Routine Fluctuation into a Problem for Science. Bulletin of the American Meteorological Society. 2015. doi: 10.1175/BAMS-D-14-00106.1.

[32] Risbey JS, Lewandowsky S, Cowtan K, Oreskes N, Rahmstorf S, Jokimäki A, et al. A fluctuation in surface temperature in historical context: reassessment and retrospective on the evidence. Environmental Research Letters. 2018;13(12):123008.

[33] Cahill N, Rahmstorf S, Parnell AC. Change points of global temperature. Environmental Research Letters. 2015; 10(8):084002.

[34] Fyfe JC, Meehl GA, England MH, Mann ME, Santer BD, Flato GM, et al. Making sense of the early-2000s warming slowdown. Nature Climate Change. 2016;6(3):224-8.

[35] Meehl GA, Hu A, Arblaster JM, Fasullo J, Trenberth KE. Externally Forced and Internally Generated Decadal Climate Variability Associated with the Interdecadal Pacific Oscillation.

Journal of Climate. 2013;26(18): 7298-310. doi: 10.1175/JCLI-D-12-00548.1.

[36] Trenberth KE. Has there been a hiatus? Science. 2015;349(6249):691-2.

[37] Foster G, Abraham J. Lack of evidence for a slowdown in global temperature. US CLIVAR. 2015:6.

[38] Vives B, Jones RN. Detection of abrupt changes in Australian decadal rainfall (1890-1989): CSIRO Atmospheric Research; 2005.

[39] White H. A heteroskedasticity-consistent covariance matrix estimator and a direct test for heteroskedasticity. Econometrica: Journal of the Econometric Society. 1980:817-38.

[40] Rodionov SN. Use of prewhitening in climate regime shift detection. Geophysical Research Letters. 2006;33 (12).

[41] Mizon GE. A simple message for autocorrelation correctors: Don't. Journal of Econometrics. 1995;69(1): 267-88.

[42] Beaulieu C, Killick R. Distinguishing trends and shifts from memory in climate data. Journal of Climate. 2018;31 (23):9519-43.

[43] Percival DB, Overland JE, Mofjeld HO. Interpretation of North Pacific variability as a short-and long-memory process. Journal of Climate. 2001;14(24):4545-59.

[44] Stock JH. Unit roots, structural breaks and trends. Handbook of econometrics. 41994. p. 2739-841.

[45] Chang Y, Kaufmann RK, Kim CS, Miller JI, Park JY, Park S. Time series analysis of global temperature distributions: Identifying and estimating persistent features in temperature anomalies. 2016.

[46] Tsonis AA, Swanson K, Kravtsov S. A new dynamical mechanism for major climate shifts. Geophysical Research Letters. 2007;34(13).

[47] Fischer JW, Walter WD, Avery ML. Brownian Bridge Movement Models to Characterize Birds' Home Ranges: Modelos de Movimiento de Puente Browniano Para Caracterizar el Rango de Hogar de las Aves. The Condor. 2013; 115(2):298-305.

[48] Dickey DA, Fuller WA. Likelihood ratio statistics for autoregressive time series with a unit root. Econometrica: Journal of the Econometric Society. 1981:1057-72.

[49] Kwiatkowski D, Phillips PC, Schmidt P, Shin Y. Testing the null hypothesis of stationarity against the alternative of a unit root: How sure are we that economic time series have a unit root? Journal of econometrics. 1992;54 (1):159-78.

[50] Zivot E, Andrews DW. Further Evidence on the Great Crash, the Oil-Price Shock, and the Unit-Root Hypothesis. Journal of Business & Economic Statistics. 1992.

[51] Fukac M. Inflation Expectations in the Czech Interbank Market. 2005.

[52] Kejriwal M, Perron P. A sequential procedure to determine the number of breaks in trend with an integrated or stationary noise component. Journal of Time Series Analysis. 2010;31(5):305-28.

[53] Harvey DI, Leybourne SJ, Taylor AR. Testing for unit roots in the possible presence of multiple trend breaks using minimum Dickey–Fuller statistics. Journal of Econometrics. 2013; 177(2):265-84.

[54] Liddle B, Messinis G. Revisiting sulfur Kuznets curves with endogenous breaks modeling: Substantial evidence of inverted-Us/Vs for individual OECD

countries. Economic Modelling. 2015; 49:278-85. doi: http://dx.doi.org/ 10.1016/j.econmod.2015.04.012.

[55] Elliott G, Rothenberg TJ, Stock JH. Efficient tests for an autoregressive unit root. National Bureau of Economic Research Cambridge, Mass., USA; 1992.

[56] Byrne JP, Perman R. Unit roots and structural breaks: a survey of the literature. Paper provided by Business School-Economics, University of Glasgow in its series Working Papers with. 2006;(2006_10).

[57] Pfaff B, Zivot E, Stigler M. Unit Root and Cointegration Tests for Time Series Data. 2016.

[58] Hacker RS. The Effectiveness of Information Criteria in Determining Unit Root and Trend Status. Royal Institute of Technology, CESIS-Centre of Excellence for Science and Innovation Studies, 2010.

[59] Kočenda E, Černý A. Elements of time series econometrics: An applied approach: Charles University in Prague, Karolinum Press; 2015.

[60] Trapletti A, Hornick K, LeBaron B. Time series analysis and computational finance. 2017.

[61] Newey WK, West KD. Automatic lag selection in covariance matrix estimation. The Review of Economic Studies. 1994;61(4):631-53.

[62] Glynn J, Perera N, Verma R. Unit root tests and structural breaks: a survey with applications. Faculty of Commerce-Papers. 2007:455.

[63] Gay-Garcia C, Estrada F, Sánchez A. Global and hemispheric temperatures revisited. Climatic Change. 2009;94 (3-4):333-49.

[64] Lumsdaine RL, Papell DH. Multiple trend breaks and the unit-root hypothesis. Review of economics and Statistics. 1997;79(2):212-8.

[65] Granger CW, Morris MJ. Time series modelling and interpretation. Journal of the Royal Statistical Society Series A (General). 1976:246-57.

[66] Jones RN, Ricketts JH. The Pacific Ocean heat engine: global climate's regulator. Earth System Dynamics (for open review). 2019.

[67] Allen MR, Smith LA. Investigating the origins and significance of low-frequency modes of climate variability. Geophysical Research Letters. 1994;21 (10):883-6.

Chapter 14

International Benchmark Activity in the Field of Sodium Fast Reactors

Domenico De Luca, Simone Di Pasquale, Marco Cherubini, Alessandro Petruzzi and Gianni Bruna

Abstract

Global interest in fast reactors has been growing since their inception in 1960 because they can provide efficient, safe, and sustainable energy. Their closed fuel cycle can support long-term nuclear power development as part of the world's future energy mix and decrease the burden of nuclear waste. In addition to current fast reactors construction projects, several countries are engaged in intense R&D and innovation programs for the development of innovative, or Generation IV, fast reactor concepts. Within this framework, NINE is very actively participating in various Coordinated Research Projects (CRPs) organized by the IAEA, aimed at improving Member States' fast reactor analytical simulation capabilities and international qualification through code-to-code comparison, as well as experimental validation on mock-up experiment results of codes currently employed in the field of fast reactors. The first CRP was focused on the benchmark analysis of Experimental Breeder Reactor II (EBR-II) Shutdown Heat Removal Test (SHRT-17), protected loss-of-flow transient, which ended in the 2017 with the publication of the IAEA-TECDOC-1819. In the framework of this project, the NINE Validation Process– developed in the framework of NEMM (NINE Evaluation Model Methodology) – has been proposed and adopted by most of the organizations to support the interpretation of the results calculated by the CRP participants and the understanding of the reasons for differences between the participants' simulation results and the experimental data. A second project regards the CRP focused on benchmark analysis of one of the unprotected passive safety demonstration tests performed at the Fast Flux Test Facility (FFTF), the Loss of Flow Without Scram (LOFWOS) Test #13, started in 2018. A detailed nodalization has been developed by NINE following its nodalization techniques and the NINE validation procedure has been adopted to validate the Simulation Model (SM) against the experimental data of the selected test. The third activity deals with the neutronics benchmark of China Experimental Fast Reactor (CEFR) Start-Up Tests, a CRP proposed by the China Institute of Atomic Energy (CIAE) launched in 2018 the main objective of which is to improve the understanding of the start-up of a SFR and to validate the fast reactor analysis computer codes against CEFR experimental data. A series of start-up tests have been analyzed in this benchmark and NINE also proposed and organized a further work package focused on the sensitivity and uncertainty analysis of the first criticality test. The present chapter intends to summarize the results achieved using the codes currently employed in the field of fast reactor in the framework of international projects and benchmarks in which NINE was involved

 IntechOpen

and emphasize how the application of developed procedures allows to validate the SM results and validate the computer codes against experimental data.

Keywords: SFR, Benchmark, EBR-II, FFTF, CEFR, M&S tools

1. Introduction

1.1 History, main features, advantages and future of sodium-cooled fast reactors

Since the very beginning of its commercial operation, immediately after the end of the Second World War, nuclear energy has been getting a significant and often increasing part in the production of safe, secure, and economic low carbon electricity.

Innovation has always been - and still is nowadays - a powerful engine for progress in the fields of regulation (also including the trans-national aspects of the emergency management), safety (with a specific care and attention paid to severe accident prevention and mitigation, e.g. through inherent and passive safety features), reliability and efficiency in design and operation (including reliability and independence of control systems), incineration of long-life by-products of the fission-conversion-breeding process, non-proliferation (uselessness of fuel materials for weapon production), environmental impact (to the air, the soil and the water, both in normal operation and in emergency), management of high and low activity wastes, and also in the very sensitive domain of the public-awareness and acceptance, which are the key-issues for the civil nuclear future [1, 2].

This trend has been even more reinforced after the Fukushima events, also accounting for the wide stress test campaign conducted worldwide, as well as the large effort for public information, participation, and inclusion carried-out by International Bodies, Governments, Constructors, Operators, Academia and Research Organizations [3].

Safety has undergone a continuous improvement effort and has been a relevant driving force for progress, improvement as well as research and development in different fields of endeavor for the current GEN III (Generation III) and GEN III+ (Generation III plus) reactor designs, but also for the advanced concept-designs both inside the GIF (Generation IV International Forum) framework and outside, and in complement to it, e.g. with the ever growing interest for the SMRs (Small Modular Reactors), small compact elementary modules, generally sizing from 10MWe to 300MWe, which are designed and engineered along with a modular construction approach enabling to combine them and incrementally extend the power capacity of the overall plant thus offering economy of scale and reducing both capital costs and construction time [4]. Designs with power outputs smaller than 10MWe, often designed for semiautonomous operation, have been referred to as Micro Modular Reactors (MMRs).

Today, facing the high investment needs and the ever increasing costs, the large delays in the licensing process and the construction, the highly expensive financing modes as well as the low public acceptance and sometimes even the fierce opposition of a majority of the population, some developed countries (mainly in the Western Europe, even if Europe in a whole lasts hosting the largest nuclear capacity of the world), have decided to either step out or phase out nuclear energy in a short-medium term.

Nevertheless, nuclear energy still enjoys an increasing and dynamic trend. The Year 2018 has even been a hit as for the installed new nuclear capacity, mainly because the interest for nuclear reactors has widely moved from developed to emergent - developing countries. This trend is to continue and even expand as, according to current estimations, the installed nuclear capacity should double in the emergent economies within the next 20years. Relying upon a robust industrial capacity, the Russian Federation is today by far the larger exporter/provider of nuclear technology worldwide, and the People Republic of China is on the way to become a future leader in the nuclear field.

In order to allow nuclear power contribute effectively to the solution of the global warming challenge in the future, it shall be necessary: to continuously up-date and improve regulations; to enhance the safety under the guidance of proactive, transparent and independent Safety Authorities; to establish suitable roadmaps providing all the actors in the nuclear field with a medium - long term clear vision, and to reduce overall costs through continuous improvement, harmonization of practices and standardization. But, mostly, it will be worth addressing and providing a long-lasting and sustainable solution to the crucial problem of the long-lived wastes.

Today, the installed nuclear capacity is by far from GEN III reactors, only a few of them belonging to the GEN III+ generation (which includes, e.g., French EPR, American AP-1000, Russian VVER-1000 ...), and even less to other concepts. During their operation, such reactors produce, as by-products of the fission-conversion-breeding process, a large quantity of long-lived isotopes, quoted as Actinides and/or Minor Actinides, depending on their features and nature, which contribute to the activity of the spent fuel for thousands of years.

The build-up of such by-products turns-out a major challenge both from the non-proliferation and the waste management viewpoint. Their recycling in the reactor fuel as well as their incineration through suitable strategies will contribute to "close the cycle", i.e., at least theoretically, to bring the spent fuel activity back to a level comparable to the natural earth radiation. The acceptance of the public of further installations of nuclear power plants will strongly depend in the future on this crucial problem.

Fast reactors closed fuel cycle can efficiently and effectively contribute to the solution of the problem decreasing the burden of nuclear waste and supporting long-term nuclear power development as part of the world's future energy mix [5].

Global interest in fast reactors has been growing since their inception in 1960's because they can provide efficient, safe and sustainable energy. In addition to the current fast reactor construction projects underway, several countries are engaged in intense research and development programs for the development of innovative, Gen IV, fast reactor concepts, as proposed by the GIF. They include three fast neutrons concepts: the SFR - Sodium Fast Reactor -, the LFR - Lead (or Lead-Bismuth) Fast Reactor - and the GFR (Gas Fast Reactor), as well as the MSR (Molten Salt Reactor) which can be declined both in a thermal and a fast neutrons version.

Moreover, current developments of SMRs include, among the more than 100 versions under study, development and/or licensing, several fast neutrons concepts, even though the most mature ones are undoubtedly based on LWR (Light Water Reactor) technology. The fast SMRs, in addition to their efficient use of the fuel, are flexible because they can operate either as breeders, to produce fissile material, or as burners of Plutonium and/or long-lived Minor Actinides. Combining this capability with the benefits in terms of power generation flexibility, SMRs could turn-out quite attracting.

The SFR is, by far, the fast reactor technology most widely spread-out worldwide. It enjoys an acknowledged maturity due to the numerous constructions and because it underwent many years of operation in several countries, from the late '60 prototypes up to the development and deployment of the industrial French fleet (including RAPSODIE, Phénix and Superphénix - the biggest fast reactor ever built, now decommissioned -, and the project ASTRID, now delayed), and other reactors now either in operation or under construction in Russia, India, China and Japan (see **Figure 1**).

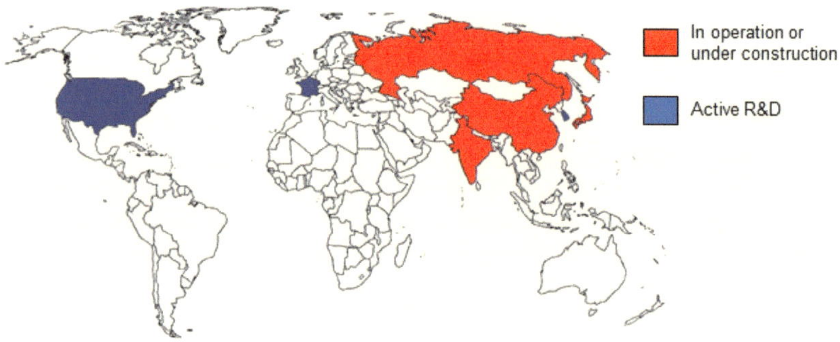

Figure 1.
World Sodium Fast Reactor Status.

Design and operation of such reactors are demanding extended computation capacity, to assess their safety, security, and economics [6], which justifies the organization under IAEA's umbrella of Coordinated Research Projects (CRPs) aimed at improving Member States' fast reactor analytical simulation capabilities and international qualification through code-to-code comparison, as well as experimental validation on mock-up experiment results of codes currently employed in the field of fast reactors. NINE is very actively participating in these exercises, and sometimes conducting them.

The present chapter summarizes the results and discusses the main outcomes of the above-mentioned benchmark exercises, in the aim at underlying the wide convergence among the computational tools adopted by the participants, as well as detecting the main discrepancies and seeking for their common origin and trend, whether and whenever existing. That should enable defining a mid-term vision for further development of the computer codes in the field of fast reactors, whatever their features and nature, and identifying new needs for their extended validation against either available or expected experimental data.

1.2 NINE involvement/interest in sodium-cooled fast reactors

Starting from the considerations above regarding the deployment of fast reactors and the maturity gained by the SFR, NINE joints the effort of International community to assess the actual computational capabilities in modeling SFRs features. Taking advantage of reactor data gathered in full scale reactor demonstrators, NINE participated, and is still doing, in several International benchmarks aiming at demonstrating the applicability of its modeling methodology to Fast Reactor design and, in particular, to SFRs; to evaluate the level of assessment of computer codes available at NINE in respect to SFR specific features; to check the applicability of the NINE Validation Process – which is part of the more general framework of NEMM (NINE Evaluation Model Methodology)[1] - with particular focus on the quantification of accuracies of the Thermal–Hydraulic (TH) simulations by means of Fast Fourier Transform Based Method (FFTBM) and finally to perform independent validation of the Serpent code.

All the analysis presented hereafter have been performed following a best estimate approach which requires, among the other things, a high-fidelity Simulation

[1] Within the NEMM (NINE Evaluation Model Methodology), NINE adopts the conventional and internationally acknowledged process to achieve the validation of computation tools and define the inherent uncertainties. It includes three steps: the analytical compliance test - the verification -, the qualification through code-to-code comparisons and benchmarks, the actual validation – supported by scaling analysis – on experimental data originating from mock-up experiments and the outcomes of the operating experience, including downgraded operation and emergency.

Model (SM), i.e., a SM that represents with a high level of detail the hardware subject of the analysis to avoid the introduction of inaccuracies due to rough approximations and assumptions.

2. Analysis and validation of EBR-II SHRT-17

The IAEA CRP "Benchmark Analyses of EBR-II Shutdown Heat Removal Tests" was initiated in 2012 [7] with the objective of improving the state-of-the-art SFR codes by extending their validation to include comparisons against whole-plant data recorded during landmark Shutdown Heat Removal Tests (SHRT) conducted at Argonne's Experimental Breeder Reactor II (EBR-II) in the 1980's.

The EBR-II plant was a uranium metal-alloy-fuelled liquid-metal-cooled fast reactor designed and operated by Argonne National Laboratory (ANL) for the U.S. Department of Energy at the Argonne-West site.

Several loss of flow tests were conducted in the facility between 1984 and 1986, as a part of SHRT series [8]. SHRT-17, protected loss of flow transient, was one of the mentioned tests to demonstrate the inherent safety of LMR type reactors. At the beginning of the test, the primary pumps were tripped and at the same time a scram was actuated through a full control rod insertion. The effectiveness of natural circulation cooling capability of the reactor, which makes them inherently safe under described accident conditions, was successfully demonstrated by this test.

2.1 The NEMM validation process

2.1.1 Overview of SCCRED methodology

A key feature of the activities performed in the field of nuclear reactor safety is the need to demonstrate the validation level of each tool adopted within an assigned process and of each step of the concerned process. Therefore, the validation of best estimate codes, models, "best modeling practices" and uncertainty methods must be considered of great importance to ensure the validity of performed Best Estimate and Uncertainty analysis. A consistent code assessment supported by a qualified experimental database is an important step for developing a solid ground for the uncertainty evaluation in the frame of Best Estimate and Uncertainty approach. Thus, the SCCRED (Standardized and Consolidated Calculated & Reference Experimental Database) methodology [9], embedded in NEMM [10], has been developed to generate a series of documents and tools to set up a qualified experimental and calculated database for Verification and Validation (V&V) purposes of Best Estimate and Uncertainty applications.

Figure 2 depicts the SCCRED diagram: the information contained in the experimental reports together with the code input nodalization are the sources to be elaborated in a systematic way by a qualified database made up of the following documents:

- The Reference Data Set for the selected facility or test (RDS-facility and RDS-test) containing the information (geometrical data of the facility and boundary and initial conditions of the selected test, respectively) needed for the code input development;

- The Validation Report (VR), which collects the results of the Validation Process of the performed code calculation;

- The Engineering Handbook (EH), that describes the code input file and provides the engineering justifications of the code-user choices.

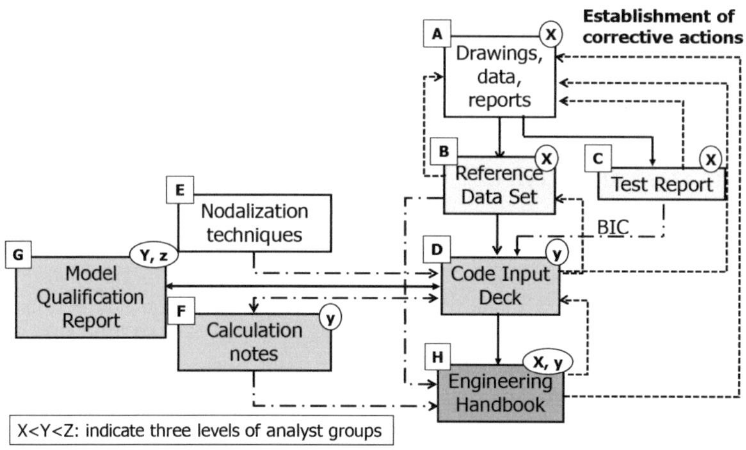

Figure 2.
SCCRED Flow Chart.

The flow-chart linking the RDS, the Input deck, the VR and the EH is highlighted in **Figure 2**. The solid lines show the time sequence of the activities, the dotted lines indicate the feedback for the review and the dashed lines are the necessary input to develop the input deck and the EH.

2.1.2 The validation procedures

The Validation Process of a thermal–hydraulic system code calculation has the goal to demonstrate that the code results (obtained by the application of the code with the developed nodalization) constitute a realistic approximation of the reference plant behavior (a full-size Nuclear Power Plant or a facility). The flow chart of the adopted Validation Process is given in **Figure 3**.

A SM representing an actual system (ITF or NPP) is qualified when:

- It enjoys a large geometrical fidelity with the involved system;
- It reproduces the measured nominal steady state condition of the system;
- It shows a satisfactory behavior in time dependent conditions.

Based on this, three main phases of the Validation Process can be distinguished:

1. The demonstration of the geometrical fidelity of the developed nodalization;
2. The demonstration of the steady state achievement;
3. The “on-transient” Validation.

In relation to the first step of the methodology it is worth demonstrating that the discrepancies between relevant geometrical parameters of the plant and the data implemented into the nodalization are within acceptable values.

The second step of the Validation Process deals with the achievement of the steady state. A set of significant parameters is identified to demonstrate that the discrepancies between calculated and measured data available from nominal stationary conditions are within acceptability thresholds.

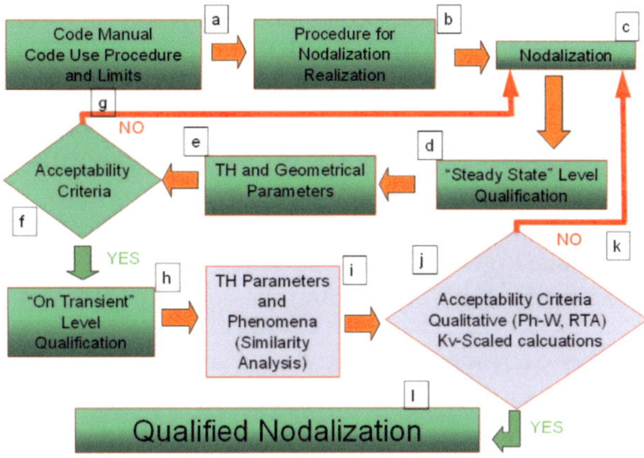

Figure 3.
Flow Chart of the Validation Procedures for the SM.

The third step of the Validation Process is the "On-transient" validation, a very complex step requiring several different sub-steps which include qualitative and quantitative accuracy evaluations performed to evaluate the acceptability of the calculation on "transient level". If the qualitative accuracy evaluation is acceptable, the accuracy of the code calculation can be quantified utilizing the Fast Fourier Transform Based Method (FFTBM) [11].

2.2 EBR-II plant and the developed SM

The EBR-II plant, located in Idaho, was operated by ANL for the U.S. Department of Energy from the beginning of 1964 until 1994. EBR-II rated thermal power was 62.5 MW, with electric output of 20MW. EBR-II was a sodium-cooled reactor fueled with uranium metal alloy fuel, with a pool type primary system. **Figure 4** shows the configuration of the main components in the EBR-II primary system [12] together with the developed RELAP5 SM.

All major primary system components were submerged in the primary tank. Two primary pumps drew sodium from the pool and provided sodium to the two inlet plena for the core, through high pressure and low-pressure pipes. The reactor

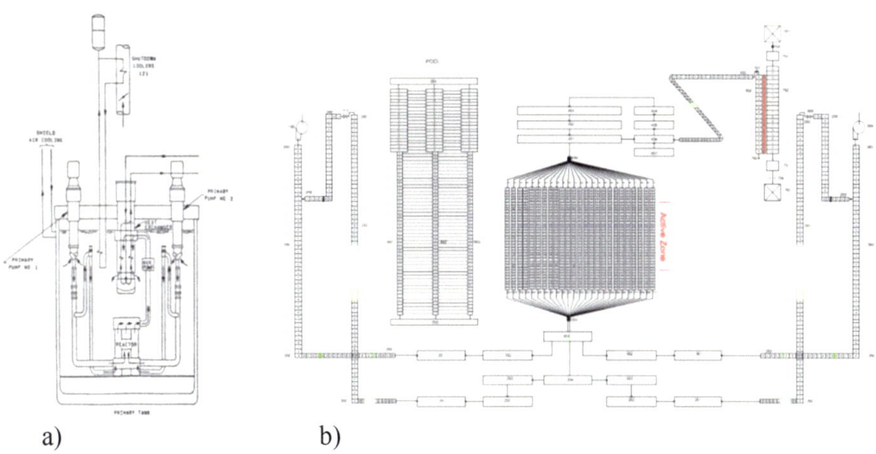

Figure 4.
(a) EBR-II Primary System and (b) The adopted Nodalization.

vessel accommodated 637 hexagonal subassemblies divided in three regions: central core (up to row 5), inner blanket (rows 6 and 7) and outer blanket (up to row 16). Hot sodium exited the subassemblies into a common upper plenum where it mixed before passing through the reactor outlet pipe ("Z-pipe") into the Intermediate Heat Exchanger (IHX). Sodium then exited the IHX into the primary sodium tank before entering sodium primary pumps.

The EBR-II benchmark specifications [12] were used to develop a detailed thermal hydraulic model (see **Figure 4b**) of the reactor. The RELAP5 system thermal hydraulic code was used for both performing the nodalization and running the calculations.

The whole reactor core consisted of 96 channels representing all 10 types of subassemblies used in the reactor, and two bypass flow paths. The reactor vessel was first subdivided into 16 rows. The subassemblies in the first 6 rows have been modeled separately (1 by 1) with 81 channels, except for safety/control rods that have been merged into one channel. Rows 7 to 16 made of reflector and blanket subassemblies have been modeled with one channel per type of subassembly in each row. One heat structure component has been used to simulate the active part of the fuel pins for each subassembly in the central core region, assuming a flat and constant power profile along all the active length. The pool was modeled with a cylindrical multi-dimensional component having 3 azimuthal meshes, two of which were thermally linked to the pumps and the third one to the IHX. The heat exchanger was of counter-current flow type. The primary side of IHX has been modeled with a pipe which takes hot sodium flowing out from the "Z-pipe" and discharges the cold sodium directly into the pool. The intermediate side of IHX has also been modeled with a pipe equivalent to 3026 secondary tubes through which the intermediate sodium flows. The boundary conditions for the intermediate side were imposed by the time-dependent volume and time-dependent junction components.

2.3 Transient results and sensitivity analysis of EBR-II SM

2.3.1 Reference results

The transient was initiated by a trip of primary and intermediate pumps, which instantaneously scrammed the reactor. While the coast-down shapes for SHRT-17 were designed to be identical for the two primary pumps, intrinsic differences between the two pump drive units caused a difference in the stop times.

The transient calculation was performed after the achievement of acceptable steady-state conditions. Starting from full power and flow, both the primary loop and the intermediate loop coolant pumps were simultaneously tripped, and the reactor was scrammed to simulate a protected loss-of-flow accident. Therefore, in the early stage of the pump coast-down (up to about 10 s) the cladding and the outlet coolant temperature decreased. During the transition from the forced to natural circulation (between 10 and 100 s) the mass flow rates decreased rapidly and the unbalance between the total core power and the energy removed from the primary coolant caused a rapid increase of the cladding temperatures and a slower increase of the coolant temperatures. When the natural circulation is fully established (after about 100 s) the total core power is efficiently removed in all subassemblies and the coolant and cladding temperatures decrease.

During the pump coast-down the mass flow rate in the instrumented subassembly XX09 remained a little bit higher than the experimental data (see **Figure 5**), which affected the coolant and cladding temperatures in the whole subassembly. Indeed, both the coolant temperatures below (**Figure 6**) and above (**Figure 7**) the active core region and the cladding temperatures at the middle and at the top of

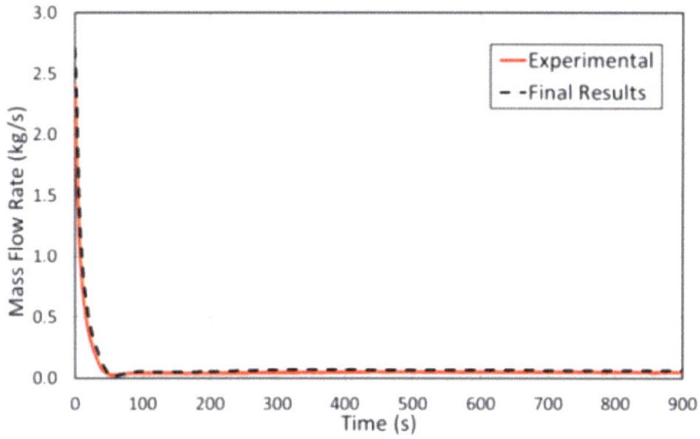

Figure 5.
XX09 Mass Flow Rate.

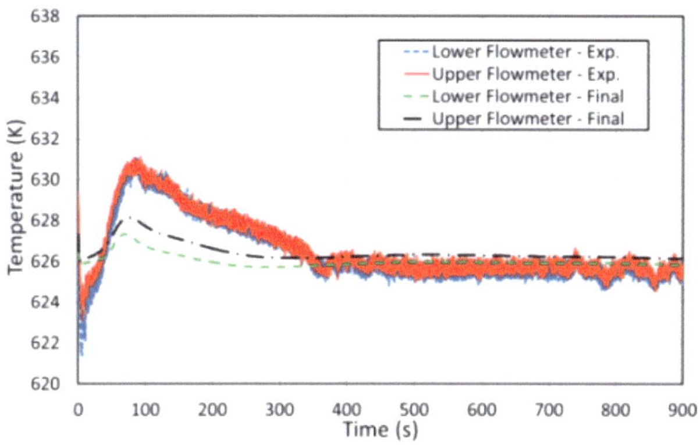

Figure 6.
XX09 Lower and Upper Flowmeter Coolant Temperatures.

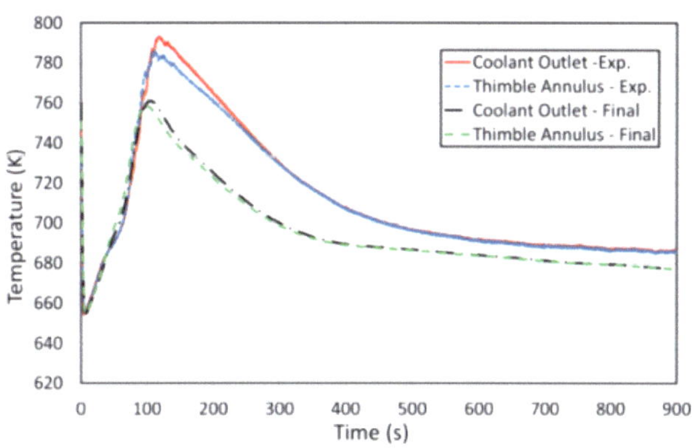

Figure 7.
XX09 Outlet and Thimble Annulus Coolant Temperatures.

the core (**Figure 8**) were slightly lower than the experimental data. These small differences became negligible during the long-term cooling because the mass flow rate reached the correct value. It should be noted that the flowmeter temperatures, where the gamma heating occurs, were qualitatively correctly predicted by the simulation.

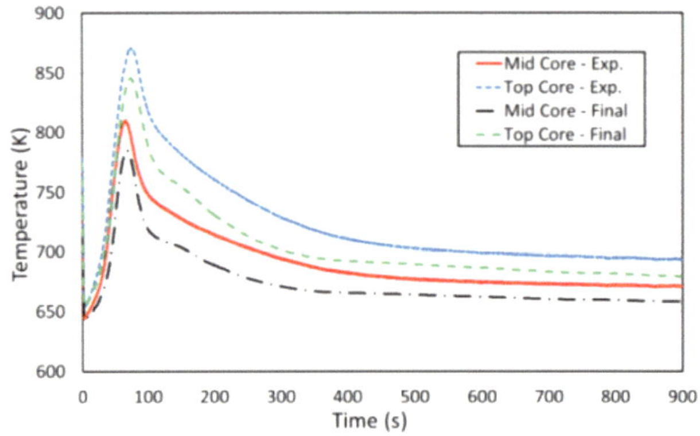

Figure 8.
XX09 Clad Temperatures.

2.3.2 Sensitivity analysis

During the phase 2 of the benchmark, a sensitivity analysis on the gamma heating was performed aimed at understanding the experimental behavior of the coolant temperature at the inlet and outlet of instrumented subassembly (in particular, the instrumented subassembly XX09).

To perform the sensitivity analysis a simple model (see **Figure 9**) of the instrumented subassembly XX09 was developed considering only the subassembly channel and the guide thimble annulus channel, thermally connected with a passive heat structure simulating the subassembly walls. The heat structure simulating the guide thimble wall has been isolated. Regarding the active heat structure, in addition to the flat power profile adopted in the early stage of the phase 2 of benchmark (Phase-2A), four different axial power distribution (see **Figure 9**) have been implemented:

1. Power supplied also below the active part of the core;

2. Power supplied also above the active part of the core;

3. Power supplied also above and below the active part of the core;

4. Axial power distribution as in SHRT-45.

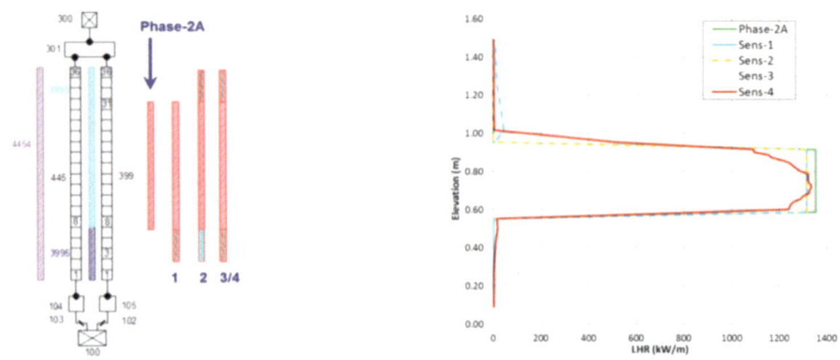

Figure 9.
XX09 Model and Axial LHR used in the Sensitivity Analysis.

DOI: http://dx.doi.org/10.5772/intechopen.97812

The axial power distribution below the BAF has been calculated to match the experimental steady state values of the coolant temperature at the lower and upper flowmeter thermocouples. It can be noted that the power supplied below the active part of the core (sensitivity #1, 3 and 4) positively affects the temperature trends at the upper flowmeter thermocouples (see **Figure 10**).

On the contrary, the power supplied above the active part of the core (sensitivity #2, 3 and 4) results in minor effect on the temperature trends. In particular, the coolant outlet temperature (see **Figure 11**) shows a light delay in the temperature increase after the pump coastdown compared to the experimental data and to the other sensitivity cases.

2.4 Validation process of the EBR-II SM

In the framework of the benchmark, a simplified version of the Validation Process was adopted selecting a smaller set of parameters to carry-out the demonstration of the geometrical fidelity and the demonstration of the steady state achievement (see §2.1.2). In addition, only a quantitative analysis by the FFTBM was carried-out, without performing the qualitative analysis (which is instead a mandatory step for a full application of the Validation Process) due to limited project's recourses. The main goal of the quantitative evaluation, as well as the analysis carried out, was to support the interpretation of the results calculated by the CRP participants, i.e., to provide quantitative measures of the discrepancies between the assumptions made by the participants and the reference specification data. These discrepancies can provide a support to understand the reasons for the differences between the participants' results and the experimental data. Results from the application of the Validation Process are available in [7].

First, a list of more than 50 parameters was selected to perform the geometrical fidelity between the EBR-II hardware and the developed nodalization.

For the achievement of the steady state, a set of significant parameters was identified to demonstrate that the discrepancies between calculated and measured experimental data were within acceptability thresholds.

Regarding the third step of the Validation Process, the "On-transient" Validation, the focus was only on the so called "Quantitative Accuracy Evaluation" that is performed by the FFTBM. A list of about 50 parameters was selected,

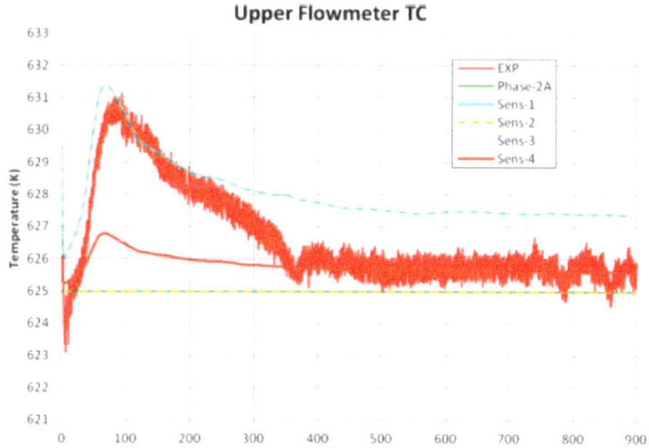

Figure 10.
Upper Flowmeter Temperature.

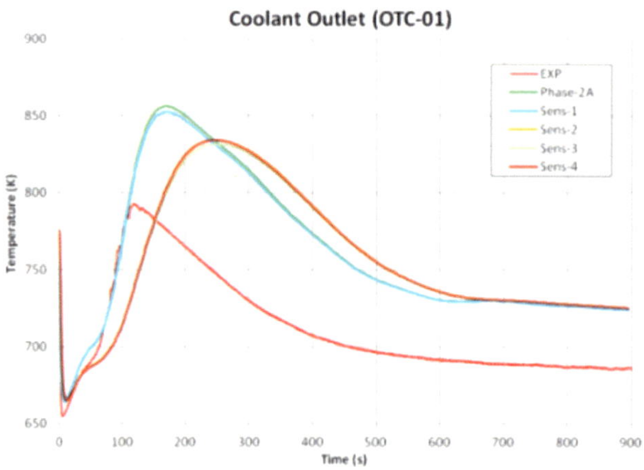

Figure 11.
Coolant Outlet Temperature.

varying from power, absolute pressures, velocity and mass flow rates, fluid temperatures, rod surface temperatures, pressure drops and mass inventory. In the case of parameters for which no reference or measured value was available a code-to-code comparison was performed.

3. Analysis of FFTF LOFWOS Test #13

The IAEA CRP focused on benchmark analysis of one of the unprotected passive safety demonstration tests performed at the Fast Flux Test Facility (FFTF) was launched in 2018 to support collaborative efforts within international partnerships on the validation of simulation tools and models in the area of sodium fast reactor passive safety.

The Fast Flux Test Facility was a 400 MW-thermal loop type SFR prototype with mixed oxide fuel, built to assist development and testing of advanced fuels and materials for fast breeder reactors. It was located at the Hanford site in Washington and designed by the Westinghouse Electric Corporation for the U.S. Department of Energy (DOE). FFTF reached criticality in 1980 and has been operating until 1992 [13].

The loss of flow without scram (LOFWOS) Test #13 was performed on July 18, 1986 as part of the Passive Safety Testing (PST) program with the aim of confirming the safety margins of FFTF as a liquid metal reactor, providing data for computer code validation, and demonstrating the inherent and passive safety benefits of its specific design features. One of the passive reactivity control devices are the Gas Expansion Modules (GEMs) located at the periphery of the FFTF core. GEMs are hollow tubes sealed at the top and open on the bottom with Argon cover gas trapped inside. During normal operation, the pressure head of the primary pumps compresses the gas to a level above the active part of the core, filling the GEMs with sodium. Following a pump trip and a corresponding decrease in the sodium pressure, the trapped gas would expand and displace sodium, increasing the neutron leakage from the core and decreasing the core reactivity.

Starting from 50% power and 100% flow, the Test #13 was initiated when the three primary sodium pumps were simultaneously tripped. The secondary loop sodium pumps remained operational throughout the whole test.

3.1 Overview of FFTF and of the developed SM

An overview of the FFTF coolant system is shown in **Figure 12**, where three main parts can be distinguished: the reactor vessel, the primary loop, and the secondary loop.

Regarding the reactor vessel, cold sodium was discharged from the three primary loop inlet pipes into an inlet plenum at the bottom of the reactor vessel. Sodium was then drawn up into the core support structure and distributed to the core assemblies and radial shields, as well as leakage and bypass flow paths. Sodium discharged from these flow paths was mixed above a horizontal baffle plate in a common outlet plenum before exiting the reactor vessel through one of three primary loop outlet pipes. The outlet plenum was bounded at the top by a region of Argon cover gas.

The IHX was vertically mounted counterflow shell and tube designs and separated activated sodium coolant in the primary loops from nonradioactive sodium in the secondary loops. Within each secondary loop, hot leg piping ran from the IHX outlet to a Dump Heat Exchanger (DHX) unit, which discharged heat to the environment. Each DHX unit contained four individual sodium-to-air dump heat exchanger modules. The cold leg sodium ran from the DHX unit to a sodium pump, and back to the IHX.

The FFTF core was loaded with199 hexagonal assemblies, that could be grouped in 91 core locations, from row 1 to row 6, including 7 different types of driver fuel assemblies, control and safety rods and test locations, 60 internal reflector assemblies, in rows 7 and 8A, and 48 external reflector assemblies, in rows 8B and 9.

Starting from the benchmark specifications [13], a detailed SM reproducing each component depicted in **Figure 12** was developed following the NINE nodalization techniques, except for the DHXs, that were replaced by boundary conditions.

The reactor vessel has been modeled with a cylindrical multi-dimensional component having three radial meshes: the innermost region represents the area occupied by the core basket and the leakage flow that passes around the fuel assemblies and reflector assemblies (up to row 8A), the intermediate zone models the annular plenum and the flow around reflector assemblies (rows 8B and 9) and through radial shields, and the outermost region simulates the peripheral plenum and the in-vessel storage region. The flow through the assemblies in the reactor core was modeled with 18 channels, 16 pipe components simulating the sixteen assembly flow zones representing the different types of assemblies and 2 pipe components to simulate separately the instrumented assemblies, the Row 2 fast response Proximity

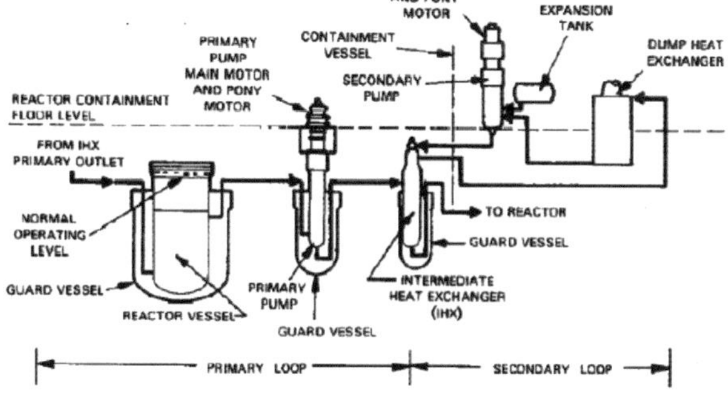

Figure 12.
FFTF Coolant System Overview.

Instrumented Open Test Assemblies (PIOTA) and the Row 6 fast response PIOTA. As for the hydraulic part, one heat structure was inserted in each channel to simulate the active part of the assemblies. A flat axial power profile was imposed along the active length of all the assemblies.

The three primary loops and secondary loops were modeled separately, each one having the same number of hydraulic components. Regarding the secondary loops, two time-dependent components were inserted to provide the proper boundary conditions, one component at the exit of the hot leg piping to set the secondary side pressure, and one component at the beginning of the cold leg piping, downstream the DHXs, to set the appropriate sodium temperature.

3.2 Reference results and sensitivity analysis of the FFTF SM

After imposing the boundary conditions (i.e., pumps speed, core power and secondary loops flow conditions) provided by the benchmark team, acceptable steady-state conditions were achieved before performing the transient simulation. The RELAP5 system thermal hydraulic code was used to make the analysis.

The FFTF LOFWOS Test #13 was initiated when the primary pumps tripped simultaneously. The row 2 PIOTA outlet temperature shown in **Figure 13** can be observed to describe the behavior of the FFTF core during the transient.

The initial rapid rise of the PIOTA outlet temperature was caused by the increasing core power-to-flow ratio following the pump trips. The outlet temperature peaked at around 10 seconds when the power-to-flow ratio reached its maximum value. Then, the increase in core temperatures together with the drop of the GEM sodium level introduced a large negative reactivity feedback, so power decreased faster than the primary flow rate. The drop in reactor power was quick enough to compensate for the reduced flow rate in the primary loop and the sodium temperature started to decrease. As the GEM sodium level approached the bottom of the core, the negative reactivity insertion slowed down. The core outlet temperature began to rise again, and a second peak occurred when the natural circulation was established. Natural circulation was maintained while power continued to decrease resulting in a decrease of the core outlet temperature until the end of the test.

Figure 13 shows the comparison of the PIOTA outlet temperature predicted by the SM against the measurement (in this and subsequent figures the solid line shows the experimental data while the dashed line displays the SM results). The SM results are in good agreement with the experimental data for the entire duration of

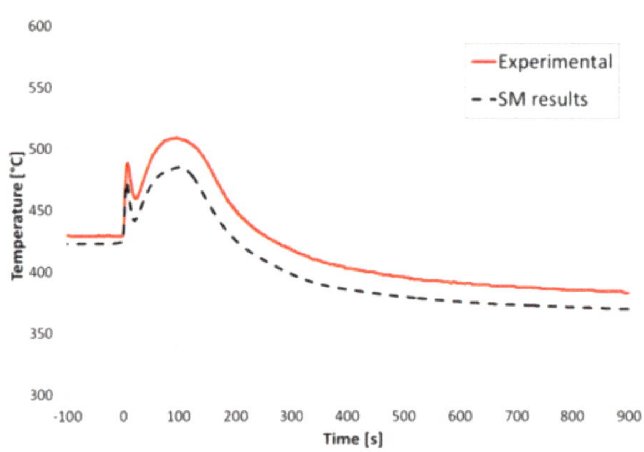

Figure 13.
Row2 PIOTA Outlet Temperature.

the transient. In particular, the time of occurrence of the two peaks is captured very well by the simulation.

The cold leg and hot leg temperatures in one of the primary loops are shown in **Figure 14**. The hot leg fluid temperature has been quite well predicted by the SM, showing a trend slightly oscillating around the experimental value. That may be due to a.

different prediction of the sodium mixing and thermal stratification phenomena in the outlet plenum of the reactor vessel during the natural circulation phase that are difficult to simulate by a system thermal–hydraulic code. The calculated cold leg fluid temperature showed a faster rise at the beginning of the transient following the increase in the DHX sodium outlet temperatures, it reached a higher peak value and it decreased faster compared to the experimental trend. In addition, it can be noted that the oscillations shown in the calculated time trend occurred about 30 seconds earlier and with a greater amplitude than the experimental data. Similar behavior can be observed in **Figure 15**, where the cold leg and the hot leg temperatures in one of the secondary loops are displayed. The cold leg temperature followed the time trends of the DHX sodium outlet temperatures, which had been specified as boundary conditions. The hot leg temperatures decreased quickly at the beginning of the transient reaching the cold leg temperature in about 200 seconds due to the reduction in heat transferred from the primary to the secondary systems across the IHXs, as the primary loop flow rates decreased, and the secondary pumps remained at full speed. It can be noted that, also in this case, the fluctuations of the SM results occur earlier than the measured data.

A sensitivity simulation to account for the thermal inertia of the temperature instrumentation has been performed to investigate the origin of the discrepancies between the SM results and the experimental data. Beyond the reactor vessel, sodium temperatures were measured in the hot and cold legs of all primary and secondary loops by the Resistance Temperature Detectors (RTDs). The RTDs were spring loaded against the bottom of a thermowell to provide a short response time. Unfortunately, there is no information in the benchmark specifications on the geometry of the temperature detectors, so, after a quick search in the literature and some assumptions, a cylindrical heat structure of 5 cm in diameter was inserted at each RTD location. The sodium temperatures detected by the heat structures are shown in **Figure 14** and **Figure 15** (dotted lines) for the primary loop and secondary loop, respectively. When considering the thermal inertia of the RTDs, the SM

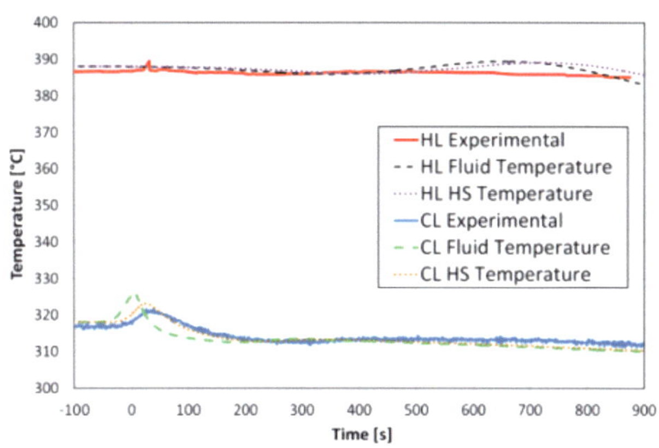

Figure 14.
Primary Loop Hot and Cold Leg Temperatures.

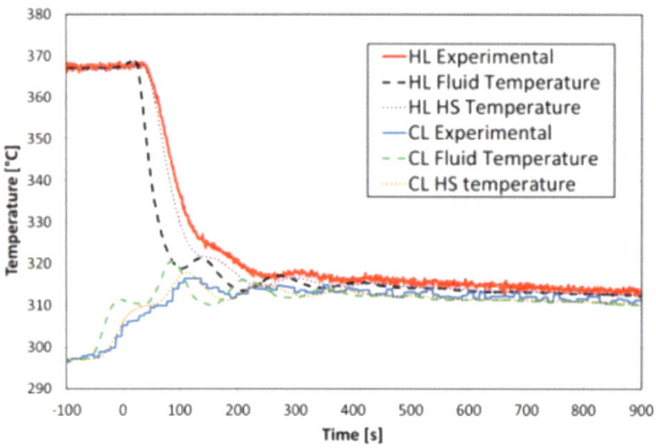

Figure 15.
Secondary Loop Hot and Cold Leg Temperatures.

results are in a very good agreement with the experimental data, improving both the timing and the amplitude of the temperature oscillations.

Then, an additional sensitivity simulation was performed to study the effect of the modeling choice made for the reactor vessel outlet plenum and its impact on sodium mixing. As mentioned before, in the reference case, the reactor vessel and the upper plenum were modeled with a cylindrical multi-dimensional component having three radial meshes. In the two sensitivity simulations, the 3D volumes of the upper plenum were replaced by a 1D vertical pipe component in the first simulation and by a 1D single-volume component in the second simulation. **Figure 16** shows the comparison of the sodium temperature in the hot leg primary loop #1 among the three different reactor vessel outlet plenum modeling choices and with the experimental data. In the reference case (3D) thermal stratification occurs with the hot sodium that tends to go upwards (it should be remembered that the hot leg connection is at the bottom of the outlet plenum, just above the core outlet). In the first sensitivity (1D PIPE), no thermal stratification is observed and the hot sodium exiting the core does not mix with the upper cold sodium at the triggering of natural circulation. In the second sensitivity (1D Single Volume) the hot sodium that would be deposited on top of the reactor completely mixes with the core outlet flow, thus resulting in a slightly higher sodium temperature in the hot leg after the beginning of the transient, compared to the reference case, and which continues to gradually increase even during natural circulation.

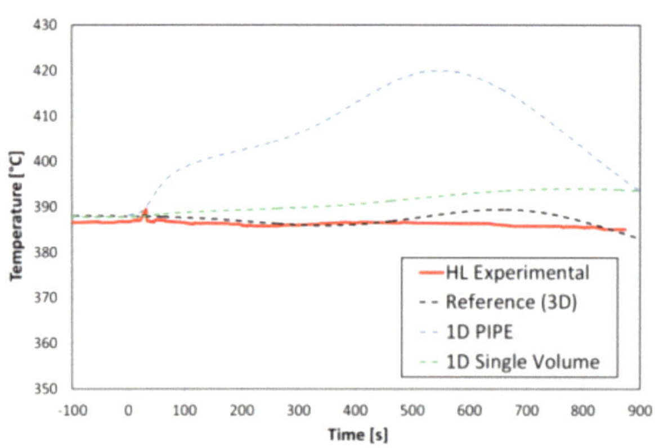

Figure 16.
Sensitivity on Upper Plenum Modeling: Comparison of Primary Loop Hot Leg Temperature.

4. Simulation of the CEFR Start-UP Tests with the Serpent Code

The Neutronics Benchmark of CEFR Start-Up Tests is a CRP proposed by the China Institute of Atomic Energy (CIAE), under the direction and support from IAEA. The CRP was launched in 2018. The main objective of this benchmark is to improve the understanding of the start-up of a SFR and validate the fast reactor analysis computer codes against experimental data obtained at the China Experimental Fast Reactor (CEFR). The CEFR is the first Chinese fast reactor; it is a pool-type sodium cooled reactor, with a nominal thermal power of 65 MWth [14].

NINE, in collaboration with University of Pisa, participated in all the proposed work packages and, in turn, proposed and organized a work package focused on sensitivity and uncertainty analysis of the first criticality test, that, for the time being, has just begun.

The tests included in this benchmark were part of the reactor start-up tests, which included both the fuel loading and the first criticality, the control rod worth measurement, the reactivity coefficients measurement, and the foil activation analysis. All the details of these tests are reported in the Technical specifications [14]. In this chapter, only a sub-set of these tests are analyzed, and their results are compared to experimental measurements.

The first test described is the one referred to the "Fuel loading and criticality" (here and after called work-package 1, WP1). It is focused on the analysis of the first criticality achievement; the tests performed are composed by ten sub-critical steps, with different number of fuel Sub-Assemblies (SAs) loaded, and 3 super-critical steps, which have different RE2 control SA insertion levels. The critical RE2 position was found and reported in the Technical Specification through an extrapolation of the experimental super-critical steps. In this work the results for the super-critical steps, as well as the critical ones, are compared with the experimental measurement; a comparison between two different nuclear libraries is also reported, to exploit the dependence on the nuclear data of the effective multiplication factor.

The second test analyzed is the "Control rod worth measurement", (work-package 2, WP2). The main goal of this test is to evaluate the control rod worth of each control rod SA and of different group of control rod SAs. Also, in this case a comparison with the experimental measurement is reported in the following sections.

The last test presented is the "Foil activation measurements" (work-package 6, WP6): it concerns the foil activation analysis made through the irradiation of different material samples inside the reactor core in both the axial and radial directions.

4.1 The simulation models developed for CEFR

The core geometry has been modeled in its full 3D configuration keeping the heterogeneity of most of the present structures (e.g., the hollowed pellet geometry has been modeled). The SAs model starts from the region above the nozzle (the nozzle is not modeled) and reaches the corresponding head. Only few regions of the core, considered less relevant for the simulations, have been homogenized for the sake of simplification (e.g., SAs Handling head, Spring, etc.). The spacer wires have been homogenized with the corresponding cladding, to guarantee the conservation of the stainless-steel mass. **Figure 17(a)** shows the horizontal section of the core taken at a height of 105.1 cm from the bottom; the operation layout is shown, with all the fuel SAs already loaded in the core. **Figure 17(b)** shows a vertical section crossing the center of the core, along the x axis: the homogenized components are noticeable, mostly on the upper part of each assembly.

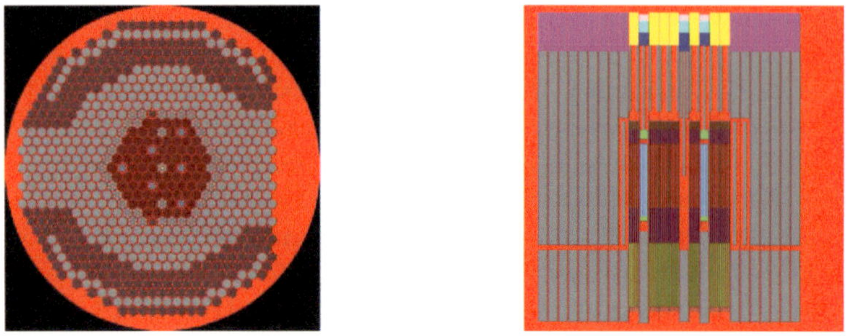

(a) (b)

Figure 17.
Serpent Geometrical Core Model: (a) Horizontal Section; (b) Vertical Section.

The Serpent 2.1.31 code [15] was used for all simulations. The ACE format cross-section libraries used have been processed at different temperatures, to match the specification of the various experiment. For the Work Packages (WPs) analysed in this work the data made available by SCK-CEN were used. All the used libraries were based on ENDF/B-VIII.0 [16]; a comparison with the library based on ENDF/B-VII.1 [17] has been performed for one test case of WP1 and has been reported in the following section.

For each experiment, both whole core geometry and material densities have been adjusted to the experimental temperatures, to consider expansion effects, and, therefore, the leakage variation. For most of the materials involved, the temperature adjusted parameters have been determined making use of the linear thermal expansion coefficients. For the sodium coolant density, the correlation provided in the technical specification has been used, while for the Helium gas, the density has been evaluated dividing the mass of gas at cold condition (at 20°C) by the fuel rods free volume available after the expansion (at hot zero power temperature) of the surrounding materials.

For all the experiments, which require a multiplication factor, the Serpent implicit k effective has been considered. In **Table 1** the cycle population, as well as the number of active and inactive cycles are reported.

For each experiment, both whole core geometry and material densities have been adjusted to the experimental temperatures, to consider expansion effects, and, therefore, the leakage variation. For most of the materials involved, the temperature adjusted parameters have been determined making use of the linear thermal expansion coefficients. For the sodium coolant density, the correlation provided in the technical specification has been used, while for the Helium gas, the density has been evaluated dividing the mass of gas at cold condition (at 20 °C) by the fuel

WP	Test cases	Particles number per cycle	Active cycle	Inactive cycle
1	All	5.0E+05	500	50
2	SH and SA	5.0E+05	500	50
	RE	1.0E+06	500	50
6	Axial: U-238, Al-27, U-235, Np-237, Ni-58	2.0E+06	500	50
	Axial: Au-197	4.0E+06		

Table 1.
Simulations Set-up.

rods free volume available after the expansion (at hot zero power temperature) of the surrounding materials. For all the experiments, which require a multiplication factor, the Serpent implicit k effective has been considered. In **Table 1** the cycle population, as well as the number of active and inactive cycles are reported.

4.2 Simulation results and preliminary interpretation of the CEFR SMs

The results presented here are taken from [18]. First, the results of WP1 are analysed. **Table 2** shows the comparison between the calculated and measured values of the effective multiplication factor for the three supercritical steps and the critical position. The relative errors are small and practically constants for all the cases when ENDF/B-VIII.0 nuclear data are used, that suggests a systematic error originating from the nuclear data library can hold. Confirming this, a much better agreement is obtained when use is made of the ENDF/B-VII.1 nuclear data, as the error drops from 0.19% to 0.03% in the case of the RE2 positioned at 190 mm.

Secondly the results of WP2 are presented. **Figure 18** shows the comparison between the calculated and measured integral worth of different control rod SAs, of the two-shutdown system (groups of control rod SAs, considered both fully

# of fuel SAs loaded	Rod position	Serpent output		Experimental	Relative error
	RE2 [mm]	keff	Std. Dev.	keff,exp	$\left\|k_{eff,exp} - k_{eff}\right\| / k_{eff,exp} * 100$
72	70	0.99817	6.00E-05	1.000000	0.18%
72	151	0.99837	6.10E-05	1.000245	0.19%
72	170	0.99848	6.50E-05	1.000335	0.19%
72	190	0.99854	6.00E-05	1.000395	0.19%
72	190 (ENDF/B-VII.1)	1.00072	6.30E-05		0.03%

Table 2.
WP1, Comparison with Experimental Data.

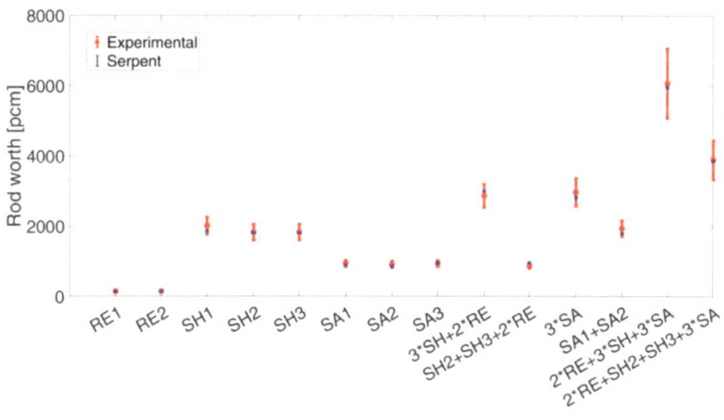

Figure 18.
WP2, Comparison with Experimental Data, Rod Worth.

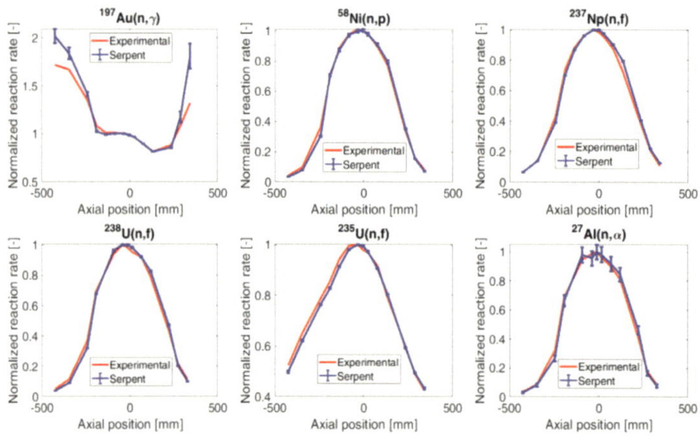

Figure 19.
WP6, Comparison with Experimental Measurements, Axial Reaction Rates.

operating or with one SA stuck) and of all control rods together. The experimental values are obtained through a rod drop experiment. It appears from the figure that a noticeable good agreement between results and measurements has been achieved.

The last results belong to WP6. In **Figure 19** is reported the comparison between the reaction rates axial distributions evaluated with Serpent and the measured values. The agreement between measurements and simulations is generally quite good. Only the case of ^{197}Au(n,γ) shows a noticeable difference particularly for the positions at the top and bottom of the core. Further investigating and understanding the origin and nature of this discrepancy would contribute to a better understanding of the of the measured activity.

5. Challenges and opportunities for enhanced computation capacity and extended investigation in the sodium-cooled fast reactors field

The core of SFR is subdivided into several hexagonal fuel bundles. Unlike LWR squared subassemblies, these hexagonal subassemblies are compact in size leading to higher power density. Higher power densities and higher coolant temperature can lead to coolant boiling. Besides the difference between the boiling points of the two coolants, a relevant distinguishing feature of sodium is its higher thermal conductivity. While the conduction in water is usually neglected when modeling LWRs, the heat conduction in sodium cannot be ignored in SFRs, mainly when natural circulation occurs. The difference in opacity between the two coolants, also implies adoption of different surveillance methods and the differences in the activation products lead to specific features of shielding (these topics are only mentioned here without deepening because they fall out the scope of this chapter).

Moreover, while LWRs systematically use uranium-oxide fuel (UO2), SFRs can use both oxide and metallic fuel, depending on the design features. The oxide fuel has the advantage of having a high melting temperature relative to the metallic. It is also less ductile and have higher strength. The prediction of the temperature distribution in fuel assemblies should be accurate to assure the safety and reliability of the reactor operation. The focus of the new codes and modeling will be on enhancements to the thermal hydraulics modeling aspects of the SFR and the modeling of metallic fuel.

Another key difference between the designs of Pressurized Water Reactors (PWRs) and SFRs is the structural support of the fuel. While the PWR fuel is supported by grid spacers that are located at specific heights, the SFR fuel is supported by wire-wrappings that extend along the whole length of the fuel. The wire wrapping brings the advantage of a better coolant flow mixing and a lower temperature gradient across the subassemblies, but that comes at the expense of higher pressure-losses along the of the fuel height.

Accounting for the above-mentioned general considerations and trends as well as of the outcomes of the investigations carried-out in the framework of the Benchmark exercises described here above, in paragraphs 2 to 4, the Participants were able to drive some general conclusions on the expected new features of the computation tools as well as on the up-dated methodologies to be adopted for the design and the safety assessment of the Sodium Fast Reactors.

These conclusions identify the main trends for improvement. They can either contribute to further and proficiently expanding the computation capacity of existing tools (and even considering the development of new ones), as well as adapting and/or updating the methodologies adopted so far in the studies.

Moreover, the importance of the representativeness, exhaustiveness and comprehensiveness of the data stored in the data base adopted for validation of the computation tools have been once more and even farther pointed-out. Accordingly, it is considered worth complementing and/or implementing the existing data base with the results of ad-hoc experimental programs, accurately designed, and engineered to match some specific validation needs, thus addressing, and filling the main and more crucial knowledge gaps. Definition of such experimental programs should be made relying upon accurate investigation adopting, e.g., Kriging-like methodologies to avoid duplication and dispersions.

The main topics found-out to be potential levers for further improvement in the SFR computation capacity and reliability are the following:

- A major need for the SFR M&S (Modelling and Simulation) appears to be the improvement of the calculation of inlet-plenum flow distribution and of lateral mixing of coolant flows in the assembly to reduce conservatism in the estimation of the peak fuel and cladding temperatures [19]. The high-fidelity Computational Fluid Dynamics (CFD) models should be developed, tested, and adapted for application to liquid sodium for the relevant flow characteristics and geometric configurations, especially wire-wrapped pin bundles contained within the hexagonal assembly boxes. Additional near term needs for the SFR M&S are related to the thermal-mechanical modeling capabilities. Recent advances in modeling of oxide fuels behavior in thermal reactors (e.g., fuel and clad conductivity and gap conductance modeling) should also be adapted for fast reactor applications.

- The pressure drop and flow distribution estimation during the transients can be significantly improved when using suitable empiric correlations to model friction losses in the wire-wrapped fuel bundle region, as proposed by Cheng & Todreas [20] and Pontier [21].

- The sodium mixing and the thermal stratification phenomena play a crucial role during the transients, mainly during the natural circulation phase. They cannot be accurately predicted by the current existing SYS-TH (System Thermal–Hydraulic) codes. In addition, these phenomena are also sensitive to the nodalization scheme adopted in the calculations. It is suggested to address this major issue for design and operation through either suitable sensitivity analyses or more in-depth investigations (i.e., adopting CFD codes).

- SFRs use fuel pins tightly packed in a hexagonal duct. As said, to maintain a gap for coolant to flow through, fuel pins are separated with a metallic helical wire spacer wrapped around the pin entire length. Additionally, these wraps mitigate vortex-induced vibration and increase convective heat transfer by enhancing sub-channel mixing. Modeling the flow in wire-wrapped rod bundles is still a challenging problem [22]. Large uncertainties exist in the treatment of wire spacers and drag models used for momentum transfer in current low- resolution (lumped parameter) models. Sub-channel codes apply "forcing functions" to model wrap induced flow mixing. However, these approaches are limited to conditions submitted to a specific validation (flow regime, channel geometry, or operating conditions) and rely on complex coefficients which were derived from fitting the experimental databases the models are based on. High fidelity tools such as CFD can simulate wire-wrapped rod bundles with more detailed resolution. However, CFD simulations are still limited in their capability to characterize long-term transients or large system simulations. They are also bounded by the quality and resolution of the experimental data these models are benchmarked against.

- The axial power profile and the gamma heating outside the core region, below and above the active fuel zone, affect the coolant temperature distribution. That way, they can have a significant impact on the behavior of some transients. In addition, approximations adopted in the calculation of the radial power distribution can engender large computation vs. measurement discrepancies for both dummy and reflector subassemblies (non-fuelled SAs). Accurately account for the gamma heating contribution to the power both in the active and non-active reactor regions can help to correctly address the issue, thus improving the quality of computation results.

- The axial conduction heat transfer modelling is generally either non available or quite poor in the current SYS-TH codes. Even if the contribution of the axial heat conduction with respect to the other heat transfer mechanisms is always negligible in fast systems, accounting for the phenomenon in the heat structures improves the simulation results for the coolant temperature distribution outside the core region, while it does not provide any measurable gain for the evaluation of the cladding temperature, and even, sometimes it turns-out damaging. No specific action is recommended on this issue.

- The correct and comprehensive simulation of the heat transfer between adjacent subassemblies is mandatory to improve the agreement between computation results and measured data. Accounting for radial heat transfer with neighboring subassemblies is mandatory to obtaining good coolant temperature predictions. Accordingly, it is recommended to accurately account for the heat transfer between adjacent subassemblies and developing the suitable computation capacity to do.

- From a pure phenomenological point of view, the transport codes can catch all the phenomena included in the libraries. Nevertheless, data in the libraries have not got the same accuracy level for all phenomena, that can propagate discrepancies to the results. Accordingly, it is strongly recommended to carry-out sensitivity and uncertainty analysis on the nuclear data, when estimated necessary.

- The cross section pre-processing routine implemented in some codes, such as the Doppler-broadening pre-processor routine inside Serpent 2, are not able to adjust the temperature of the unresolved region probability tables. According to the importance of that region for the stability and operability of the core, when using such codes for fast reactor systems, it is recommended to evaluate the Doppler broadening with a suitable nuclear data processing code.

- When using Monte-Carlo codes, the calculation time to achieve a good statistic can turn-out remarkably high. This issue being in common with all Monte Carlo calculations and not being a specific problem for the Sodium Fast Reactor simulation, no specific recommendation is done.

- Many experiments have been performed to study thermal- hydraulics characteristics, primarily pressure drops, of the wire-wrapped fuel bundles. Often however, the uncertainties of pressure drop measurements associated with these experiments is high due to the geometrical complexity of the hexagonal wire-wrapped fuel bundle. Recently [23, 24], a database of pressure drops and flow-field measurements in a 61-pin wire-wrapped hexagonal fuel bundle was developed with the sole purpose to benchmark the existing correlations and validate the CFD calculations. These experiments investigate the flow characteristics in the near-wall region of the 61- pin wire-wrapped hexagon fuel bundle.

All the above-mentioned topics merit for careful consideration and further investigation in view of the definition of future R&D programs in support to the industrial development and the deployment of the SFRs. Due to their size and scope, they should mainly be addressed in the framework of international collaborations.

6. Conclusions

Sodium Fast Reactors is a branch of Fast Reactors technology developed since the early 60s, which nowadays regains large interest and attractiveness thanks to their flexibility and their potential to be operated as Actinide burners and/or breeders, thus playing a crucial role in the closure of nuclear fuel cycle and solving the burden of long-lived nuclear wastes.

Design and operation of such reactors require a noticeable computational capacity, but also specific means to assess their safety both in normal and downgraded operation as well as in emergency conditions. In this prospect, IAEA has organized several Coordinated Research Projects aimed at improving Member States' fast reactor analytical simulation capabilities.

The participation in such research programs - allowing direct comparison of computation results with measured data - contributes to increase the confidence in the capabilities of available computational tool, in the meantime highlighting the potential for improvements which could address and solve the pending issues and for the identification of new ones.

NINE has been actively involved in such research activities within the IAEA CRPs, thus catching the opportunity to independently assess and validate several commonly and widely used Thermal–Hydraulic and Reactor Physic codes. Moreover, it contributed to the comparison of results and the interpretation of discrepancy origins, identifying trends, and driving conclusions for future developments.

The NINE's Simulation Models have been developed strictly complying with the Best Estimate principle - which namely requires avoiding the introduction of inaccuracies due to rough approximations and assumptions - thus trying to represent at best the problem under investigation without adopting any major simplification. Despite this approach requires relevant and continuous computational efforts, the obtained results show-up highly satisfactory and encouraging in a whole. Moreover, as far as the computation capacity is concerned, the NEMM model methodology developed by NINE confirmed its applicability also in the case of SFR simulations.

The present chapter summarizes the activity carried-out, presents the results and discusses the main outcomes of the mentioned benchmark exercises, underlying the wide convergence among the computational tools, as well as detecting the main discrepancies and seeking for their common origin and trends, which should enable defining a mid-term vision for further development of the computer codes in the field of fast reactors and identifying new needs for their extended validation against either available or expected experimental data.

Among others, the following items have been identified as meriting careful and particular attention in the future: the need for an accurate modelling of the mixing of coolant flows in the assembly, the estimation of the pressure drop and the flow distribution during the transients which could be significantly improved using suitable empiric correlations, the sodium mixing and the thermal stratification phenomena which play a crucial role during the transients, and are sensitive to the nodalization scheme adopted and cannot be accurately predicted by the current existing SYS-TH codes, the need for a correct and comprehensive simulation of the heat transfer between adjacent subassemblies, the suitability for improvement of the calculation of inlet-plenum flow distribution.

Moreover, the importance of the representativeness, exhaustiveness and comprehensiveness of data have been once more and even farther pointed-out, claiming the need for complementing and/or implementing the existing data base with the results of experimental programs, engineered to match some specific validation needs, thus addressing, and filling the main and more crucial knowledge gaps. Definition of such programs should rely upon accurate to avoid duplication and dispersions.

Acknowledgements

The data and information presented in this chapter are part of two ongoing IAEA coordinated research project; the first one on "Benchmark Analysis of Fast Flux Test Facility (FFTF) Loss of Flow Without Scram Test – CRP-I32011", and the second one on "Neutronics Benchmark of CEFR Start-Up Tests – CRP-I31032".

Glossary

ANL	Argonne National Laboratory
CEFR	China Experimental Fast Reactor
CFD	Computational Fluid Dynamics
CIAE	China Institute of Atomic Energy
CRP	IAEA Coordinated Research Project
DOE (US)	Department of Energy
DHX	Dump Heat Exchanger
EBR-II	Experimental Breeder Reactor II
FFTBM	Fast Fourier Transform Based Method

FFTF	Fast Flux Test Facility
GEN III	Generation III
GEN IV	Generation IV
GEM	Gas Expansion Module
GFR	Gas Fast Reactor
GIF	Generation IV International Forum
IAEA	International Atomic Energy Agency
IHX	Intermediate Heat Exchanger
ITF	Integral Test Facility
LFR	Lead (or Lead-Bismuth) Fast Reactor
LOFWOS	Loss of Flow Without Scram
LWR	Light Water Reactor
MMR	Micro Modular Reactor
MSR	Molten Salt Reactor
NEMM	NINE Evaluation Model Methodology
M&S	Modelling and Simulation
NPP	Nuclear Power Plant
PST	Passive Safety Testing
PWR	Pressurized Water Reactor
R&D	Research and Development
RDS	Reference Data Set
RTD	Resistance Temperature Detector
SA	Sub-Assembly
SCCRED	Standardized and Consolidated Calculated & Reference Experimental Database
SFR	Sodium Fast Reactor
SHRT	Shutdown Heat Removal Test
SM	Simulation Model
SMR	Small Modular Reactor
SYS-TH	System Thermal–Hydraulic
PIOTA	Proximity Instrumented Open Test Assembly
TH	Thermal–Hydraulic
VR	Validation Report
WP	Work Package

Author details

Domenico De Luca[1*], Simone Di Pasquale[1], Marco Cherubini[1], Alessandro Petruzzi[1] and Gianni Bruna[1,2]

1 Nuclear and Industrial Engineering (NINE), Lucca, Italy

2 NucAdvisor, Courbevoie, France

*Address all correspondence to: d.deluca@nineeng.com

References

[1] Faudon V. Relocaliser en décarbonant grâce à l'énergie nucléaire. Fondation pour l'Innovation Politique, fondapol.org, janvier 2021.

[2] Baroni M. Energie nucléaire: la nouvelle donne international. Fondation pour l'Innovation Politique, fondapol.org, février 2021.

[3] Bruna GB, Apostolakis G., Yasui M., Diaz N., in Buongiorno J., Near - and long-term Regulatory Changes after Fukushima: Does the Accident in Japan call for a Major Overhaul of Nuclear safety Regulations, Embedded Topical Meeting, ANS Winter Meeting, San Diego, California, 11-15 November 2012.

[4] IAEA. Technical Meeting on the Benefits and Challenges of Fast Reactors of the SMR Type. Milan, Italy, 24-27 September 2019.

[5] Commissariat à l'énergie atomique et aux énergies alternatives. Les réacteurs nucléaires à caloporteur sodium. Ed. LeMoniteur, Paris, October 2014.

[6] Bruna GB, et al. Advanced Numerical Simulation and Safety Demonstration. In Numerical Simulations - Applications, Examples and Theory, INTECH, 2011.

[7] IAEA. Benchmark Analysis of EBR-II Shutdown Heat Removal Tests. IAEA-TECDOC-1819, IAEA, Vienna 2017.

[8] Planchon HP, et al. The Experimental Breeder Reactor II Inherent Shutdown Heat Removal Tests – Results and Analysis. Nuclear Engineering and Design, 91(1986) 287-296, 1985.

[9] Petruzzi A, D'Auria F. Standardized Consolidated Calculated and Reference Experimental Database (SCCRED): a Supporting Tool for V&V and Uncertainty Evaluation of Best-Estimate System Codes for Licensing Applications. Nucl. Sci. Eng., 182(1). 2016. pp. 13-53.

[10] Petruzzi A, Giannotti W, Modro M. NEMM - NINE Evaluation Model Methodology. NURETH-18, Portland OR, USNRC, 18-23 August 2019.

[11] Petruzzi A, D'Auria F. Accuracy Quantification: Description of the Fast Fourier Transform Based Method (FFTBM). 3D S.UN.COP., Barcelona, Spain, 7-25 October 2013.

[12] Sumner T, Wei TYC. Benchmark Specifications and Data Requirements for EBR-II Shutdown Heat Removal Tests SHRT-17 and SHRT-45R. ANL-ARC-226 (rev 1), May 31, 2012.

[13] Sumner T. Benchmark Analysis of Fast Flux Test Facility (FFTF) Loss of Flow Without Scram Test. IAEA CRP-I332011. ANL, USA, 2018.

[14] Huo H. Technical Specsifications for Neutronics Benchmark of CEFR Start-up Tests. IAEA CRP-I31032. CIAE, Beijing, 2019.

[15] Leppänen J, et al. The Serpent Monte Carlo code: Status, development and applications in 2013. Ann. Nucl. Energy, 82. 2015, 142-150.

[16] Brown DA, et al. ENDF/B-VIII.0: The 8th Major Release of the Nuclear Reaction Data Library with CIELO-project Cross Sections, New Standards and Thermal Scattering Data. Nuclear Data Sheets, 148. 2018, 1-142.

[17] Chadwick MB, et al. ENDF/B-VII.1 Nuclear Data for Science and Technology: Cross Sections, Covariances, Fission Product Yields and Decay Data. Nuclear Data Sheets, 112. 2011, p. 2887-2996.

[18] Di Pasquale S., Petruzzi A., Giusti V., NINE/UNIPI Final Results, Presentation at the 3rd RCM of the IAEA CRP I31032 on Neutronics Benchmark of CEFR Start-up Tests, April 2021, Virtual Meeting (2021)

[19] Khalil H. Modeling and Simulation Needs for Future Generation Reactors. Joint International Topical Meeting on Mathematics & Computation and Supercomputing in Nuclear Applications (M&C + SNA 2007) Monterey, California, April 15-19, 2007, on CD-ROM, American Nuclear Society, LaGrange Park, IL, 2007.

[20] Cheng SK, Todreas NE. Hydrodynamic models and correlations for bare and wire-wrapped hexagonal rod bundles – bundle friction factors, subchannel friction factors, and mixing parameters. Nuclear Engineering and Design 92. 1986, p. 227-251.

[21] Tenchine D, et al. Status of CATHARE code for sodium cooled fast reactors. Nuclear Engineering and Design 245. 2012, p. 140-152.

[22] Delchini MO, et al. Assessment of SFR Wire Wrap Simulation Uncertainties. ORNL/TM-2016/540, 2016.

[23] Nguyen T, et al. PIV measurements of turbulent flows in a 61-pin wire-wrapped hexagonal fuel bundle. International Journal of Heat and Fluid Flow, 65. 2017, 47-59.

[24] Vaghetto R, et al. Pressure Measurements in a Wire-Wrapped 61-Pin Hexagonal Fuel Bundle. Journal of Fluids Engineering, 140. 2018.

Chapter 15

On the Determination of Molar Heat Capacity of Transition Elements: From the Absolute Zero to the Melting Point

Ivaldo Leão Ferreira, José Adilson de Castro and Amauri Garcia

Abstract

Molar specific heat is one of the most important thermophysical properties to determine the sensible heat, heat of transformation, enthalpy, entropy, thermal conductivity, and many other physical properties present in several fields of physics, chemistry, materials science, metallurgy, and engineering. Recently, a model was proposed to calculate the Density of State by limiting the total number of modes by solid–liquid and solid–solid phase nucleation and by the entropy associated with phase transition. In this model, the new formulation of Debye's equation encompasses the phonic, electronic, and rotational energies contributions to the molar heat capacity of the solids. Anomalies observed in the molar specific heat capacity, such as thermal, magnetic, configurational transitions, and electronic, can be treated by their transitional entropies. Model predictions are compared with experimental scatter for transitional elements.

Keywords: molar heat capacity, density of state, phase transition entropies, transitional elements

1. Introduction

Einstein [1] developed the first model approach regarding the atoms in a crystalline solid as independent harmonic oscillators vibrating at the same frequency by assuming the density of state as a delta function. Debye [2–4] modeled the vibrations in a solid as normal modes of a continuous elastic body, which corroborates well for long-wavelength vibrations that do not depend on the detailed atomic character of the solid and do conform better with experimental scatters at lower temperatures. The density of state modeled by Debye failed for many materials, which present a gap in the density of state [5, 6]. The Debye model does not consider rotational, electronic, and magnetic contributions [7–11]. Ferreira et al. [12, 13] considering Gibbs–Thomson coefficients for equilibrium and non-equilibrium nucleation conditions, and the assumption that when cleaved, certain crystals exhibit surface stress that gives rise to small but detectable strains in the interior of the crystal, i.e., microscopic considerations that predict the presence of surface stress whenever a new surface is created [14], derived a model for pure elements and compounds, regarding the critical radius expressed in terms of the temperature drop employing the correlation between the solid–liquid surface tension and the bulk melting entropy by unit volume, given in terms of the Gibbs–Thomson coefficient [15, 16]. Consequently, based on the nucleation of solid–liquid or solid–solid phases, the total number of atoms in the volume and a correspondent density of n atoms limited by nucleation conditions were proposed to calculate the density of state. Ferreira et al.'s model consists of the phonic, electronic, rotations contributions and predicts magnetic anomalies, such as phase transition temperatures.

In this paper, model predictions of the molar heat capacity of transitional elements from absolute zero to the melting point are compared with the Thermo-Calc Software simulations and experimental data.

2. Modeling

The Gibbs–Thomson coefficient describes for pure elements the melting temperature depression $\Delta T_m[K]$, based on the solid–liquid interface energy $\gamma_{sl}[N.m^{-2}]$ and on the bulk melting entropy by unit volume $\Delta S_\forall[J.K^{-1}.m^{-3}]$. Let us consider an isolated solid particle of radius r in the liquid phase; the Gibbs–Thomson equation for the structural melting point depression can be expressed by [12, 13]:

$$\Gamma = \frac{\gamma_{sl}}{\Delta S_\forall} \tag{1}$$

According to Gurtin et al. [14], surface stress gives rise to detectable strains in the interior of the crystal whenever a solid surface is created. A relation of surface tension in terms of η and ζ parameters is given by:

$$\Gamma = \eta \frac{\sigma_{sl}}{\Delta S_\forall} \zeta \tag{2}$$

and,

$$\eta \cong \frac{\sigma_{sl}}{\gamma_{sl}} \tag{3}$$

By substituting (3) into (2) and making $\zeta = 1\,[m]$ provides

$$\Gamma \cong \frac{\sigma_{sl}}{\Delta S_\forall} \cong \frac{\sigma_{sl} T_m^{bulk}}{\Delta H_\forall} = \frac{2\,\Gamma}{r} \tag{4}$$

where σ_{sl} is the solid–liquid interface tension $[N.m^{-1}]$,T_m^{bulk} is the bulk melting temperature $[K]$, $\Delta H_\forall$ is the latent heat of melting per unit volume $[J.m^{-3}]$ and r is the spherical grain radius$[m]$, respectively.

For a stable nucleus, the critical radius can be expressed in terms of the temperature drop $\Delta\mathrm{T}(r)$ through the correlation between the solid–liquid surface tension σ_{sl} and the bulk melting entropy by unit volume $\Delta S_\forall$, which can be written in terms of the Gibbs–Thomson coefficient Γ.

$$\Delta\mathrm{T}(r \geq r_C) = \frac{2\,\Gamma}{r} \tag{5}$$

The density of state $D(\omega)$ for a given grain of volume $\forall$ regarding the critical nucleation radius, is defined as

$$D(\omega) = \frac{\forall \omega^2}{2\,\pi^2 \nu^3} \tag{6}$$

where ω is the frequency, ν is the speed of sound in the solid. For a total number of atoms N in the volume $\forall$ and a correspondent density of atoms n, these variables can be expressed as,

$$N = n\forall \tag{7}$$

The first Brillouin zone is exchanged by an integral over a sphere of radius k_D, containing precisely N wave vectors allowed. As a volume of space k by wave vector requires,

$$\frac{(2\pi)^3}{\forall} N = \frac{4\pi k_D^3}{3} \tag{8}$$

Then, the density of atoms n can be obtained as,

$$n = \frac{k_D^3}{6\pi^2} = \frac{1}{6\pi^2}\left(\frac{k_B \Theta_D}{\hbar\,\nu}\right)^3 \tag{9}$$

As observed in Eq. (9), the element fundamental frequency is expressed as

$$\omega_D = \frac{k_B \cdot \Theta_D}{\hbar} \tag{10}$$

where, Θ_D is the Debye‘s temperature of the element, k_B and are the constants of Boltzmann and Planck, respectively.

The electronic contribution to c_{ve} is written in terms of the phonon energy c_v^{Vib} as,

$$\frac{c_{ve}}{c_v^{Vib}} = \frac{5}{24\,\pi^3} Z \frac{\Theta_D^3}{T^2 T_m^{bulk}} \tag{11}$$

where, Z is the valence of the element, T_m^{bulk} is the melting temperature of the element $[K]$ and T is the absolute temperature $[K]$.

In 2019, Ferreira et al. [15, 16] considered the following approach for the rotational energy,

$$E_{Rot} = \frac{5}{4}\hbar^2 \frac{J(J+1)}{\overline{M} \cdot \mathrm{r}^2}\;[J] \tag{12}$$

where, J is the rotational level corresponding to integer $J = 0, 1, 2, 3, \ldots$, r and $\overline{M}$ are the atomic radius and the molar mass, respectively. The rotational contribution c_v^{Rot} to the molar heat capacity can be derived as,

$$c_v^{Rot} = \frac{5}{4} \frac{R \cdot \hbar^3 \cdot}{k_B^2 \omega_D (T + \Theta_D)^2} \frac{J(J+1)}{\overline{M} \cdot \mathrm{r}^2}\;\left[J.mol^{-1}.K^{-1}\right] \tag{13}$$

where, ω_D is the maximum admissible frequency known as Debye's frequency and, R is the universal gas constant $\left[J.mol^{-1}.K^{-1}\right]$.

Debye's temperature for transition elements is found in the literature [4]. The additions of Eq. (17) of the electronic and of Eq. (19) of the rotational contributions to c_v, provide,

$$c_v = (1.0 + D(\omega))\, 9\, N_a k_B \left(\frac{T}{\Theta_D}\right)^3 \int_0^{\frac{T}{\Theta_D}} \frac{x^4 e^x}{(e^x - 1)^2} dx\, (1 + c_{ve}) + \left(n + \frac{1}{2}\right)\left[9.0\, c_v^{Rot} + \left(1 - \sqrt{\frac{\mathrm{E} \cdot \rho_{Dia}}{E_{Dia} \cdot \rho}}\right) \frac{RT^3}{\Theta_D T_m^2}\right] \tag{14}$$

3. Results and discussion

Figure 1 presents the model predictions of molar heat capacities for pure Chromium and experimental data from absolute zero to the melting point. Debye's model predictions are presented as a reference for Ferreira's model calculations [15, 16]. Thermo-Calc equilibrium calculations were performed in the range 176 K

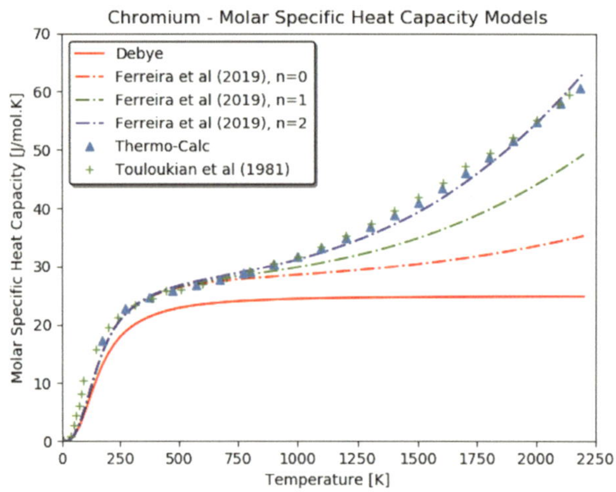

Figure 1.
Comparison of the molar heat capacity of pure chromium by applying Debye, Thermo-Calc, and Ferreira et al. [15, 16], and Touloukian et al. [17].

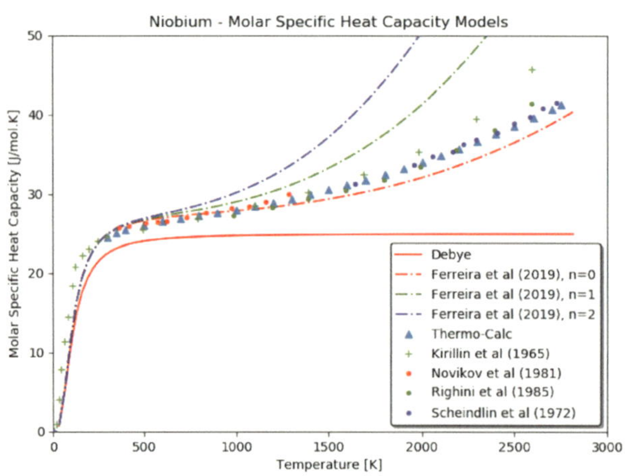

Figure 2.
Comparison of the molar heat capacity of pure niobium by applying Debye, Thermo-Calc, and Ferreira et al. [15, 16], and Kirillin et al. [18], Novikov et al. [19], Righini et al. [20] and Scheindlin et al. [21].

to 2180 K. The proposed model agrees, for low and high temperatures, with the experimental data of Touloukian et al. [17].

Figure 2 shows model calculations for Niobium compared with calculations performed with Thermo-Calc, and four experimental data sets [18–21]. In the literature, experimental values of molar heat capacity at high temperatures (for which measurements are complicated) generally overestimate the heat capacity. On the other side, at low temperatures, where measurements are difficult to control, the experimental values underestimate this property [22]. Furthermore, observations of the thermophysical properties of Nb applied in the model predictions, such as surface tension, Debye's temperature, atomic radius, the density of solid at the melting point, latent heat of fusion, among others, should be carefully compared with those from different authors, as values for the thermophysical properties found in the literature differ from author to author, and they could also be a likely source of the slight deviation observed in the predicted curve. The equivalent

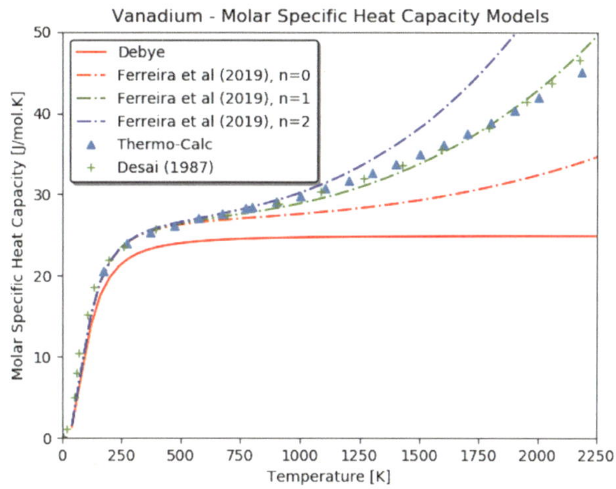

Figure 3.
Comparison of the molar heat capacity of pure Vanadium by applying Debye, Thermo-Calc, and Ferreira et al. [15], and experimental data Desai [23].

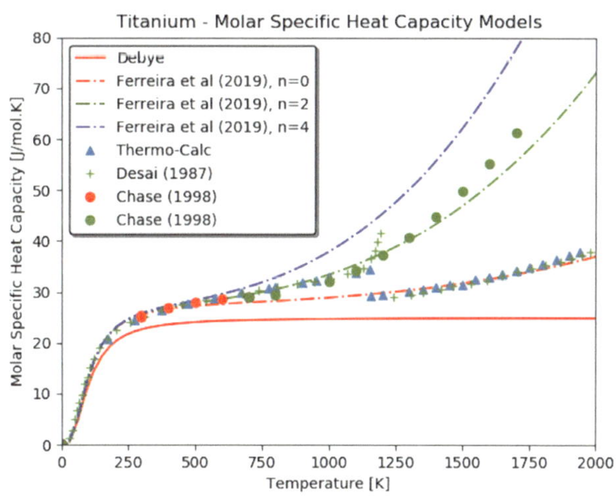

Figure 4.
Comparison of the molar heat capacity of pure titanium by applying Debye, Thermo-Calc, and Ferreira et al. [15, 16], and experimental data Desai [23] and Chase [24].

wavevectors simulated are $n = 0, 1$, and 2. The experimental data are close to the theoretical calculations for $n = 0$.

Figure 3 shows the experimental scatter for Vanadium from the absolute zero to the melting point, Thermo-Calc and Ferreira et al. model's calculations [15, 16]. The predictions for $n = 1$ agrees, for the whole temperature range, with the experimental data, and Thermo-Calc.

Figure 4 shows the molar specific heat for Titanium, the experimental data from Chase [24] and found in Desai [23]. Chase experimental data, in green, follow $n = 2$ for the whole temperature range. In this case, Chase's experiment's thermodynamic conditions allow concluding that no phase transition at $T = 1156K$ takes place, which configures a non-fundamental state specific heat. On the other hand, after the transition temperature, Desai's [23] experimental data and Thermo-Calc agree with the theoretical model for $n = 0$ from 1156 to 1941 K, configuring a fundamental state specific heat.

4. Conclusions

The model previously proposed by Ferreira et al. [15, 16] based on the critical radius of phase nucleation to determine the total numbers of modes, and consequently, the Density of State successfully predicted the molar specific heat capacity of transitional elements. In Cr and V, the experimental data follow the theoretical prediction curves with $n = 2$ and $n = 1$, respectively. Furthermore, the model's calculation for Nb agrees with the experimental data except for the set found in Kirillin et al. [18]. The thermophysical properties of Niobium at high temperatures and experimental difficulties might be the reasons responsible for the slight deviation observed between the predictions and experimental data at high temperatures. For Titanium, non-fundamental states and fundamental state molar heat capacity were predicted experimentally and theoretically, as Chase's experiments follow the model's theoretical predictions for $n = 2$.

Acknowledgements

The authors acknowledge the financial support provided by FAPERJ (The Scientific Research Foundation of the State of Rio de Janeiro), CAPES and CNPq (National Council for Scientific and Technological Development).

Author details

Ivaldo Leão Ferreira[1]*, José Adilson de Castro[2]* and Amauri Garcia[3]*

1 Faculty of Mechanical Engineering, Federal University of Pará, UFPA, Belém, PA, Brazil

2 Graduate Program in Metallurgical Engineering, Fluminense Federal University, Volta Redonda, RJ, Brazil

3 Department of Manufacturing and Materials Engineering, University of Campinas – UNICAMP, Campinas, SP, Brazil

*Address all correspondence to: ileao@ufpa.br, joseadilsoncastro@id.uff.br and amaurig@fem.unicamp.br

References

[1] S.I. Abu-Eishah; Y. Haddad; A. Solieman; A. Bajbouj, A new correlation for the specific heat of metals, metal oxides and metal fluorides as a function of temperature, Lat. Am. Appl. Res. 34 (2004) 257–265.

[2] A. Einstein. Die plancksche Theorie der Strahlung und die Theorie der Spezifischen Wärme, Ann. Phys. 22 (1907) 180–190.

[3] P. Debye. Zur Theorie der Spezifischen Wärmen, Ann. Phys. 344 (1912) 789–839.

[4] N.W. Ashcroft; N.D. Mermin. Solid State Physics, 1. ed., Cengage Learning, New York, 2011, pp. 491–598.

[5] J.M. Schliesser; B.F. Woodfield. Development of a Debye heat capacity model for vibrational modes with a gap in the density of states, J. Phys. Condens. Matter 27 (2015) 285402.

[6] E.D.M. Costa; N.H.T. Lemes; M.O. Alves; J.P. Braga. Phonon density of states from the experimental heat capacity: an improved distribution function for solid aluminum using an inverse framework, J. Mol. Model. 20 (2360) (2014) 1–6.

[7] D.V. Schroeder. An Introduction to Thermal Physics, 1st ed., Addison-Wesley Professional, New York, 1999, p. 409.

[8] U. Mizutani; A. Kamiya; T. Matsuda; K. Kishi; Takeuchi, S. Electronic specific heat measurements for quasicrystals and Frank-Kasper crystals in Mg-AI-Ag, Mg-Al-Cu, Mg-Al-Zn, Mg-Ga-Zn and Al-Li-Cu alloy systems, J. Phys. Condens. Matter 3 (1991) 3711–3718.

[9] G. Inden. Computer calculation of the free energy contribution due to chemical and/or magnetic ordering, Proc. Project Meeting CALPHAD, Dusseldorf, 1976, pp. 1–13.

[10] M. Hillert; M. Jarl. A model for alloying in ferromagnetic metals, CALPHAD 2 (1978) 227–238.

[11] Y. Chuang; R. Schmid; Y.A. Chang. Magnetic contributions to the thermodynamic functions of pure Ni, Co and Fe. Metall. Trans. A 16 (1985) 153–165.

[12] I.L. Ferreira; A. Garcia. The application of numerical and analytical approaches for the determination of thermophysical properties of Al–Si–Cu–Mg alloys. Continuum Mech. Thermodyn. 32 (2020) 1231–1244.

[13] P.A.D. Jácome; M.C. Landim; A. Garcia; A.F. Furtado; I.L. Ferreira. The application of computational thermodynamics and a numerical model for the determination of surface tension and Gibbs Thomson coefficient of aluminum-based alloys. Thermochimica Acta 523 (2011) 142–149.

[14] M. E. Gurtin, A. I. Murdoch. Surface Stress in Solids. Int. J. Solids Struct. 14 (1978) 431–440.

[15] I.L. Ferreira; J.A. de Castro; A. Garcia. Determination of heat capacity of pure metals, compounds and alloys by analytical and numerical methods. Thermochim. Acta 682 (2019) 178418.

[16] I.L. Ferreira. On the heat capacity of pure elements and phases. Materials Res (2021) in press.

[17] Y. S. Toulokian, C. Y. Ho. Properties of selected ferrous alloying elements. McGraw-Hill Book Company, 1981.

[18] V. A. Kirillin, A. E. Scheindlin, V. Ya. Chekhovskoi, I. A. Zhukova. Themodynamic Properties of Niobium from 0K to the Melting Point, 2740K. In Advances in Thermophysical Properties at Extreme Temperature and Pressures Proceedings of the Third Symposium on

Thermophysical Properties. ASME (1965) 152.

[19] I. I. Novikov, V. V. Roshchupkin, A. G. Mozgovoi, N. A. Semashko. Specific heat of nickel and niobium in the temperature interval 300-1300K. High Temperature. 19 (1981) 694.

[20] F. Righini, R. B. Roberts, A. Rosso. Measurements of thermophysical properties by a Pulse-Heating Method: Niobium in the range 1000-2500K. Int. J. Thermophys. 6 (1985) 681.

[21] A. E. Scheindlin, B. Ya. Berezin, V. Ya. Chekhovskoi. Enthalpy of niobium in the solid and liquid state. High Temp – High Press. 4 (1972) 611–619.

[22] V. Boodu, Paul Redner. Energetic Materials: Thermophysical properties, Predictions, and experimental measurements. CRC-Press – New York, 1st Ed. (2010).

[23] P. D. Desai. Thermodynamic Properties of Nickel. Int. J. Thermophys. 8 (1987) 763–780.

[24] M.W. Chase Jr. NIST-JANAF Thermochemical Tables, 4th Edition. J. Phys. Chem. Ref. Data, Monograph 9 (1998) 1–1951.